Lecture Notes in Nanoscale Science and Technology

Volume 47

Lecture Notes in Nanoscale Science and Technology (LNNST) aims to report latest developments in nanoscale science and technology research and teaching – quickly, informally and at a high level. Through publication, LNNST commits to serve the open communication of scientific and technological advances in the creation and use of objects at the nanometer scale, crossing the boundaries of physics, materials science, biology, chemistry, and engineering. Certainly, while historically the mysteries in each of the sciences have been very different, they have all required a relentless step-by-step pursuit to uncover the answer to a challenging scientific question, but recently many of the answers have brought questions that lie at the boundaries between the life sciences and the physical sciences and between what is fundamental and what is application. This is no accident since recent research in the physical and life sciences have each independently cut a path to the edge of their disciplines. As both paths intersect one may ask if transport of material in a cell is biology or is it physics? This intersection of curiosity makes us realize that nanoscience and technology crosses many if not all disciplines. It is this market that the proposed series of lecture notes targets.

Shrikaant Kulkarni
Editor

A Journey from Nanotechnology to Picotechnology

Atomic and Molecular Scale Engineering

Springer

Editor
Shrikaant Kulkarni
Faculty of Engineering and Technology
Sanjivani University
Kopargaon, Maharashtra, India

ISSN 2195-2159 ISSN 2195-2167 (electronic)
Lecture Notes in Nanoscale Science and Technology
ISBN 978-3-032-07227-6 ISBN 978-3-032-07228-3 (eBook)
https://doi.org/10.1007/978-3-032-07228-3

This Springer imprint is published by the registered company Springer Nature Switzerland AG
The registered company address is: Gewerbestrasse 11, 6330 Cham, Switzerland

Preface

Humankind has always striven to manipulate the world around it, from the macroscopic scale of monumental architecture to the microscopic intricacies of clockwork. In recent decades, our ability to engineer matter has undergone a profound revolution, plunging us into the realm of the ultra-small. This book charts a course through this exciting territory, beginning with the now-established field of nanotechnology and venturing into the nascent and potentially transformative domain of picotechnology—the engineering of matter at the atomic and molecular level.

The journey from manipulating structures measured in nanometers (billionths of a meter) to those measured in picometers (trillionths of a meter) represents not just a quantitative leap in scale, but a qualitative shift in our capabilities and understanding. At these dimensions, the very nature of matter reveals itself in its most fundamental form, governed by the intricate dance of individual atoms and the subtle forces that bind them. This opens up unprecedented opportunities for creating materials and devices with entirely new functionalities, pushing the boundaries of what we thought possible in fields ranging from electronics and medicine to energy and materials science.

This book is intended to be a guide for researchers, students, and anyone curious about the frontier of small-scale engineering. We begin by laying the groundwork in nanotechnology, exploring the principles, tools, and groundbreaking applications that have paved the way for the picoworld. We then delve into the concepts and emerging techniques that are enabling us to manipulate matter at the atomic and molecular level. We will examine the theoretical underpinnings, the experimental challenges, and the breathtaking potential of picotechnology, while also acknowledging the ethical and societal considerations that inevitably accompany such powerful capabilities.

The path from nanotechnology to picotechnology is not a simple extrapolation. It demands new ways of thinking, innovative experimental approaches, and a deep understanding of the quantum mechanical principles that govern behavior at these scales. This book aims to illuminate this path, providing a comprehensive yet accessible overview of the current state of the art and a glimpse into the exciting future that lies ahead.

Join us on this journey into the heart of matter, where the ability to sculpt the world atom by atom promises to unlock unimaginable possibilities and reshape the very fabric of our technological landscape.

Kopargaon, Maharashtra, India Shrikaant Kulkarni

Contents

Laying the Nanoscale Foundation

Fundamental Concepts in Nanotechnology 3
Rhea Vinod Nair, Bidyut Kumar Murmu, Simran Budhathoki, and Krupa S

Introduction to Nanotechnology and Picotechnology 27
P. Saranraj, K. Gayathri, Shrikaant Kulkarni, M. Manigandan, and K. Kesavardhini

A Pervasive Study on the Fabrication Techniques of Nanotechnology 45
Jincemon Cyriac, Jisty Ann Jiji, Joice K. Joseph, Dan Mathew Sabu, Ashtami R. Nair, Tina Sebastian, V. N. Sheeja, and Tom Cherian

Exploring the Realm of Nanotechnology

Advanced Techniques for Material Characterization and Analysis in Nano-Technology 93
Raja Kaliyaperumal and Karuppiah Nagaraj

Nanotechnology in Medicines and Healthcare 105
Priyanka S. Deorankar, Amol B. Kumbhar, Nilakshi V. Nerkar, and Shakila S. Shaikh

Nanotechnology in Energy and Environment 133
Jibrin Muhammad Yelwa, Joseph Usman, Ibrahim Umar, Alhagie Dremmah, Jibrin Yusuf Yahaya, Tari Joel Honda, and Haruna Musa

Nanotechnology in Computing and Communication 165
Nitiraj V. Kulkarni, Sampathi Sunitha, Jagadish V. Tawade, and Ashish Katke

Nanopores for Molecular Sequencing 191
Sunitha Sampathi and Alifeya Fidaali

Venturing into the Picoscale Frontier

Fabrication Techniques in Picotechnology 231
P. Saranraj, B. Lokeshwari, K. Gayathri, Hasanain A. J. Gharban, and M. Divya

Picotechnology in Energy and Environment 255
P. Saranraj, Ashish Kumar Nayak, Shrikaant Kulkarni, K. Kesavardhini, and E. Tharani

Picotechnology in Computing and Communication 275
Yuvalakshmi Logesan, Shrikaant Kulkarni, M. Divya, R. Nisha, and P. Saranraj

Picotechnology for Biological Science 299
K. Gayathri, P. Saranraj, Riyaz Sayyed, L. Charlie Jelura, and S. Monisha

Picotechnology: Advancing Precision in Bioimaging and Disease Diagnosis 313
Kevin Philip Sabu and Rony Rajan Paul

Picotechnology in Material Science 335
M. Manigandan, Shrikaant Kulkarni, F. Fareedhul Fahmitha, R. Yuvaraj, and P. SivaSakthivelan

The Journey Ahead

Future Trends and Challenges in Nanotechnology and Picotechnology 357
B. Lokeshwari, Shrikaant Kulkarni, U. Subhalakshmi, P. Saranraj, and N. Usha

Index 373

Contributors

Simran Budhathoki Jain (Deemed-to-be-University), School of Sciences, Bengaluru, Karnataka, India

Tom Cherian Department of Chemistry, Christ College (Autonomous) Irinjalakuda, University of Calicut, Irinjalakuda, Thrissur, Kerala, India

Jincemon Cyriac Department of Physics, Christ College (Autonomous) Irinjalakuda, University of Calicut, Thrissur, Kerala, India

Priyanka S. Deorankar Department of Pharmaceutics, Rajmata Jijau Shikshan Prasarak Mandal, College of Pharmacy, Pune, India

M. Divya PG and Research, Department of Microbiology, Sacred Heart College (Autonomous), Tirupattur, Tamil Nadu, India

Alhagie Dremmah Chemistry Unit, Division of Physical and Natural Sciences, School of Arts and Sciences, University of the Gambia, Serekunda, The Gambia

F. Fareedhul Fahmitha PG and Research Department of Microbiology, Sacred Heart College (Autonomous), Tirupattur, Tamil Nadu, India

Alifeya Fidaali School of Pharmacy, Vishwakarma University, Pune, Maharashtra, India

K. Gayathri PG and Research Department of Microbiology, Sacred Heart College (Autonomous), Tirupattur, Tamil Nadu, India

Hasanain A. J. Gharban Department of Internal and Preventive Veterinary Medicine, College of Veterinary Medicine, University of Wasit, Wasit, Iraq

Tari Joel Honda Department of Chemistry, Modibbo Adama University, Yola, Nigeria

L. Charlie Jelura PG and Research Department of Microbiology, Sacred Heart College (Autonomous), Tirupattur, Tamil Nadu, India

Jisty Ann Jiji Marian College Kuttikkanam (Autonomous), Kuttikkanam, Kerala, India

Joice K. Joseph Marian college Kuttikkanam (Autonomous), Kuttikkanam, Kerala, India

Raja Kaliyaperumal Department of Biological and Chemical Sciences, Liberal Arts and Sciences, Mohan Babu University (Erstwhile Sree Vidyanikethan Engineering College), Tirupathi, Andhra Pradesh, India

Ashish Katke School of Pharmacy, Vishwakarma University, Pune, Maharashtra, India

K. Kesavardhini PG and Research Department of Microbiology, Sacred Heart College (Autonomous), Tirupattur, Tamil Nadu, India

Krupa S Jain (Deemed-to-be-University), School of Sciences, Bengaluru, Karnataka, India

Nitiraj V. Kulkarni Faculty of Science and Technology, Vishwakarma University, Pune, Maharashtra, India

Shrikaant Kulkarni Sanjivani University, Kopargaon, India

Faculty of Business, Victorian Institute of Technology, Melbourne, VIC, Australia

Amol B. Kumbhar Department of Pharmaceutics, Rajmata Jijau Shikshan Prasarak Mandal, College of Pharmacy, Pune, India

Yuvalakshmi Logesan PG and Research, Department of Microbiology, Sacred Heart College (Autonomous), Tirupattur, Tamil Nadu, India

B. Lokeshwari PG and Research, Department of Microbiology, Sacred Heart College (Autonomous), Tirupattur, Tamil Nadu, India

M. Manigandan PG and Research Department of Microbiology, Sacred Heart College (Autonomous), Tirupattur, Tamil Nadu, India

S. Monisha PG and Research Department of Microbiology, Sacred Heart College (Autonomous), Tirupattur, Tamil Nadu, India

Bidyut Kumar Murmu Jain (Deemed-to-be-University), School of Sciences, Bengaluru, Karnataka, India

Haruna Musa Department of Pure and Industrial Chemistry, Bayero University Kano, Kano, Nigeria

Karuppiah Nagaraj Surfactant Biomedical Research Laboratory, Center for Global Health Research (CGHR), Saveetha Medical College and Hospitals, Saveetha Institute of Medical and Technical Sciences (SIMATS), Chennai, Tamil Nadu, India

Ashtami R. Nair Marian College Kuttikkanam (Autonomous), Kuttikkanam, Kerala, India

Rhea Vinod Nair Jain (Deemed-to-be-University), School of Sciences, Bengaluru, Karnataka, India

Ashish Kumar Nayak Department of Civil Engineering, Prasad V. Potluri Siddhartha Institute of Technology, Vijayawada, Andhra Pradesh, India

Nilakshi V. Nerkar Department of Pharmaceutics, Rajmata Jijau Shikshan Prasarak Mandal, College of Pharmacy, Pune, India

R. Nisha PG and Research, Department of Microbiology, Sacred Heart College (Autonomous), Tirupattur, Tamil Nadu, India

Rony Rajan Paul Department of Chemistry, CMS College, Kottayam, Kerala, India

Dan Mathew Sabu Marian College Kuttikkanam (Autonomous), Kuttikkanam, Kerala, India

Kevin Philip Sabu Centre for Immunology & Cancer Research, Faculty of Medicine, Hebrew University of Jerusalem, Lautenberg, Israel

P. Saranraj PG and Research Department of Microbiology, Sacred Heart College (Autonomous), Tirupattur, Tamil Nadu, India

Riyaz Sayyed Department of Biological Sciences and Chemistry, College of Arts and Science, University of Nizwa, Nizwa, Oman

Tina Sebastian Department of Physics, Deva Matha College, Kuravilangad, Kerala, India

Shakila S. Shaikh Department of Pharmacology, Rajmata Jijau Shikshan Prasarak Mandal, College of Pharmacy, Pune, India

V. N. Sheeja Marian College Kuttikkanam (Autonomous), Kuttikkanam, Kerala, India

P. SivaSakthivelan Agricultural Research Station, Tamil Nadu Agricultural University, Erode, Tamil Nadu, India

U. Subhalakshmi PG and Research Department of Microbiology, Sacred Heart College (Autonomous), Tirupattur, Tamil Nadu, India

Sampathi Sunitha School of Pharmacy, Vishwakarma University, Pune, Maharashtra, India

Jagadish V. Tawade Faculty of Science and Technology, Vishwakarma University, Pune, Maharashtra, India

E. Tharani PG and Research, Department of Microbiology, Sacred Heart College (Autonomous), Tirupattur, Tamil Nadu, India

Ibrahim Umar Department of Chemistry, Aliko Dangote University of Science and Technology, Wudil, Kano State, Nigeria

N. Usha PG and Research Department of Microbiology, Sacred Heart College (Autonomous), Tirupattur, Tamil Nadu, India

Joseph Usman Department of Electrical and Electronics Engineering, Abubakar Tafawa Balewa University, Bauchi, Nigeria

Jibrin Yusuf Yahaya Scientific and Industrial Research Department, National Research Institute for Chemical Technology, Zaria, Kaduna State, Nigeria

Jibrin Muhammad Yelwa Scientific and Industrial Research Department, National Research Institute for Chemical Technology, Zaria, Kaduna State, Nigeria

Department of Pure and Industrial Chemistry, Bayero University Kano, Kano, Kano State, Nigeria

R. Yuvaraj PG and Research Department of Microbiology, Sacred Heart College (Autonomous), Tirupattur, Tamil Nadu, India

Abbreviations

2DOBM	Two-dimensional organic biomaterial
5G	Fifth generation
5mC	5-Methylcytosine
6G	Sixth generation
AAAS	American Association for the Advancement of Science
AAM	Anodized alumina membrane
AAO	Anodic aluminum oxide
AC-TEM	Aberration-corrected transmission electron microscopy
AD	Alzheimer's disease
AD3D-MIL	Attention-based deep 3D multi-instance learner
ADA	Adenosine deaminase
AFM	Atomic force microscopy
AI	Artificial intelligence
ALD	Atomic layer deposition
ALP	Alkaline phosphatase
AM	Additive manufacturing
AMR	Antimicrobial resistance
AuNPs	Gold nanoparticles
BBB	Blood-brain barrier
BCP	Block copolymer
BIAcore	Surface plasmon resonance biosensor
BS	Base station
CAR-T	Chimeric antigen receptor T-cell
CBP	Customs and Border Protection
CCS	Circular consensus sequencing
CFTR	Cystic fibrosis transmembrane conductance regulator
CHO	Chinese hamster ovary
CIT	California Institute of Technology
CMC	Critical micelle concentration
CNNs	Convolutional neural networks
CNT	Carbon nanotubes

CNT-PDMS	Carbon nanotube–poly(dimethylsiloxane)
COFs	Covalent organic frameworks
COG-UK	COVID-19 Genomics UK
CRISPR	Clustered Regularly Interspaced Short Palindromic Repeats
CRISPR-Cas9	Clustered Regularly Interspaced Short Palindromic Repeats
CSA	Cooperative self-assembly
CsgG	A lipoprotein
CSNPs	Chitosan-based nanoparticles
CT	Computed tomography
CVD	Chemical vapor deposition
DBD	Dielectric barrier discharge
DC-PECVD	Direct current plasma-enhanced CVD
DFT	Density functional theory
DLS	Dynamic light scattering
DNA	Deoxyribonucleic acid
DNI	Dynamic nano-inscribing
DRIE	Deep reactive ion etching
dsDNA	Double-stranded DNA
DSL	Digital subscriber line
EA	Electroacupuncture
EBL	Electron-beam lithography
EC	Edge computing
ECG	Electrocardiogram
ECM	Extracellular matrix
EDS	Energy-dispersive X-ray spectroscopy
EE	Energy efficiency
EEG	Electroencephalogram
EELS	Electron energy loss spectroscopy
EHD	Electrohydrodynamic printing
EISA	Evaporation-induced self-assembly
ELC	Electronic load controller
ELISA	Enzyme-linked immunosorbent assay
EMG	Electromyography
EPM	Electrophoretic mobility
EPR	Enhanced permeation and retention
EUV	Extreme ultraviolet (lithography)
EXAFS	Near edge X-ray absorption fine structure
FBS	Functional biomimetic synthesis
FeO_3	Iron oxide
FESEM	Field emission scanning electron microscopy
FIB	Focused ion beam
FIBID	Focused ion beam-induced deposition
FIBM	Focused ion beam machining
FinFET	Fin field-effect transistor
FPGA	Field-programmable gate array

FTIR	Fourier Transform infrared spectroscopy
FUs	Function units
GAA	Gate-all-around (transistor)
GB	Gigabyte
GFP	Green fluorescent protein
GO	Graphene oxide
GO-Gel	GO-gelatin
GPUs	Graphics processing units
GQDs	Graphene quantum dots
GSR	Galvanic skin response
HA	Hydroxyapatite
h-BN	Hexagonal boron nitride
HIV	Human immunodeficiency virus
HMMs	Hidden Markov Models
HR-TEM	High-resolution transmission electron microscopy
IBSM	International Business Machine Corporation
ICP	Inductively coupled plasma
IF	Intermediate Filaments
II	Iteration interval
IMs	Internal memories
IoMT	Internet of Medical Things
IONPs	Iron oxide nanoparticles
IoT	Internet of Things
ITRS	International Technology Roadmap for Semiconductor materials
LEDs	Light-emitting diodes
LIFT	Laser-induced forward transfer
m6A or 6mA	N6-methyladenosine
Mb	Megabyte
MBE	Molecular beam epitaxy
MD	Molecular Dynamics
MDA	Mass drug administration
MEMS	Microelectromechanical systems
MIMO	Multiple input multiple output
M-IoTs	Medical Internet of Things
ML	Machine Learning
MLD	Molecular layer deposition
MnO	Manganese oxide
MNPs	Magnetic nanoparticles
MOFs	Metal organic frameworks
MoS_2	Molybdenum disulfide
MOSFET	Metal oxide semiconductor field effect transistors
MR	Magnetic resonance
MRAM	Magnetic random-access memory
MRFM	Magnetic resonance force microscopy
MRI	Magnetic resonance imaging

MS	Molecular sequencing
MspA	*Mycobacterium smegmatis* porin A
MTs	Microtubules
MW-PECVD	Microwave plasma-enhanced CVD
nc-AFM	Non-contact atomic force microscopy
NCL	Nanochannel-guided lithography
NEMS	Nanoelectromechanical systems
NF-H	Neurofilament-Heavy
NF-L	Neurofilament-Light
NF-M	Neurofilament-Medium
NGS	Next-generation sequencing
NIH	National Institutes of Health
NIL	Nanoimprint Lithography
NM	Nanomaterial
NP	Nanopore
NPA	Non-programmable accelerator
NPS	Nanoparticle synthesis
NPS	Nanopore sequencing
NSOM	Near-field scanning optical microscopy
NTA	Nanoparticles tracking analysis
NV	Nitrogen-vacancy
OCT	Optical coherence tomography
OLEDs	Organic light-emitting diodes
ONT	Oxford Nanopore Technologies
Otfts	Organic thin-film transistors
PAMAM	Poly(amidoamine)
PBS	Process biomimetic synthesis
PC	Polycarbonate
PCA	Principal component analysis
PCR	Polymerase chain reaction
PDMS	Polydimethylsiloxane
PECVD	Plasma-enhanced CVD
PET	Positron emission tomography
PLD	Pulsed laser deposition
PLGA	Poly(lactic-co-glycolic acid)
PM	Picomolecules
PMMA	Polymethyl methacrylate
PNF	Peptide nanofiber
PNIPAM	Poly(N-isopropylacrylamide)
PSA	Polystyrene-acrylate
PTMs	Post-translational modifications
PTT	Pulse transit time
PVD	Physical vapor deposition
QCD	Quantum chromo dynamics
QDSCs	Quantum dot solar cells

QED	Quantum electrodynamics
QKD	Quantum key distribution
Qubits	Quantum bits
RCD	Residual current device
ReRAM	Resistive random-access memory
RF-PECVD	Radio-frequency plasma-enhanced CVD
RHEED	Reflection high-energy electron diffraction
RIE	Reactive-ion etching
RNA	Ribonucleic acid
RTL	Register transfer level
SA	Self-assembly
SACs	Single-atom catalysts
SAMs	Self-assembled monolayers
SARS-CoV-2	Severe acute respiratory syndrome coronavirus 2
SAXS	Small angle X-ray scattering
SCID	Severe combined immunodeficiency
SERS	Surface-enhanced Raman spectroscopy
Si3N4	Silicon nitride
SiO2	Silicon dioxide
siRNA	Small interfering ribonucleic acid
SLI	System-level integration
SMA	Spinal muscular atrophy
SMN1	Survival of motor neuron 1
SPM	Scanning probe microscopy
SQUIDs	Superconducting quantum interference devices
ssDNA	Single-stranded DNA
STM	Scanning tunneling microscopy
TB	Tuberculosis
TEM	Transmission electron microscopy
TFET	Tunnel field-effect transistor
TiO_2	Titanium dioxide
TLCT	True liquid crystal templating
TMDs	Transition metal dichalcogenides
TMS	Transcranial magnetic stimulation
TPU	Tensor processing unit
UDN	Ultra-dense networks
UHV	Ultrahigh vacuum
US NIST	United States National Institute of Standards and Technology
USB	Universal serial bus
UV	Ultraviolet
VDP	Vapor deposition polymerization
VEGF	Vascular endothelial growth factor
VHDL	VHSIC Hardware Description Language
VLIW	Very Long Instruction Word
VOCs	Volatile organic compounds

VR	Virtual reality
XANES	X-ray absorption near-edge structure
XAS	X-ray absorption spectroscopy
XR	Extended reality
XRD	X-ray diffraction
ZnO	Zinc oxide

Laying the Nanoscale Foundation

Fundamental Concepts in Nanotechnology

Rhea Vinod Nair, Bidyut Kumar Murmu, Simran Budhathoki, and Krupa S

1 Introduction

Nanotechnology is the science and engineering of managing and modifying materials at the nanoscale, generally from 1 to 100 nm. Materials exhibit distinct biological, chemical, and physical characteristics, distinguishing them from the bulk material at this size scale. Along with design, characterization, manufacturing, and application, the technology includes exact control over size and shape at the nanoscale (Ratner & Ratner, 2003). The notion of nanotechnology is not new; its roots may be seen in the ground-breaking lecture of 1959 "There's Plenty of Room at the Bottom," in which Richard Feynman envisioned manipulating individual molecules and atoms (Feynman, 2002). Although Feynman's lecture established the conceptual foundation, the term "nanotechnology" was initially used by Norio Taniguchi in 1974 as precision machining with tolerances in the order of nanometers (Taniguchi, 1974). The development of nanotechnology is another field that marks great historical milestones, thus it becomes even more significant since it's the most fast-paced area. The history of nanoscale engineering begins with the fourth-century A.D. Roman Lycurgus Cup., showing dichroic properties because of gold and silver nanoparticles embedded within (Ahire et al., 2022). Later on, Michael Faraday's study in 1857 of colloidal gold proved for the first time that nanoscale particles show optical effects Mustafa.

However, the new nanotechnology field started gaining momentum during the 1980s when it saw significant breakthroughs in enabling technologies. A key advancement in nanoscale research was made in 1981 when the scanning tunneling microscope (STM) was developed, enabling researchers to view and even work with

R. V. Nair · B. K. Murmu · S. Budhathoki · Krupa S (✉)
Jain (Deemed-to-be-University), School of Sciences, Bengaluru, Karnataka, India

S. Kulkarni (ed.), *A Journey from Nanotechnology to Picotechnology*, Lecture Notes in Nanoscale Science and Technology 47,
https://doi.org/10.1007/978-3-032-07228-3_1

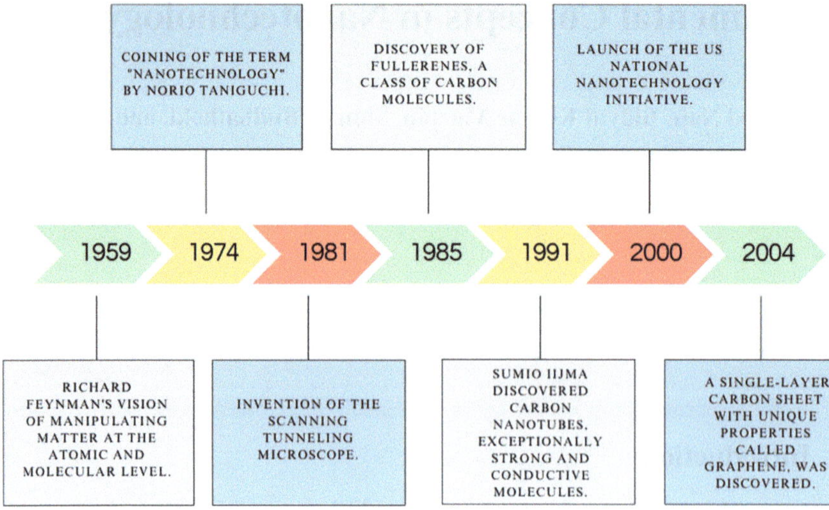

Fig. 1 A timeline of key discoveries and events in nanotechnology

individual atoms (Binnig et al., 1982). In addition, fullerenes (C60), a new form of carbon at the nanoscale, were found in 1985. This development opened several doors toward "nanomaterials"research (Kroto et al., 1985). This led to the burgeoning of interdisciplinary study and established nanotechnology as a well-defined scientific discipline.

The integration of nanotechnology across multiple domains, including environmental science, electronics, medicine, and energy, has been made easier by the special qualities and capabilities of nanoscale materials (Fig. 1). Because it tackles intricate issues at the atomic and molecular level, nanotechnology has remained essential to the advancement of science in the twenty-first century.

2 Nanomaterials and their Unique Properties

Nanomaterials are a broad category of materials that are characterized by having at least one dimension in the nanoscale, usually between 1 and 100 nanometers. Compared to their bulk versions, these tiny particles have distinct physical, chemical, and biological characteristics. The remarkable traits arise from two main elements:

1. *Increased surface area-to-volume ratio:* At the nanoscale, the area of a material's surface increases with its volume in a direct proportion. Due to the improved environment-material interaction, the mechanical properties of the material are greatly improved compared to regular-sized materials, and the catalytic efficiency and reactivity are also increased (Salata, 2004).

2. *Quantum effects:* As materials are reduced in size, quantum mechanical phenomena become more noticeable. These effects, like quantum confinement and size quantization, affect the electronic and optical characteristics of nanomaterials. For example, the bandgap of semiconductor nanocrystals (quantum dots) can be adjusted by manipulating their size, which leads to size-dependent optical emissions (Alivisatos, 1996a; Kumar Sahu et al., 2017).

The distinctive characteristics of nanomaterials allow nanoparticles and nanotubes to demonstrate remarkable versatility in a wide range of applications. Nanoparticles, which are zero-dimensional structures, are restricted to the nanoscale in all dimensions. Their small size and outstanding surface area-to-volume ratio greatly improve their usefulness in areas like drug delivery, sensing, and catalysis (Salata, 2004). As noted by Dreaden et al. (2012), for instance, gold nanoparticles are useful for imaging and biosensing applications because of their distinctive optical characteristics connected to their localized surface plasmon resonance. Nanotubes are one-dimensional structures that are prized for their exceptional thermal properties, outstanding electrical conductivity, and remarkable mechanical strength (Iijima, 1991). As noted by Dresselhaus et al. (1996), their distinct structural characteristics make them ideal for use in energy storage, electronics, and composite materials. Nanowires are one-dimensional structures with high aspect ratios that display electrical and optical characteristics that change with size (Dippong, 2024; Xia et al., 2003). Applications of these materials in sensors, energy harvesting devices, and nanoelectronics are being studied in great detail (Duan et al., 2001). Quantum dots are semiconductor nanoparticles that emit light in a range of hues based on their dimensions, exhibiting size-tunable spectrum properties (Bruchez et al., 1998; Kumar Sahu et al., 2017). These materials possess special optical qualities which make them advantageous in displays, solar cell applications, and bioimaging (Klimov et al., 2000).

These nanomaterial's distinctive characteristics stem from their size, shape, and surface features at the nanoscale, which enable customized functionalities that are very different from their bulk counterparts. The combination of these nanomaterials propels fundamental developments in a number of domains, including electrical technology, environmental research, and healthcare. The uniqueness of nanomaterials, physical, chemical, optical, and mechanical properties is attributed mainly to their high surface area-to-volume ratio and quantum confinement effects. High surface area enhances the chemical reactivity of nanomaterials and makes them very good catalysts (Rao et al., 2005). Electronic band structures are altered by quantum confinement, which emerges when the material size approaches that of the de Broglie wavelength of electrons. Quantum dots show size-dependent fluorescence, while unique optical properties include the capability of altering electronic band structures as seen in the case of nanomaterials (Alivisatos, 1996b). Mechanically, nanomaterials are much stronger and stiffer than their bulk counterparts because of the absence of defects and changes in dislocation dynamics (Ashby et al., 2007). Such properties make nanomaterials very appealing for a variety of applications, ranging from catalysis and sensing to structural reinforcement and optoelectronics.

The surface area-to-volume ratio is a critical factor that the nanomaterials' properties and behavior most vary. Since materials can be reduced to the nanoscale, their surface area tends to increase relative to their volumes, which dictates the higher ratio that improves their reactivity and interaction with the environment. The surface area A of a sphere was provided using the formula $A = 4\pi r^2$ whereas its volume V as; $V = (4/3)\ \pi r^3$. This results in a surface area-to-volume ratio that comes as $A/V = 3/r$, which tells that with a smaller radius r, the ratio will be larger. As such, nanoparticles possess a substantially higher surface area per unit volume compared to larger particles, which means they have heightened chemical reactivity. This is clearly so especially because, with the particle being just 3 nm in diameter, there is the possibility of as many as 50% of the atoms to be on the surface. A particle this large would exhibit a much lower percent of surface atoms. This kind of characteristic enables nanoparticles to be involved in reactions that, in bulk form, are inert. Hence, the fields that benefit from them are vast: catalysis, drug delivery, and sensor technology, to mention a few (Mekuye & Abera, 2023). The impact of this feature serves to remind how nanomaterials revolutionize scientific and industrial uses.

Nanomaterials are distinguished by having a surface area-to-volume ratio that is significantly greater than that of bulk material. This is due to their nanoscale size wherein the majority of atoms are present on the surface and less within the volume. When the size of the particles decreases, the surface area dramatically increases while the volume decreases; hence, it gives rise to this high ratio (Rao et al., 2005). This huge surface area profoundly influences many other properties. It makes chemical reactivity improved because more atoms on the surface are used to interact with other substances, making nanomaterials a very effective catalyst (Bond, 1987). The process of adsorption and absorption is also affected because the molecules that may be bound or absorbed on the surface increase with nanomaterials (Langmuir, 1916). Additionally, the mechanical, electrical, and optical properties will have a significant correlation with the high surface area.

3 Scale and Units in Nanotechnology

Dimensions for materials lie on a scale which runs between 1 and 100 nm—the nanoscale where, often significantly differently compared with bulk versions, they gain novel physical or chemical properties that materialists possess. A nanometer is defined as one billionth of a meter, or 10^{-9} m, making it hard to visualize because of its very small size; to put this into perspective, a human hair is about 80,000 to 100,000 nm in width. At this scale, quantum effects dominate and have a controlling influence on the melting point, electrical conductivity, and chemical reactivity. This high surface area-to-volume ratio at the nanoscale also makes it highly reactive and interactive with the environment, which has highly applied purposes in catalysis and drug delivery. The importance of the nanoscale is also shown in the progress of technologies like quantum dots, used in high-resolution displays and medical imaging. Understanding the dimensions of nanoscale is fundamental to the harnessing of unique properties of nanomaterials in various scientific and industrial applications.

The nanoscale, that scale of dimensions from 1 to 100 nm, is the only one in which matter exhibits properties very different from those seen in bulk (Ratner & Ratner, 2003). A billionth of a meter (10^{-9} m) is equal to one nanometer. That's really too small to see. Put in perspective, a human hair is about 80,000–100,000 nm in diameter (Silverman, 2017). When scaled down, the surface area-to-volume ratio really increases. Quantum mechanical effects dominate and will give rise to novel biological, chemical, and physical properties, as pointed out by Bhushan (2017). A great understanding of manipulating matter on the nanoscale is also crucial for new developments in such fields as medicine, electronics, and materials science.

Accurate characterizations of matter and its manipulation at the nanoscale require measurement techniques and tools that are able to resolve dimensions on the order of nanometers and even picometers. For perspective, the nanometer is 10^{-9} meters while a picometer is even smaller: 10^{-12} m (Bhushan, 2017). It is difficult to visualize or measure features of this scale without the use of methods such as Atomic Force Microscopy (AFM) and Scanning Tunneling Microscopy (STM). STM, developed by Binnig and Rohrer in 1982, allows for atomic resolution imaging by measuring the variation of tunneling current with a sharp tip poking into the surface of materials. AFM, developed subsequently in 1986 (Binnig et al., 1986), extends this to measuring the force between the tip and sample, enabling an even wider range of material studies than was possible with STM, including insulators. The Electron Microscopy techniques, including SEM and TEM, allow for very high-resolution imaging of nanomaterials; TEM can further provide higher magnification and even atomic structure resolution (Williams & Carter, 2009). Some of these and other tools are critical to nanotechnology advancement through the precise characterization and manipulation of nanomaterials.

Techniques and tools for measurements are critical to characterize nanomaterials in the nanoscale, that is, in the range of 1–100 nm. Measurement methods at this scale lack the needed resolution and sensitivity. Transmission Electron Microscopy (TEM) is one of the most commonly used techniques, where high-resolution images can be directly measured for particle size, morphology, and distribution by transmitting a beam of electrons through a sample. Dynamic Light Scattering (DLS) is the other important method, which monitors light scattering arising as a result of Brownian motion, for deducing the nanoscale size distribution in liquid suspensions. Atomic Force Microscopy is also a method in which it images three-dimensional topographies of the surface of nanostructures with resolutions down to individual atoms by laterally scanning a sharp tip against the sample surface. Techniques like Inductively Coupled Plasma Mass Spectrometry are employed to provide elemental analysis at trace levels to measure the purity and composition of nanomaterials. The precision of such measuring tools plays a very key role in enhancing nanotechnology applications in material science, electronics, and medicine, among others. These methods collectively demonstrate the significance of precise nanoscale material measurement.

Scanning tunneling microscopy and atomic force microscopy are the major tools in nanotechnology, and they allow scientists to visualize and manipulate matter at atomic and nanoscales. STM was discovered by Binnig and Rohrer in 1982; it works by the principle of quantum tunneling, where the electrons tunnel from a sharp conducting tip to the sample surface. STM creates a real-space, atomic resolution image of the surface topography through the scanning of the tip during the measurement of the tunneling current. This image shows individual atoms and their relative arrangement (Chen, 2008). Atomic Force Microscopy (AFM) allows this capability to expand when measuring the forces between a sample and sharp tip developed some years after the invention of STM (Binnig et al., 1986). This enables AFM to image conductive and nonconductive materials, which yields information on the surface topography, mechanical properties, and biological structures (Garcia, 2010). STM and AFM are now indispensable in nanotechnology and have enabled research scientists to be able to probe and manipulate nanomaterials, characterize properties, and manufacture nanoscale devices.

STM scans the surface by making use of a sharp conductive tip that hovers just close to a sample, hence making it possible to measure the tunneling current between the tip and the sample. This current is very sensitive to the distance between them, so STM can actually achieve atomic resolution and image electronic properties of surfaces (Binnig & Rohrer, 1982). On the other hand, AFM makes use of a mechanical probe that interacts with the surface through intermolecular forces and can image an extensive range of materials such as insulators and biological samples (Binnig et al., 1986). The ability to work in air and liquids increases the applicability of AFM over STM, which requires vacuum or controlled atmospheres for it to work to its best extent. While both techniques support high-resolution imaging, they can also be manipulated at the nanoscale level, making these two techniques irreplaceable instruments for researchers involved in nanoscience and materials engineering. STM and AFM have, together, revolutionized our perception of surface science; further applications of nanotechnology are still in the process of evolving.

4 Principles of Nanoscience

At the nanoscale, the properties of matter are determined by quantum mechanics (Heinrich et al., 2021). However, this is significantly different from the field of physics that governs macroscopic objects, such as the possibility of particles existing in multiple states at the same time and their ability to move through barriers. Quantum effects, such as quantum confinement, are very significant here. Dimensions of the material are reduced to the nanoscale, generally through methods like etching, lithography, or self-assembly. Such confinement reduces the possible motion of the electrons, allowing for enhanced characteristics. For example, in a quantum dot, the dimensions of the crystal itself determine the energy levels available for electron transitions to occur (Osypiw et al., 2022). When the crystal size shrinks, the energy levels also become more discrete, leading to quantum confinement which results in the emission of light at specific wavelengths, which can be controlled by changing the size of nanocrystals. This confinement alters the electronic energy levels of the material, leading to unique optical, electrical, and magnetic properties (Heinrich et al., 2021).

As material dimensions shrink, the ratio of surface area to volume tends to increase. This is because, with a reduction in material size, the volume also decreases much faster than the surface area (Kumari & Yadav, 2024). For example, if a rectangular block has a volume of 1 m^3 with dimensions $2 \times 1 \times 0.5$ m. Its initial surface area is 7 m^2. If this block is split into eight equal little blocks, its mass remains the same, but the surface area is increased to 14 m^2. This explains how the surface area increases rapidly when the material is divided into small pieces while the total volume remains uniform. Such an increased surface area can have a lot of advantages such as an increase in surface interactions like adsorption and reactivity (Roy et al., 2021). Highly effective catalysts speeding up chemical reactions can also be regarded as high surface area nanoparticles. Hence, the main effects of a greater surface area are on the physicochemical properties of a given material.

One of the main concepts in nanoscience is self-assembly. It refers to the quick assembly of molecules into complex structures without any external interference. This can be generally achieved through some intermolecular forces and other reactions that can help particles bind specifically to one another and form the desired nanostructures (Li et al., 2020). The most common example of this is how hydrogen bonds between water molecules come together to form ice crystals. Self-assembly can be initiated by the careful preparation of molecular building blocks and precise control over the surrounding environmental conditions to create desired configurations with better control over their dimensions and functionality (Rao et al., 2022).

Amphiphilic polymer-based nanoparticles primarily depend on self-assembly (Choi et al., 2022). This category of polymers has both hydrophobic and hydrophilic domains and often tends to self-organize in an aqueous medium spontaneously. Self-assembly through hydrophobic interactions serves to reduce the free energy associated with interfacial surfaces (Sun, 2022). Here, the hydrophobic portion is directed inward into the core of the nanoparticle, whereas hydrophilic moieties mostly lie outward, at the surface. This example shows how micelles can be formed with the help of intermolecular forces. A main application of this concept is in drug delivery, where self-assembled nanoparticles can encapsulate and protect therapeutic molecules like proteins and peptides from degradation in the gastrointestinal tract while also facilitating their safe transport into the systemic circulation (Yang et al., 2023).

Interfacial phenomena in nanoparticles refer to the several interactions that take place at the boundary that exists within the nanomaterial and its ambient surroundings. These interactions are heavily influenced by the Nano crystal's high surface area per unit volume, adsorption, wetting, electrostatic forces, etc. Surfactants or polymers are generally used for surface modification to control these interfacial properties of the nanoparticles and influence dispersion, aggregation, and other related interactions with surrounding molecules (Miyazawa et al., 2021). For example, electrostatic interactions between the nanoparticle surface charge and nearby ions may affect their stability and aggregation behavior. Adsorption of molecules from the environment onto the nanoparticle surface forms an interfacial layer, significantly altering its interactions with other species. Wetting behavior, which is determined by the angle of contact between the nanoparticle and its surrounding fluid, influences dispersion and other interactions. Capillary forces, such as those appearing in nanoparticles, influence foam stabilization and liquid penetration into porous media.

In biology, the adsorption of proteins on the nanoparticle surface can lead to the formation of a "protein corona," which can influence how they interact with cells and the fate of nanoparticles (Bilardo et al., 2022). In material sciences, nanoparticles can act as Pickering stabilizers (Pinto et al., 2025), allowing for adsorption onto the oil–water interface and stopping droplet coalescence in emulsions. Interfacial forces also influence the stability of nanoparticle dispersions which generally depend on attractive or repulsive forces. The interfacial behavior of nanoparticles can affect their interaction with other materials, as demonstrated by studies on the influence of nano-silica on the behavior of water and alkali solutions. These studies indicate how the surface properties of nanoparticles may modify the properties of bound water, leading to different clustering and hydrogen bond network behaviors (Gun'ko et al., 2023).

5 Synthesis of Nanomaterials

Nanoparticles are those materials that fall in the range of less than 100 nm (Khan, Asmatulu, et al., 2025). They have distinct physical and chemical properties due to their nanoscale dimensions. Synthesis of these particles is achieved by carefully controlling the nucleation and growth processes involved in achieving the desired size, shape, and surface properties. Synthesis should be done accurately to ensure consistent productivity and reproducibility in different applications, including drug delivery and catalysis. Here are some of the most common approaches used in its synthesis:

The top-down method generally involves downsizing of bulk material to its nanodimensions (Chen et al., 2023), which is attained by techniques like ball milling and sputtering. These methods are usually conducted through physical or chemical processing, which typically breaks up a large structure into smaller parts (Dash & Panda, 2024). However, such an approach may introduce defects or other unwanted impurities in our nanomaterial due to the aggressive nature of the size reduction process. For example, ball milling is done based on high-energy impacts which cause our materials to be smashed into small-sized particles (Joy et al., 2022). Thus, this method can introduce defects in the crystal structure of nanoparticles. Similarly, laser ablation can cause localized heating and melting, leading to variations in the physical aspects of the particles.

Nanolithography is a top-down approach wherein nanoscale structures are created by directly manipulating materials on the surface. The most prominent technique among this is soft lithography which utilizes elastomeric stamps and transfers materials onto a surface or substrate whereas in dip-pen nanolithography, atomic force microscope tips are used for depositing molecules (Li et al., 2021). These methods allow for accurate placement and patterning of nanoparticles on surfaces along with control over their arrangement, thus allowing for the fabrication of functional nanostructures with customizable properties. This controlled state is necessary to form complex nanostructures with specific functionalities. This allows for the development of arrays of nanoparticles with specific sizes and shapes. These can then be used for various concerns such as enhancement of sensitivity in biosensors (An et al., 2022) or efficiency of catalysts.

On the contrary, the bottom-up strategy entails building nanostructures from smaller units, like atoms and other molecules (Abid et al., 2022). Some commonly used techniques under this are chemical vapor and atomic layer deposition (Zhou & Yu, 2021). Nanostructures can either be made using controlled chemical reactions or even self-assembly, to create particles with effective influence over their form. Although this approach helps form an ultra-fine nanoparticle with fewer defects, it often involves the use of extremely sensitive chemical processes and requires strict control over surrounding environmental factors. For example, in a chemical vapor deposition process; the pressure, flow rate, etc., of reactant gases need to be maintained to produce our desired nanostructures (Yang et al., 2021). Atomic layer deposition requires the alternate pulses of different precursor gases, with precise timing and control to reach the desired thickness and uniformity of the nanostructure (Oke & Jen, 2022).

Chemical vapor deposition is a bottom-up approach that involves coatings and nanostructures being deposited onto substrates such as glass or metal surfaces with the help of chemical reactions. Generally, in this method, precursor gases are introduced into a reactor wherein chemical processes occur at high temperatures. Reaction products are then deposited onto hot surfaces, thereby creating the desired material. Among the several benefits of this method, it allows the creation of materials of higher quality with specific properties. It has acquired great relevance in the fabrication of graphene (Liu et al., 2022) as well as in other nanomaterials due to their scalability, accommodation to industrial conditions, and minimal environmental hazards.

The sol gel technique is a very flexible way to produce nanoparticles, especially metal oxides (Bokov et al., 2021). It involves the dissolution of a molecular precursor in a solvent, followed by hydrolysis and condensation reactions that eventually lead to the formation of a colloidal suspension, called the sol. The subsequent gelation stage involves the formation of a three-dimensional network due to the interconnection of particles. The solvent is then removed through drying and the acquired gel can then be processed through methods like grinding to obtain our required nanoparticle (Bokov et al., 2021). This technique can also be applied in many fields such as ceramics and electronics (Zanurin et al., 2022).The green synthesis of nanomaterials provides new solutions to current sustainable development challenges. Microorganisms, such as bacteria, fungi, and other plant extracts, are involved in green synthesis due to their natural ability to reduce metal ions to nanoparticles

(Jeevanandam et al., 2022). In the case of microbial synthesis, the microbial enzymes and metabolites reduce and stabilize metal ions (Bahrulolum et al., 2021), while the reducing and stabilizing agents in plant-mediated synthesis are flavonoids, terpenoids, and alkaloids (Munir et al., 2021). This process allows for precise control over the size and shape of nanoparticles. Redox reactions in which biological compounds donate electrons to metal ions, allow for their reduction and aggregation into nanoparticles. Green-synthesized nanoparticles have enhanced catalytic properties, helping to remove pollutants during environmental remediation and wastewater treatment. In medicine, silver nanoparticles improve therapeutic efficacy, reduce side effects, and provide targeted cancer therapy (Gomes et al., 2021). Green synthesis is an environmental-friendly and biocompatible technique as compared to other chemical synthesis approaches (Fig. 2).

6 Characterization Techniques

The assessment of nanocrystals is very essential for interpreting and optimizing their attributes for different applications. These materials possess unique size-dependent features like enhanced chemical activity, improved electric conductivity and other properties that have a significant effect on the interactions of nanomaterials in various settings through changes in their material properties with other biosystems. Accurate characterization is a must for maintaining structure property ties, consistent and improved performance, and advancement of nanotechnology (Jayawardena et al., 2021). The following are a few important methods for characterizing nanoparticles:

Scanning electron microscopy is a method that is used for identifying and understanding the general structure of nanoparticles (Vijayaram et al., 2023). It provides high-quality, black-and-white visuals by directing a narrow beam of electrons across the sample exterior, revealing details such as their respective shape and size. In nanoparticle characterization, this method requires careful sample preparation (Ghomrasni et al., 2020) since nanoparticles is often fragile and sensitive to the high-energy electron beam being used. Nanoparticles are usually mounted on a conductive substrate such as aluminum to provide a stable and electrically conductive platform for the nanoparticles to prevent damage to them and ensure accurate imaging (Gutiérrez et al., 2022). This substrate can also help dissipate charge buildup that often happens during the scanning of electrons and, in this way, enhance image quality and prevent artifacts. It runs under a high vacuum environment. This form of microscopy also gives details of the particle's composition and orientation. However, this approach may have certain limitations such as potentially causing damage to sensitive samples and the inability to image wet or nonconductive samples.

Transmission Electron Microscopy is a powerful analysis method used to image nanoparticles at the nanoscale. Unlike optical microscopy, this requires a high-

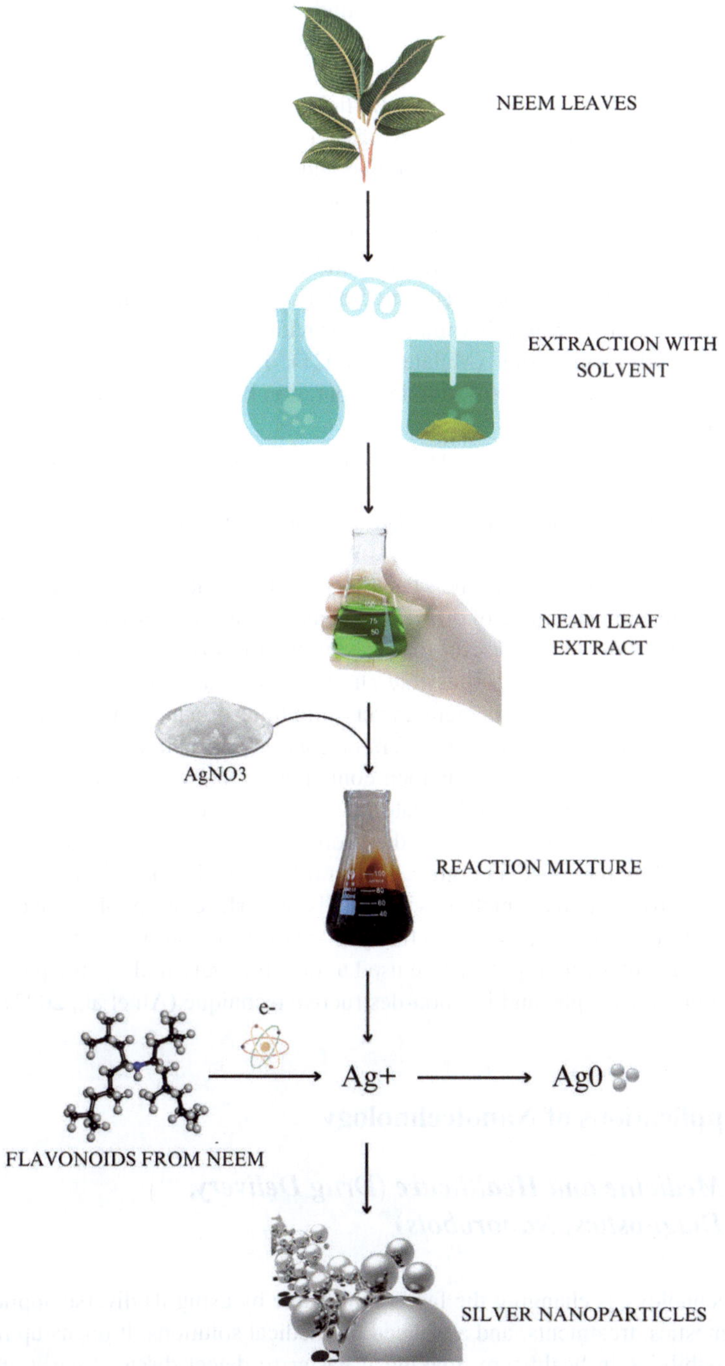

Fig. 2 Process for biosynthesis of silver nanoparticles

energy electron beam which produces higher resolution images for the detailed visualization of nanoparticle size, shape, and morphology as well as its internal nanostructure (Malatesta, 2021). The main principle of TEM includes accelerating electrons from an electromagnetic field to a far shorter wavelength as compared to light. It allows for the passing of an electron ray through a fine sample, where the visual generated is magnified and projected onto a detector. Hence, it is useful for imaging a variety of nanomaterials in material sciences and biological macromolecules in life sciences. High-definition images of nanoparticles allow for the observation of fine structural details like crystal lattice structures, defects, and other interfaces. This method also allows us to understand the interaction of particles with other materials, providing information on their behavior and related applications (Malatesta, 2021). However, transmission electron microscopy requires careful sample preparation because it needs to be very thin for charge flow to occur.

X-ray diffraction is another approach for characterizing nanoparticles and provides useful information about the crystal properties. It involves the analysis of interference patterns generated when X-rays engage with the crystal arrangement of nanoparticles (Thakur et al., 2025). When X-rays interface with the smaller atoms in the nanoparticle, they generally tend to scatter in all directions. The alignment and magnitude of diffraction peaks reveals details about the material's crystal structure and are useful for determining the size of crystallites and in the assessment of other properties like strain, texture, and defects (Pandey et al., 2021). Diffraction analysis requires very thin films of nanoparticles on appropriate substrates, usually, silicon wafers or glass slides, and subsequently diffraction data is collected, which is then compared to known diffraction patterns in online databases to identify the material's atomic arrangement. This technique is important in order to understand the properties of nanoparticles since they are influenced by their size and other structural features. Unlike scanning and other microscopy methods, which provide details on surface morphology and internal structure of nanocrystals, respectively, X-ray diffraction has many advantages such as its relatively rapid, can be used to quantitatively find out the phase composition of a sample, and is a non-destructive technique (Ali et al., 2022).

7 Applications of Nanotechnology

7.1 Medicine and Healthcare (Drug Delivery, Diagnostics, Nanorobots)

Nanotechnology is changing the face of medicine by using its diverse applications in diagnostics, treatments, and advanced biomedical solutions. It opens up incredible possibilities in healthcare, making it easier to detect diseases early, improve imaging, and deliver treatments directly to where they're needed (Malik et al.,

2023a), for example, tiny particles can carry drugs, heat, or light directly to diseased cells, ensuring healthier cells are left untouched. This precise targeting not only reduces side effects but also helps catch illnesses in their early stages, making treatments more effective and less invasive. The field of nanotechnology and nanomedicine has made several diverse and wide-ranging discoveries. Significant progress in nanomedicine has elevated the drug to a new level with astounding therapeutic outcomes (Di Stefano, 2023). Nanoparticles in those devices can function as molecular imaging agents and input genetic alterations associated with cancer and tumor cell activity (Liu & Grodzinski, 2021). Due to their nanoscale size, nanoparticles have a high surface-area-to-volume ratio and fast blood circulation, which allows them to absorb huge doses of medications. Because of their larger surface area, they have special qualities that make them appropriate for a wider range of medical applications, including enhanced mechanical, magnetic, optical, and catalytic capabilities... According to their chemical makeup, nanoparticles are divided into three categories: carbon-based, inorganic, and organic (Haleem et al., 2023), for example, drug combinations that are considered chemotherapy are directly delivered to cancer cells using nanoparticles which reduce the side effects associated with treatments, especially cancer therapy. Real-time health monitoring can be achieved through wearable devices powered by nanotechnology (Pokrajac et al., 2019).

7.2 Electronics (Semiconductors, Transistors, and Sensors)

Nanotechnology plays a pivotal role in advancing electronic devices, drawing significant research interest from both developed and developing nations. Nanoelectronics, an interdisciplinary field merging nanotechnology and electronics, focuses on the manipulation, design, characterization, and fabrication of electronic devices at the nanoscale. Approach: Shrinking material dimensions alters their electronic properties, amplifies inter-atomic interactions, and brings quantum effects to the forefront (Payal & Pandey, 2022). Numerous electronic uses of nanotechnology, especially in semiconductors, transistors, and sensors, can result in notable improvements in efficiency and performance. For example, by absorbing a wider spectrum of the sun's light, including ultraviolet and infrared wavelengths, quantum dots improve the efficiency of solar cells. In a similar vein, carbon nanotube field-effect transistors (CNTFETs), which offer better performance and lower power consumption than traditional MOSFETs, are showing promise as a substitute for silicon-based transistors(Zhang et al., 2024).

Furthermore, because of its remarkable qualities, graphene is essential to many electrical components, including sensors, biosensors, field-effect transistors (GFETs), photovoltaic devices, and supercapacitors. In dye-sensitized solar cells, nanofibers—especially electro-spun ones—produce photoanodes with improved light-harvesting and electron-transport capacities. These developments based on nanomaterials are revolutionizing electronics (Mhetre et al., 2021).

7.3 Environmental Impact (Water Purification, Pollution Control)

One of the largest problems facing civilization today is environmental pollution, which has prompted the creation of innovative technologies to combat contamination of the air, water, and soil. Particulate matter, heavy metals, pesticides, fertilizers, herbicides, oil spills, hazardous gases, industrial effluents, sewage, and organic compounds are among the most prevalent contaminants (Roy et al., 2021). Since the majority of research focuses on the removal of heavy metals and chlorinated organic contaminants from water, metal-based nanomaterials hold the most promise among the options offered. Iron, titanium dioxide, silica, zinc oxide, carbon nanotubes, dendrimers, and polymers are examples of nanomaterials utilized for air purification, water treatment, and soil health restoration (Choi & Lee, 2022). Beyond environmental remediation, nanotechnology is also advancing renewable energy by making it more efficient and affordable, fostering a more sustainable future Metal and metal oxide nanomaterials, in particular, stand out as highly effective adsorbents due to their rapid kinetics and high adsorption capacity (Serrano & García-Martínez, 2009). Additionally, because of their adaptability, they can be widely used in aquatic systems applications both in situ and ex situ (Das et al., 2015). Numerous human activities, industrial processes, and other agricultural operations can result in the arbitrary release of chemicals and hazardous materials into the environment. At high concentrations, these pollutants disrupt ecosystems and pose significant threats to both humans and other organisms. To fight environmental contamination, numerous remediation technologies have been developed, with bioremediation standing out as a more effective and dependable solution due to its eco-friendly nature (Babatunde et al., 2020).

7.4 Food Industry

Nanotechnology has left its presence in almost all industries and grabbed the attention of people across the globe by providing endless applications. Such a transforming field is food and beverages, wherein nanotechnology is revolutionizing the packaging, testing, processing, and preservation of food (Biswas et al., 2022). This helps to decrease the chances of contamination risks along with preserving the chemical, physical, physiological, and mechanical properties of food products. There have been various applications of nanomaterials used to produce bio-based active packaging to help preserve food items. In addition, intelligent packaging systems, in combination with sensors, enable the detection of moisture, temperature, gases, and other food-related characteristics, contributing further toward enhanced food safety and quality (Choi et al., 2022). The bioactive compounds including the phenolic compounds, flavonoids, tannins, carotenoids, terpenoids, and probiotics have a complex nature and, thus, these are the challenges for incorporating them

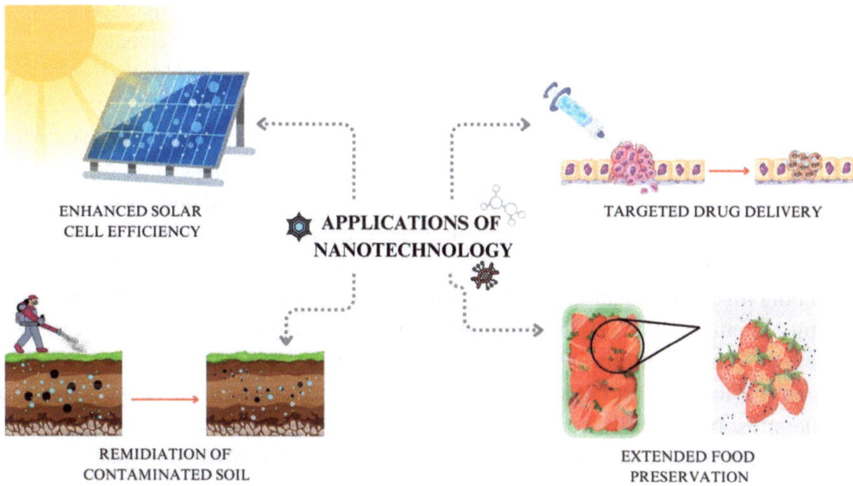

Fig. 3 Diverse applications of nanotechnology

into food, due to their low stability with the interaction of food matrices, bioavailability, phase separation, homogeneity, etc. (Bora et al., 2025). Furthermore, because of their versatility, they can be widely used in aquatic system applications both in situ and ex situ (Das et al., 2015). Chemicals and other dangerous elements are randomly released into the environment as a result of a variety of human activities, industrial processes, and other agricultural operations (Khan, Ezati, et al., 2025). Silica has been studied as an edible packaging material due to its non-toxic, non-irritant quality and its biocompatibility with food.

Nano-coatings have also been used to enhance food product packaging and increase shelf life (Fig. 3). A more advanced, long-term application involves the development and integration of smart nanosensors in plants, enabling them to detect abiotic stress through the targeted delivery of nanomaterials, potentially allowing plants to adapt to changing conditions (Shah et al., 2024).

8 Ethical and Safety Considerations

Nanotechnology holds incredible promise for transforming our lives, but we need to proceed thoughtfully. It brings up some big questions about safety and ethical implications. We need to consider these carefully now, so we can ensure nanotechnology truly benefits humanity and avoids unintended harm (Yunus et al., 2012). The science of manipulating matter at the atomic or molecular scale is a fast-growing field. This has brought new possibilities to science and technology. While these advances have produced many revolutionary breakthroughs, there are also very important ethical and safety concerns that have been raised (Srinivas et al., 2010). These issues include environmental and health risks of nanomaterials, since their potential toxicity and long-term impact on humans and ecosystems remain uncertain. Comprehensive regulatory frameworks and adherence to international standards make it even more complex for safe and responsible nanotechnology (Ray et al., 2009). Public perception and ethical considerations take precedence, because the acceptance of nanotechnology and its products within society rests on transparency and trust. Nano ethics highlights how it is essential to understand effects on humans, the environment, education, and society, considering the ethical and societal impact of nanotechnology. The need for well-informed guidelines for educating and safeguarding scientists, engineers, students, employees, legislators, consumers, and regulators involved in nanotechnology and related fields is emphasized in this paper (Khan, Asmatulu, et al., 2025).When developing and using nanotechnology, ethical and safety considerations are crucial. Risks to the environment and human health, such as the toxicity and long-term impacts of nanomaterials on ecosystems and people, must be carefully considered. Strong regulatory frameworks and adherence to international standards are necessary to overcome these issues and guarantee the appropriate and safe application of nanotechnology. Many nanomaterials can be harmful to the human body due to their exceptional qualities, mostly because of the way they are created and designed at the near-atomic size. Materials of this sort are mostly manufactured through advanced techniques, and specific regulations and guidelines to govern their manufacturing processes while ensuring the safety of workers, scientists, and engineers are not put in place yet (Asmatulu et al., 2013).

9 Future Perspectives in Nanotechnology

Nanotechnology is constantly evolving and bringing up ground-breaking discoveries in various domains, such as environmental science, energy, nanobiotechnology, and medicine. It has the power to transform several sectors of the economy and offer innovative solutions to some of the most significant issues confronting humanity (Malik et al., 2023a). While advancements like tailored drug delivery, nanoscale imaging, and sophisticated diagnostic tools are extremely promising, so are the new trends in nanomedicine. Similar to this, precision-based crop management and nanosensors are how nanobiotechnology transforms agriculture and food safety (Hua & Wu, 2018). It demonstrates the enormous potential of nanotechnology to address global problems Given the great potential of nanotechnology, a number of sustainability-related issues have surfaced, including the complexity of procedures and the impact of nano-waste contamination following the use of nanomaterials Scientists are investigating a range of green approaches to nanotechnology, including ecologically benign ways for synthesizing nanoparticles and biodegradable nanomaterials, to assure sustainable development (Srinivas et al., 2025). The creation of smart drug delivery systems, which are transforming the healthcare sector by overcoming the drawbacks of traditional drug administration techniques, is strongly linked to these developments (Liu et al., 2016). These systems provide increased safety, better patient compliance, and increased therapeutic efficacy by utilizing sustainable nanotechnology. This chapter explores a variety of intelligent medication delivery technologies, emphasizing how they could revolutionize healthcare while keeping an eye on environmentally responsible innovation developments in nanotechnology, microfluidics, bioelectronics, 3D printing, and systems analysis: stimuli-responsive, targeted delivery, and implanted devices (Boppana et al., 2024). In addition to the ethical, legal, and design issues that need to be resolved for broad adoption, there is promise for cost savings, improved patient outcomes, and individualized therapy. Moreover, optimization of these systems through AI and machine learning integration highlights their potential to completely transform the provision of healthcare. Also, nanotechnology holds promise for producing sophisticated sensors and strong, lightweight materials(Serrano et al., 2024). However, issues like ethical dilemmas, legal restrictions, and the possible health and environmental hazards connected to nanomaterials must be addressed if its transformative potential is to be completely realized. Combining artificial intelligence with nanomaterials enables the creation of "smart" materials. These cutting-edge materials have the rare capacity to react and adjust dynamically to changes in their surroundings, creating opportunities for various uses (Mhetre et al., 2021).

10 Conclusion

Nanotechnology refers to the handling of materials on the nanometer scale, with a size of between 1 and 100 nm. This chapter then explains the origin of nanotechnology. It has been traced from conceptual origins up to Richard Feynman's lecture in 1959 up to its more formal definition in 1974 by Norio Taniguchi. The innovation of scanning tunneling microscopy was in the 1980s and allowed atomic-level manipulation. Nanomaterials have distinct properties, such as a much higher surface area-to-volume ratio compared to their bulk form counterparts and quantum effects, which distinguish them from their "bulk" counterparts and make them highly applicable in wide fields. Nanomaterials of the class nanoparticles, nanotubes, nanowires, and quantum dots possess their unique structural properties and functionality. This chapter guides the reader toward thc basic principles of nanoscale science; it helps with the issues concerning quantum confinement, self-assembly, and the interfacial phenomena critical for the determination of electronic, optical, and mechanical properties of nanomaterials. Overall, this synthesis of nanomaterials can be categorized under two wide categories: top-down and bottom-up categories and several other methods, such as lithography, chemical vapor deposition, and green synthesis. For the desired properties to be achieved, characterization techniques such as SEM, TEM, and XRD become necessary for the examination of nanostructures with precise accuracy. Applications of nanotechnology are vast and include medicine, electronics, environmental remediation, and the food industry. Nanotechnology, although offering transformative improvements, is associated with ethical and safety concerns such as toxicity, environmental impact, and regulatory challenges. It concludes the chapter by putting forward the promises of nanotechnology, such as revolutionizing entire industries while underscoring responsible innovation.

References

Abid, N., Khan, A. M., Shujait, S., Chaudhary, K., Ikram, M., Imran, M., Haider, J., Khan, M., Khan, Q., & Maqbool, M. (2022). Synthesis of nanomaterials using various top-down and bottom-up approaches, influencing factors, advantages, and disadvantages: A review. *Advances in Colloid and Interface Science, 300*, 102597. https://doi.org/10.1016/j.cis.2021.102597

Ahire, S. A., Bachhav, A. A., Pawar, T. B., Jagdale, B. S., Patil, A. V., & Koli, P. B. (2022). The augmentation of nanotechnology era: A concise review on fundamental concepts of nanotechnology and applications in material science and technology. *Results in Chemistry, 4*, 100633. https://doi.org/10.1016/j.rechem.2022.100633

Ali, A., Chiang, Y. W., & Santos, R. M. (2022). X-ray diffraction techniques for mineral characterization: A review for engineers of the fundamentals, applications, and research directions. *Minerals, 12*(2), 205. https://doi.org/10.3390/min12020205

Alivisatos, A. P. (1996a). Semiconductor clusters, nanocrystals, and quantum dots. *Science, 271*(5251), 933–937. https://doi.org/10.1126/science.271.5251.933

Alivisatos, A. P. (1996b). Semiconductor nanocrystals: Shape and electronic properties. *Science, 271*(5251), 933–937.

An, T., Wen, J., Dong, Z., Zhang, Y., Zhang, J., Qin, F., Wang, Y., & Zhao, X. (2022). Plasmonic biosensors with nanostructure for healthcare monitoring and diseases diagnosis. *Sensors, 23*(1), 445. https://doi.org/10.3390/s23010445

Ashby, M. F., Evans, A. G., Fleck, N. A., & Hutchinson, J. W. (2007). *Mechanics of materials*. Butterworth-Heinemann.

Asmatulu, R., Zhang, B., & Asmatulu, E. (2013). Safety and ethics of nanotechnology. In *Nanotechnology safety* (pp. 31–41). Elsevier. https://doi.org/10.1016/B978-0-444-59438-9.00003-5

Babatunde, E., Denwigwe, I. H., Babatunde, O. M., Gbadamosi, S. L., Babalola, I. P., & Agboola, O. (2020). Environmental and societal impact of nanotechnology. *IEEE Access, 8*, 4640–4667. https://doi.org/10.1109/ACCESS.2019.2961513

Bahrulolum, H., Nooraei, S., Javanshir, N., Tarrahimofrad, H., Mirbagheri, V. S., Easton, A. J., & Ahmadian, G. (2021). Green synthesis of metal nanoparticles using microorganisms and their application in the agrifood sector. *Journal of Nanobiotechnology, 19*(1), 86. https://doi.org/10.1186/s12951-021-00834-3

Bhushan, B. (2017). *Introduction to nanotechnology*. John Wiley & Sons.

Bilardo, R., Traldi, F., Vdovchenko, A., & Resmini, M. (2022). Influence of surface chemistry and morphology of nanoparticles on protein corona formation. *Wiley Interdisciplinary Reviews Nanomedicine and Nanobiotechnology, 14*(4), 1788. https://doi.org/10.1002/wnan.1788

Binnig, G., Quate, C. F., & Gerber, C. (1986). Atomic force microscope. *Physical Review Letters, 56*(9), 930.

Binnig, G., & Rohrer, H. (1982). Scanning tunneling microscopy. *Helvetica Physica Acta, 55*(6), 726–735.

Binnig, G., Rohrer, H., Gerber, C., & Weibel, E. (1982). Surface studies by scanning Tunneling microscopy. *Physical Review Letters, 49*(1), 57–61. https://doi.org/10.1103/PhysRevLett.49.57

Biswas, R., Alam, M., Sarkar, A., Haque, M. I., Hasan, M. M., & Hoque, M. (2022). Application of nanotechnology in food: Processing, preservation, packaging and safety assessment. *Heliyon, 8*(11), e11795. https://doi.org/10.1016/j.heliyon.2022.e11795

Bokov, D., Jalil, A. T., Chupradit, S., Suksatan, W., Ansari, M. J., Shewael, I. H., Valiev, G. H., & Kianfar, E. (2021). Nanomaterial by sol-gel method: Synthesis and application. *Advances in Materials Science and Engineering*. https://doi.org/10.1155/2021/5102014

Bond, G. C. (1987). *Heterogeneous catalysis: Principles and Applications*. Clarendon Press.

Boppana, S. H., et al. (2024). Current approaches in smart nano-inspired drug delivery: A narrative review. *Health Science Reports, 7*(4), e2065. https://doi.org/10.1002/hsr2.2065

Bora, S., Revathi, V., Das, K., Katiyar, V., & Ghosh, T. (2025). Nanotechnology in functional food developments. In *Advancements in nanotechnology for food and packaging* (pp. 133–157). Elsevier. https://doi.org/10.1016/B978-0-443-21428-8.00006-0

Bruchez, M., Moronne, M., Gin, P., Weiss, S., & Alivisatos, A. P. (1998). Semiconductor nanocrystals as fluorescent biological labels. *Science, 281*(5385), 2013–2016. https://doi.org/10.1126/science.281.5385.2013

Chaudhary, P., & Janmeda, P. (2025). Industrial application of bio-nanomaterial in food industry. In N. Sharma, R. Sharma, & T. Dakal (Eds.), *Bio-nanomaterials in environmental remediation* (1st ed., pp. 191–224). Wiley. https://doi.org/10.1002/9783527848546.ch8

Chen, C. J. (2008). *Scanning tunneling microscopy: From the atom to the molecule*. Oxford University Press.

Chen, Z., Zhao, C., Zhou, X., Xiao, L., Li, Z., & Zhang, Y. (2023). A review of top-down strategies for the production of quantum-sized materials. *Small Science, 3*(12), 2300086. https://doi.org/10.1002/smsc.202300086

Choi, W. S., & Lee, H.-J. (2022). Nanostructured materials for water purification: Adsorption of heavy metal ions and organic dyes. *Polymers, 14*(11), 2183. https://doi.org/10.3390/polym14112183

Choi, Y. S., Cho, H., Choi, W.-G., Lee, S. S., Huh, K. M., Shim, M. S., Park, I. S., Cho, Y.-Y., Lee, J. Y., Lee, H. S., & Kang, H. C. (2022). Beyond hydrophilic polymers in amphiphilic polymer-based self-assembled NanoCarriers: Small hydrophilic carboxylate-capped disulfide drug delivery system and its multifunctionality and multispatial targetability. *Biomaterials, 280*, 121307. https://doi.org/10.1016/j.biomaterials.2021.121307

Christian Heyn, A., Radu, J. A., Vinasco, D., Laroze, R. L., Restrepo, V. T., Hieu, N. N., Phuc, H. V., Mora-Ramos, M. E., Ojeda, J. H., Morales, A. L., & Duque, C. A. (2021). Exciton states in conical quantum dots under applied electric and magnetic fields. *Optics & Laser Technology, 139*, 106953. https://doi.org/10.1016/j.optlastec.2021.106953

Corporativa, I. (n.d.). *Nanotechnology: A small solution to big problems*. Iberdrola. Retrieved January 29, 2025, from https://www.iberdrola.com/innovation/nanotechnology-applications

Das, S., Sen, B., & Debnath, N. (2015). Recent trends in nanomaterials applications in environmental monitoring and remediation. *Environmental Science and Pollution Research, 22*(23), 18333–18344. https://doi.org/10.1007/s11356-015-5491-6

Dash, P., & Panda, P. K. (2024). *Top-down strategies synthesis of 2D nanomaterial*. Wiley Online Library. https://doi.org/10.1002/9781394167883.ch1

Di Stefano, A. (2023). Nanotechnology in targeted drug delivery. *International Journal of Molecular Sciences, 24*(9), 8194. https://doi.org/10.3390/ijms24098194

Dippong, T. (2024). Innovative nanomaterial properties and applications in chemistry, physics, medicine, or environment. *Nanomaterials, 14*(2), 145. https://doi.org/10.3390/nano14020145

Dreaden, E. C., Alkilany, A. M., Huang, X., Murphy, C. J., & El-Sayed, M. A. (2012). The golden age: Gold nanoparticles for biomedicine. *Chemical Society Reviews, 41*(7), 2740–2779. https://doi.org/10.1039/C1CS15237H

Dresselhaus, M. S., Dresselhaus, G., & Eklund, P. C. (1996). *Science of fullerenes and carbon nanotubes*. Academic press.

Duan, X., Huang, Y., Cui, Y., Wang, J., & Lieber, C. M. (2001). Indium phosphide nanowires as building blocks for nanoscale electronic and optoelectronic devices. *Nature, 409*(6816), 66–69. https://doi.org/10.1038/35051047

Feynman, R. (2002). There's plenty of room at the bottom. In *Feynman and computation*. CRC Press.

Garcia, R. (2010). *Scanning probe microscopy: Principles and applications*. John Wiley & Sons.
Ghomrasni, N. B., Chivas-Joly, C., Devoille, L., Hochepied, J.-F., & Feltin, N. (2020). Challenges in sample preparation for measuring nanoparticles size by scanning electron microscopy from suspensions, powder form and complex media. *Powder Technology, 359*, 226–237. https://doi.org/10.1016/j.powtec.2019.10.022
Gomes, H. I. O., Martins, C. S. M., & Prior, J. A. V. (2021). Silver nanoparticles as carriers of anticancer drugs for efficient target treatment of cancer cells. *Nanomaterials, 11*(4), 964. https://doi.org/10.3390/nano11040964
Gun'ko, V. M., Turov, V. V., Krupska, T. V., & Borysenko, M. V. (2023). Surroundings effects on the interfacial and temperature behaviors of NaOH/water bound to hydrophilic and hydrophobic nanosilicas. *Journal of Colloid and Interface Science, 634*, 93–109., ISSN 0021-9797,. https://doi.org/10.1016/j.jcis.2022.12.027
Gutiérrez, J. L., Barrera, I. A., & Gómez, N. T. (2022). Nanoparticle detection on SEM images using a neural network and semi-synthetic training data. *Nanomaterials, 12*(11), 1818. https://doi.org/10.3390/nano12111818
Haleem, M., Javaid, R. P., Singh, S. R., & Suman, R. (2023). Applications of nanotechnology in medical field: A brief review. *Global Health Journal, 7*(2), 70–77. https://doi.org/10.1016/j.glohj.2023.02.008
Heinrich, A. J., Oliver, W. D., Vandersypen, L. M. K., Ardavan, A., Sessoli, R., Loss, D., Jayich, A. B., Fernandez-Rossier, J., Laucht, A., & Morello, A. (2021). Quantum-coherent nanoscience. *Nature Nanotechnology, 16*(12), 1318–1329. https://doi.org/10.1038/s41565-021-00994-1
Hua, S., & Wu, S. Y. (2018). Editorial: Advances and challenges in nanomedicine. *Frontiers in Pharmacology, 9*, 1397. https://doi.org/10.3389/fphar.2018.01397
Iijima, S. (1991). Helical microtubules of graphitic carbon. *Nature, 354*(6348), 56–58. https://doi.org/10.1038/354056a0
Jayawardena, H. S. N., Liyanage, S. H., Rathnayake, K., Patel, U., & Yan, M. (2021). Analytical methods for characterization of nanomaterial surfaces. *Analytical Chemistry, 93*(4), 1889–1911. https://doi.org/10.1021/acs.analchem.0c05208
Jeevanandam, J., Kiew, S. F., Boakye-Ansah, S., Lau, S. Y., Barhoum, A., Danquah, M. K., & Rodrigues, J. (2022). Green approaches for the synthesis of metal and metal oxide nanoparticles using microbial and plant extracts. *Nanoscale, 14*(7), 2534–2571. https://doi.org/10.1039/d1nr08144f
Joy, J., Krishnamoorthy, A., Tanna, A., Kamathe, V., Nagar, R., & Srinivasan, S. (2022). Recent developments on the synthesis of nanocomposite materials via ball milling approach for energy storage applications. *Applied Sciences, 12*(18), 9312. https://doi.org/10.3390/app12189312
Ke-Jun, W., Tse, E. C. M., Shang, C., & Guo, Z. (2022). Nucleation and growth in solution synthesis of nanostructures – From fundamentals to advanced applications. *Progress in Materials Science, 123*, 100821. https://doi.org/10.1016/j.pmatsci.2021.100821
Kensel, R., Sathish, N., & Saha, A. (2024). Chapter 3 - Importance of soft lithography. In P. V. Mohanan (Ed.), *Human organs-on-a-chip technology* (pp. 43–61). Academic Press. https://doi.org/10.1016/B978-0-443-13782-2.00015-2
Khan, P., Ezati, S. K., Tammina, Z. R., & Priyadarshi, R. (2025). Carbon-based nanomaterials in food packaging and preservation. In M. Alle, R. Priyadarshi, & S. D. Purohit (Eds.), *Functional nanomaterials and nanocomposites for biodegradable food packaging: Smart nanomaterials technology* (pp. 109–128). Springer. https://doi.org/10.1007/978-981-97-9084-5_5
Khan, W. S., Asmatulu, E., Zhang, B., & Asmatulu, R. (2025). Safety and ethics of nanotechnology. In *Nanotechnology safety* (pp. 43–57). Elsevier. https://doi.org/10.1016/B978-0-443-15904-6.00009-5
Klimov, V. I., Mikhailovsky, A. A., Xu, S., Malko, A., Hollingsworth, J. A., Leatherdale, C. A., Eisler, H.-J., & Bawendi, M. G. (2000). Optical gain and stimulated emission in nanocrystal quantum dots. *Science, 290*(5490), 314–317. https://doi.org/10.1126/science.290.5490.314
Kroto, W., Heath, J. R., O'Brien, S. C., Curl, R. F., & Smalley, R. E. (1985). C60: Buckminsterfullerene. *Nature, 318*(6042), 162–163. https://doi.org/10.1038/318162a0

Kumar Sahu, A., Katageri, I. S., Jadhav, M. P., & Vamadevaiah, H. M. (2017). A simple, rapid, and effective protocol for extraction of total plant proteins from cotton leaf. *International Journal of Current Microbiology and Applied Sciences, 6*(12), 2968–2975. https://doi.org/10.20546/ijcmas.2017.612.346

Kumari, A., & Yadav, S. K. (2024). Moving towards sustainable nanoengineered building materials with less energy consumption, energy and buildings. *International Standard Serial Number, 318*, 114475. https://doi.org/10.1016/j.enbuild.2024.114475

Langmuir, I. (1916). The constitution and fundamental properties of solids and liquids. Part I. Solids. *Journal of the American Chemical Society, 38*(11), 2221–2295. https://doi.org/10.1021/ja02268a002

Li, X., Liu, X., & Liu, X. (2020). Self-assembly of colloidal inorganic nanocrystals: Nanoscale forces, emergent properties and applications. *Chemical Society Reviews, 50*(3), 2074–2101. https://doi.org/10.1039/d0cs00436g

Li, Z., Guo, P., & Zhou, Y. (2021). Dip-pen nanolithography enabled functional nanomaterials and their applications. *Advanced Materials Technologies, 6*(4), 2000897. https://doi.org/10.1002/admt.202000897

Liu, D., Yang, F., Xiong, F., & Gu, N. (2016). The smart drug delivery system and its clinical potential. *Theranostics, 6*(9), 1306–1323. https://doi.org/10.7150/thno.1485

Liu, F., Li, P., An, H., Peng, P., McLean, B., & Ding, F. (2022). Achievements and challenges of graphene chemical vapor deposition growth. *Advanced Functional Materials, 32*(42), 2203191. https://doi.org/10.1002/adfm.202203191

Liu, H., & Grodzinski, P. (2021). Nanotechnology for cancer imaging: Advances, challenges, and clinical opportunities. *Radiology: Imaging Cancer, 3*(3), e200052. https://doi.org/10.1148/rycan.2021200052

Mahendra Kumar Yadav, A. S. (2023). Nanomaterials and nanotechnology: A comprehensive study of synthesis, characterization and multifaceted applications. *Biochemical and Cellular Archives, 23*(S1), 1641–1647. https://doi.org/10.51470/BCA.2023.23.S1.1641

Malatesta, M. (2021). Transmission electron microscopy as a powerful tool to investigate the interaction of nanoparticles with subcellular structures. *International Journal of Molecular Sciences, 22*(23), 12789. https://doi.org/10.3390/ijms222312789

Malik, S., Muhammad, K., & Waheed, Y. (2023a). Emerging applications of nanotechnology in healthcare and medicine. *Molecules, 28*(18), 6624. https://doi.org/10.3390/molecules28186624

Malik, S., Muhammad, K., & Waheed, Y. (2023b). Nanotechnology: A revolution in modern industry. *Molecules, 28*(2), 661. https://doi.org/10.3390/molecules28020661

Mekuye, B., & Abera, B. (2023). Nanomaterials: An overview of synthesis, classification, characterization, and applications. *Nano Select, 4*(8), 486–501. https://doi.org/10.1002/nano.202300038

Mhetre, H. V., Kanse, Y. K., & Patil, S. S. (2021). Nanomaterials: Applications in electronics. *The International Journal of Advanced Engineering and Nano Technology, 4*(6), 7–19. https://doi.org/10.35940/ijaent.D0464.094621

Miyazawa, T., Itaya, M., Burdeos, G. C., Nakagawa, K., & Miyazawa, T. (2021). A critical review of the use of surfactant-coated nanoparticles in nanomedicine and food nanotechnology. *International Journal of Nanomedicine, 16*, 3937–3999. https://doi.org/10.2147/ijn.s298606

Munir, H., Bilal, M., Mulla, S. I., Khan, H. A., & Iqbal, H. M. N. (2021). Plant-mediated green synthesis of nanoparticles. In *Advances in science, technology & innovation/advances in science, technology & innovation* (pp. 75–89). Springer. https://doi.org/10.1007/978-3-030-67884-5_4

Oke, J. A., & Jen, T.-C. (2022). Atomic layer deposition and other thin film deposition techniques: From principles to film properties. *Journal of Materials Research and Technology, 21*, 2481–2514. https://doi.org/10.1016/j.jmrt.2022.10.064

Osypiw, A. R. C., Lee, S., Jung, S., Leoni, S., Smowton, P. M., Hou, B., Kim, J. M., & Amaratunga, G. A. J. (2022). Solution-processed colloidal quantum dots for light emission. *Materials Advances, 3*(17), 6773–6790. https://doi.org/10.1039/d2ma00375a

Pandey, A., Dalal, S., Dutta, S., & Dixit, A. (2021). Structural characterization of polycrystalline thin films by X-ray diffraction techniques. *Journal of Materials Science Materials in Electronics, 32*(2), 1341–1368. https://doi.org/10.1007/s10854-020-04998-w

Patil, M., Mehta, D., & Guvva, S. (2008). Future impact of nanotechnology on medicine and dentistry. *Journal of Indian Society of Periodontology, 12*(2), 34. https://doi.org/10.4103/0972-124X.44088

Payal, & Pandey, P. (2022). Role of nanotechnology in electronics: A review of recent Developmentsand patents. *Recent Patents on Nanotechnology, 16*(1), 45–66. https://doi.org/10.2174/1872210515666210120114504

Pinto, R. T., Feu, K. S., Dalmaschio, C. J., Nascimento, A., & Lacerda, V. (2025). Oil recovery improvements based on Pickering emulsions stabilized by cellulose nanoparticles and their underlying mechanisms: A review. *ACS Omega, 10*, 3262–3282. https://doi.org/10.1021/acsomega.4c08428

Pokrajac, L., Nazar, L., Chen, Z., & Mitra, S. (2019). The waterloo institute for nanotechnology: Societal impact and a sustainable future. *ACS Nano, 13*(11), 12247–12253. https://doi.org/10.1021/acsnano.9b08356

Rao, A., Roy, S., Jain, V., & Pillai, P. P. (2022). Nanoparticle self-assembly: From design principles to complex matter to functional materials. *ACS Applied Materials & Interfaces, 15*(21), 25248–25274. https://doi.org/10.1021/acsami.2c05378

Rao, C. N. R., Kulkarni, G. U., Govindaraj, A., & Patnaik, S. K. (2005). *Nanomaterials*. Alpha Science Int'l Ltd..

Ratner, M. A., & Ratner, D. (2003). *Nanotechnology: A gentle introduction to the next big idea.* Prentice Hall Professional.

Ray, P. C., Yu, H., & Fu, P. P. (2009). Toxicity and environmental risks of nanomaterials: Challenges and future needs. *Journal of Environmental Science and Health, Part C, 27*(1), 1–35. https://doi.org/10.1080/10590500802708267

Roy, A., Sharma, A., Yadav, S., Jule, L. T., & Krishnaraj, R. (2021). Nanomaterials for remediation of environmental pollutants. *Bioinorganic Chemistry and Applications, 2021*, 1–16. https://doi.org/10.1155/2021/1764647

Saadbin, K., & Khalid Hossain, M. (2022). Classification and properties of nanoparticles. In S. M. Rangappa, T. G. Jyotishkumar Parameswaranpillai, Y. Gowda, M. Suchart Siengchin, & O. Seydibeyoglu (Eds.), *Woodhead Publishing Series in composites science and engineering, nanoparticle-based polymer composites* (pp. 15–54). Woodhead Publishing. https://doi.org/10.1016/B978-0-12-824272-8.00009-9

Salata, O. (2004). Applications of nanoparticles in biology and medicine. *Journal of Nanobiotechnology, 2*(1), 3. https://doi.org/10.1186/1477-3155-2-3

Serrano, G. R., & García-Martínez, J. (2009). Nanotechnology for sustainable energy. *Renewable and Sustainable Energy Reviews, 13*(9), 2373–2384. https://doi.org/10.1016/j.rser.2009.06.003

Serrano, R., et al. (2024). Artificial intelligence (AI) applications in drug discovery and drug delivery: Revolutionizing personalized medicine. *Pharmaceutics, 16*(10), 1328. https://doi.org/10.3390/pharmaceutics16101328

Shah, M. A., et al. (2024). Application of nanotechnology in the agricultural and food processing industries: A review. *Sustainable Materials and Technologies, 39*, e00809. https://doi.org/10.1016/j.susmat.2023.e00809

Silverman, M. (2017). *The basics of nanoscience and nanotechnology*. Springer.

Srinivas, K. R., Chandran, S. N., & Kumar, D. (2025). Future aspects in the field of bio-nanomaterials toward environmental assessment. In N. Sharma, R. Sharma, & T. Dakal (Eds.), *Bio-nanomaterials in environmental remediation* (1st ed., pp. 295–317). Wiley. https://doi.org/10.1002/9783527848546.ch12

Srinivas, P. R., et al. (2010). Nanotechnology research: Applications in nutritional sciences. *The Journal of Nutrition, 140*(1), 119–124. https://doi.org/10.3945/jn.109.115048

Sun, Q. (2022). The hydrophobic effects: Our current understanding. *Molecules, 27*(20), 7009. https://doi.org/10.3390/molecules27207009

Taniguchi, N. (1974). On the basic concept of 'nano-technology'. In *Proc Intl Conf Prod Eng Tokyo Part II*. Retrieved Jan 26, 2025, from https://cir.nii.ac.jp/crid/1572261550373135488

Thakur, V., Gautam, S., Gupta, M., & Chaudhary, R. (2025). X-ray tools for characterizing nanostructures: Computational aspects. In S. Gautam, J. P. Singh, D. P. Rai, A. Kumar, & A. Mishra (Eds.), *Micro and nano technologies, advances in nanostructures* (pp. 405–424). Elsevier. https://doi.org/10.1016/B978-0-443-13819-5.00015-X

Vaseashta, M., Vaclavikova, S., Vaseashta, G., Gallios, P. R., & Pummakarnchana, O. (2007). Nanostructures in environmental pollution detection, monitoring, and remediation. *Science and Technology of Advanced Materials, 8*(1-2), 47–59. https://doi.org/10.1016/j.stam.2006.11.003

Vijayaram, S., Razafindralambo, H., Sun, Y., Vasantharaj, S., Ghafarifarsani, H., Hoseinifar, S. H., & Raeeszadeh, M. (2023). Applications of green synthesized metal nanoparticles — a review. *Biological Trace Element Research, 202*(1), 360–386. https://doi.org/10.1007/s12011-023-03645-9

Williams, D. B., & Carter, C. B. (2009). *Transmission electron microscopy: A textbook for materials science* (p. 215). Springer.

Xia, Y., Yang, P., Sun, Y., Wu, Y., Mayers, B., Gates, B., Yin, Y., Kim, F., & Yan, H. (2003). One-dimensional nanostructures: Synthesis, characterization, and applications. *Advanced Materials, 15*(5), 353–389. https://doi.org/10.1002/adma.200390087

Yang, B., Gao, L., Xue, M., Wang, H., Hou, Y., Luo, Y., Xiao, H., Hu, H., Cui, C., Wang, H., Zhang, J., Li, Y., Xie, G., Tong, X., & Xie, Y. (2021). Experimental and simulation research on the preparation of carbon Nano-materials by chemical vapor deposition. *Materials, 14*(23), 7356. https://doi.org/10.3390/ma14237356

Yang, X., Yang, Y., Yu, H., & Zhou, Y. (2023). Self-assembled polymers for gastrointestinal tract targeted delivery through the Oral route: An update. *Polymers, 15*(17), 3538. https://doi.org/10.3390/polym15173538

Yunus, S., Harwin, A., Kurniawan, D. A., & Indarto, A. (2012). Nanotechnologies in water and air pollution treatment. *Environmental Technology Reviews, 1*(1), 136–148. https://doi.org/10.1080/21622515.2012.733966

Zanurin, N. A., Johari, J., Alias, H., Mas Ayu, N., & Redzuan, S. (2022). Izman, research progress of sol-gel ceramic coating: A review. *Materials Today: Proceedings, 48*(6), 1849–1854. https://doi.org/10.1016/j.matpr.2021.09.203

Zhang, Z., Wang, W., Rao, H., Pan, Z., & Zhong, X. (2024). Improving the efficiency of quantum dot-sensitized solar cells by increasing the QD loading amount. *Chemical Science, 15*(15), 5482–5495. https://doi.org/10.1039/D3SC06911G

Zhou, X., & Yu, G. (2021). Preparation engineering of two-dimensional heterostructures via bottom-up growth for device applications. *ACS Nano, 15*(7), 11040–11065. https://doi.org/10.1021/acsnano.1c02985

Introduction to Nanotechnology and Picotechnology

P. Saranraj, K. Gayathri, Shrikaant Kulkarni, M. Manigandan, and K. Kesavardhini

1 Introduction

Nanomaterials (NMs), Nanochemistry, Nanotechnology, and Nanoscience are just a few of the new terms in the nanotechnology field that are appearing in journals of materials science, engineering, and medicine, as well as popular books and newspapers. They are becoming familiar to a broad audience, including laypeople. NMs are materials with one or more nanometric dimensions (internal or external) (i.e., 1–100 nm) (Khan et al., 2019). NMs can be divided into different groups in the function of their dimensionality (0–3D), morphology (low- and high-aspect ratio), porosity (macro-, nano-, mesoporous), composition (carbon, inorganic, organic, and hybrids), origin (natural, incidental, engineered, bioinspired), phase (single phase, multiphase), and dispersion state (dispersed or aggregated) (Meftahi et al., 2022). NM synthesis methods tend to be divided into three main categories: (1) Liquid-Phase Synthesis Techniques (the most important), which involve biological or biochemical processes that occur in solution; (2) Gas-Phase Synthesis methods, which involve NM nucleation process growth, and deposition under vacuum or plasma phase limitations; and (3) Solid Phase Synthesizing, that includes grain formation and alloying. The nanoparticles are synthesis methods that can be further classified into three main types, physical, chemical, and biological processes that have been a vast improvement over time to include physiochemical and mechanochemical processes (Ijaz et al., 2020).

P. Saranraj (✉) · K. Gayathri · M. Manigandan · K. Kesavardhini
PG and Research Department of Microbiology, Sacred Heart College (Autonomous), Tirupattur, Tamil Nadu, India

S. Kulkarni
Sanjivani University, Kopargaon, India

Faculty of Business, Victorian Institute of Technology, Melbourne, VIC, Australia

S. Kulkarni (ed.), *A Journey from Nanotechnology to Picotechnology*, Lecture Notes in Nanoscale Science and Technology 47,
https://doi.org/10.1007/978-3-032-07228-3_2

Humans have always created and employed nanomaterials and nano-composites. According to recent research by Barhoum et al. (2019), NMs have evolved throughout history. Since the discovery of fire, Homo sapiens has produced carbon-based NMs. Carbon NMs (composite of fat, charcoal, and pigments) were discovered in hand stencils in Sulawesi caves (Indonesia) from 40,000 BCE (according to carbon dating results) (Barhoum et al., 2019). Ancient Egyptians (4000 years ago) used chemically synthesized PbS nanoparticles (~5 nm) to make hair colors. Inorganic NMs were first produced in Egypt and Mesopotamia in the fourteenth and fifteenth centuries BC, and other inorganic materials were used to make glassware (Barhoum et al., 2022). In 1959, Feynman first introduced the modern concepts of nanotechnology, nanoscience, and nanomaterials and their significance ("There is Plenty of Room at the Bottom") (Gaur et al., 2021). Nanoscience takes us further by focusing on studying and debating scientific phenomena and the fundamental properties of molecules and substances smaller than 100 nm. To date, there are several important applications of nanomaterials, such as water treatment, hydrogen production, fuel cells, batteries, sensors, diagnosis, and drug delivery (Gaur et al., 2021).

The development of numerous spectroscopic techniques accelerated research and innovation in nanotechnology. IBM researchers developed Scanning Tunneling Microscopy (STM) in 1982, and with STM, it became feasible to attain images of single atoms on "flat" (i.e., not a tip) surfaces (Nasrollahzadeh et al., 2019). Atomic Force Microscopy (AFM) was introduced in 1986 and has become the most significant Scanning Probe Microscopy technology. The aim to build hard discs with high storage density led to the measurements of electrostatic and magnetic forces. This led to the development of Kelvin-probe, Electrostatic, and Magnetic-force Microscopy (Butt et al., 2005). Nanotechnology is rapidly expanding and is currently integrated into nearly every field of materials chemistry. Nanotechnology continually evolves, and robust characterization and synthesis technologies are now available to create nanoparticles with more exact dimensions. Nanotechnology is an excellent instance of newly developed technology, with customized nanoparticles that can significantly improve product performance. Currently, nanomaterials find commercial roles in scratch-free paints, surface coatings, electronics, cosmetics, environmental remediation, sports equipment, sensors, and energy storage devices (Marques et al., 2021).

An emerging and much theoretical field is Picotechnology, the science of manipulating matter at the architecture of picometers (1 pm = 10^{-12} m), which is a thousand times smaller than nanometers, the unit in which nanotechnology works. At this resolution, control is shifted from atoms and molecules to subatomic particles, such as electrons and quarks. Picotechnology can realize even more abstract resolution, manipulating the position and energy states of electrons within atoms to modify material properties with extreme precision. It might even be possible to develop new material classes and invent devices with functionalities entirely governed by such configuration at the quantum level (Feynman, 1960). As Picotechnology essentially belongs to theoretical physics and advanced quantum mechanics, it operates on Nanotechnology and Atomic-level research elements. Advances in quasi-experimental techniques like Scanning tunneling, Quantum tunneling, and Atomic force microscopy have given scientists a foretaste for manipulating and visualizing matter with high precision—perhaps hinting at the potential possibility of such manipulations at the micrometer scale. It is an enticing thought that Picotechnology may someday be harnessed for advanced purposes, such as quantum processors with maximum efficiency, drug delivery at the subatomic scale, and advanced sensing and data storage techniques (Drexler, 1992).

Picotechnology is a highly challenging field for practical realization. Current tools and techniques contend with quantum uncertainty, thermal noise, and the fundamental laws of physics. As theoretical models develop further and experimental technologies ameliorate, what was once assigned to the domain of science fiction becomes ever more scientifically realistic. Picotechnology is, therefore, a probable and distant step into the realm of possibilities for the ultimate control of matter, which includes the idea that design and engineering at the level of the subatomic could bring about advance throughout the entire spectrum of scientific fields, from materials science to medicine and computing (Roco & Bainbridge, 2001).

2 Nanotechnology

2.1 Types of Nanomaterials

Nanomaterials (NMs) are divided into four basic categories regardless of location or chemical features.

2.1.1 Zero-Dimensional (Nanoparticles)

Zero-dimensional (0D) nanomaterials include quantum dots (carbon, graphene, and inorganic) and various spherical NMs (noble metallic substances, fullerenes, polymer compounds, metals organic framework, and up-and-down-conversion nanoparticles). Due to their chemical inertness, biocompatibility, optical stability, cell permeability, and wavelength-dependent photoluminescence, they are interesting for biomedical and optoelectronic applications (Sannino, 2021).

2.1.2 One-Dimensional (Nanowires, Nanotubes)

One-dimensional (1D) NMs (one dimension over 100 nm). Nanotubes, as well nanorods, nanowires, and nanofibers, are made of polymers, the element carbon, metals, and metal oxides, and thus, they emit electrons efficiently in low electric fields. Veils, matting, and nonwovens made of polymer nanofibers are examples of other 1D nanomaterials. Due to their important surface-to-volume ratio and tiny pores, they are used for filtration, decontamination, and catalysis and as scaffolds and super-absorbents for wound dressing and tissue engineering (Quan et al., 2019).

2.1.3 Two-Dimensional (Graphene, Nanofilms)

Any nanomaterial with a length and breadth in the order of 1–100 nm is defined as a two-dimensional (2D) nanomaterial, which, according to this definition, includes planar types of graphene (graphene oxide and reduced graphene oxide), transition metal dichalcogenides, metal oxides, silicates, graphitic carbon nitride, layered double hydroxides, black phosphorus, tin telluride nanosheets, antimonite, hexagonal boron nitride, Boron nanosheets, and other sheet-like nanomaterials. Their physicochemical, biological, and optical properties explain their uniform shape, surface charge, and high surface-to-volume ratio (Zhang, 2015).

2.1.4 Three-Dimensional (Bulk Nanomaterials)

Nanoporous powders, nanowire bundles, nanotube bundles, nanolayers, and nanostructured electrodes are three-dimensional (3D) nanomaterials (no dimensions are in the nanoscale range). Much research has gone into designing, manufacturing, and testing 3D nanomaterials for storage devices (supercapacitors and batteries) within applications such as wastewater treatment and electrochemical conversion. These novel nanomaterials are used extensively in biomedical devices, solar cells, microelectromechanical systems, and robotic technologies. The 3D printing of NMs will allow the development of Architectures with improved functional integration (Paramasivam et al., 2021).

2.2 *Synthesis and Fabrication Techniques*

2.2.1 Top-Down

Mechanical and chemical fabrication techniques are among the top-down approaches to NM synthesis. Top-down methods to produce NM involve the disaggregation of bulk materials to nanoparticles, which is usually done by attrition, milling, and etching. Top-down methods are more appropriate for thin film production and nanomaterials larger than 100 nm. They are used to fabricate well-integrated and connected electrical circuits. Integrated circuit fabrication can be classified as a top-down nanotechnology in which, in a tiny chip, small mechanical components (like levers, springs, and fluid channels) are placed. A significant disadvantage of current top-down methods is their limit in resolution (Ranjan et al., 2016).

2.2.2 Bottom-Up

Bottom-up methods create NMs with fewer defects than top-down methods. Atoms, molecules, and nanoparticles are starting materials in these processes to construct complex nanostructures. Ionic and molecular self-assembly is a bottom-up approach in which chemical or physical forces are used to gather individual blocks/molecules into larger structures through noncovalent bonds, such as hydrogen and ionic bonds, van der Waals forces, and water-mediated hydrogen bonding (Gherbi et al., 2022). Bottom-up techniques provide spatial control over individual blocks' features and composition. The required attributes determine the size of each building block. Wet chemical techniques (e.g., sol-gel, microemulsion) and co-precipitation are bottom-up approaches (Iqbal et al., 2012).

2.2.3 Biological and Green Synthesis Techniques

Biological synthesis methods include microorganism-assisted, biotemplate-assisted, and plant extract-assisted biogenesis, which are green synthesis approaches for nanoparticle synthesis (Jeevanandam et al., 2016). Plants use a variety of biomolecules, such as enzymes, vitamins, polysaccharides, organic acids, amino acids, and proteins, to synthesize nanoparticles in a metal-rich media. The first phase involves creating leaf tinctures using plant extracts and microorganisms, including fungi, bacteria, and lactobacilli. Plant extracts contain many active chemicals, including alkaloids, phenols, terpenoids, quinines, amides, flavonoids, proteins, and alcohols. Some contribute to the green synthesis of metallic nanoparticles by reducing metal cations to nanoparticles (i.e., reducing agents, such as flavonoids and phenols) and performing stabilization functions to avoid nanoparticle aggregation (Narayanan & Sakthivel, 2010).

Extraction is a separation method to isolate specific components of plants whereby the plant components (solid) are dissolved and entrapped into the solvent (liquid). The concentration of the plant extracts, pH, temperature, and contact duration influence the size and shape of nanoparticles and their yield. Compared with the other previously described methods, the use of plant extracts for nanoparticle synthesis is exciting because it is less expensive, more environmentally friendly, can be scaled up for industrial applications, and can be performed without high pressure, energy, temperature, or toxic chemicals (Mittal et al., 2013). Synthesis of collagen-based nanoparticles by a simple method involves mixing plant extracts with a metal salt solution at room temperature for a few minutes. The plant part for nanoparticle synthesis may be leaves, stems, stalks, or flowers. Some leaf extract examples include *Cinnamomum camphora*, *Alfalfa*, *Persimmon*, *Magnolia*, *Cycas*, *Geranium* leaves, *Aloe vera*, *Holarrhena antidysenterica*, etc., for the synthesis of Ag NPs (Rai & Yadav, 2013). Leaf extracts act safely as reducing and stabilizing agents. Future studies should investigate whether biomass or plant extracts can produce noble metals such as silver, gold, platinum, and palladium nanoparticles in particular shapes and sizes. Indeed, due to the incredible plant diversity, it is unknown whether they could help produce such nanoparticles (Thakkar et al., 2010).

The possible way bacteria and microorganisms can even be made nanofactories for green methods for making nanoparticles of diverse medical utility, like antibacterial application, antiproliferative, antioxidants, anticancer drugs, and delivery drug systems. For instance, it could include fungi, bacteria, and viruses that can produce metal and metal oxide nanoparticles using several methods, including metalliferous ion trapping outside or within microbial cells, enzymatic reduction, and capping. Since different microbes can mediate nanoparticle synthesis, it holds the utmost promise in the replacement of conventional methods of synthesis. For instance, the capacity of fungi to produce and release the specific enzymes needed for nanoparticle synthesis can be stimulated with a growth medium that contains the specific substrate for the fungal enzyme that will degrade silver and produce nanoparticles (Ovais et al., 2018).

Yeast cells bearing metal nanoparticles are "Green" hierarchical particle systems that do not require chemical stabilizers or dispersants (Li et al., 2018). Easy, non-toxic, and environmentally friendly approaches may be employed in the formation of viruses of nanoparticles. Plant viruses and Bacteriophages would qualify as promising candidates for the production of nanoparticles mainly due to their different geometries and homogeneous size distributions. Plant viral proteins called capsid proteins may be employed to create novel NMs that can assemble themselves into viruses with varying coat components of protein or as internal and exterior diameters. Particularly, the nanomaterial is enveloped by the virus and thus is protected (Iravani & Zolfaghari, 2021).

2.2.4 Template-Based Methods

The template technique has received substantial attention in the synthesis of different nanostructures. Sol-gel technology can easily produce mesoporous oxides with well-defined and organized porous structures. Another recent example is the fabrication of poly (Nisopropylacrylamide) nanogels using Au nanoparticles as templates. However, template-based manufacturing can be uniformly applied to make free-standing, non-oriented, or oriented nanowires, nanorods, or nanotubes, which are also called arrays of nanorods, nanowires, and nanotubes. The most commonly used and commercially available templates are Anodized Alumina Membrane (AAM) and radiation track-etched Polycarbonate (PC) membranes (Singh & Lyon, 2007). Other membranes under this category include nanochannel arrays on glass, radiation track-etched mica particles, mesoporous materials, porous silicon formed by electrochemical etching silicon wafer, zeolites, and carbon nanotubes. Bio-templates are also considered for nm-scale wires and nanotubes: Cu, Ni, Co, and Au nanowires. Huge alumina membranes with vertical parallel pores are produced on anodized aluminum sheets in sulfuric, oxalic, or phosphoric acid solutions. These holes can be consolidated into perfect hexagonal lattices, resulting in 10^{11} pores/cm^2

pore densities. The pore sizes range from 10 nm to 100 μm. PC membranes are made by blasting a nonporous polycarbonate sheet 6–20 μm thick with fissionable nuclear fragments to create damage tracks: these tracks are chemically etched into porosity. The pores formed by radiation track etching membranes are uniform in size, as small at 10 nm, and are dispersed randomly relative to one another.

Other membranes under this title include nanochannel arrays on glass, radiation track-etched mica particles, mesoporous material, porous silicon made by electrochemically etching a silicon wafer, zeolites, and carbon nanotubes. Bio-templates are also considered for making nm-scale wires and nanotubes: Cu, Ni, Co, and Au nanowires. These giant alumina membranes with vertical pores were made by anodizing aluminum sheets in sulfuric, oxalic, or phosphoric acid solutions. These holes may be compacted into perfect hexagonal lattices yielding 10^{11} pores/cm^2 pore densities. The pore sizes may range from 10 nm up to 100 μm. Blasting a nonporous polycarbonate sheet of 6–20 μm thickness with fissionable nuclear fragments would provide damage tracks that could then be chemically etched into porosity. The pores created are tiny (as small as 10 nm), uniformly sized, and randomly distributed within these radiation track etch membranes. Pore densities can reach up to 10^9 pores/cm^2 (Ajayan et al., 1995).

2.3 *Properties of Nanomaterials*

2.3.1 Mechanical Properties

Mechanical properties like that of a nanoparticle will be the mechanical properties of these nanostructures in terms of their response to various conditions and applied external loads. The maximum values obtained on different mechanical properties for metal nanoparticles as, maximum the degree of hardness, fracture toughness, fractured strength, ultimate tensile strength, and impact strength are given. Nanomaterials with metal nanoparticles are considered to have greater strength and fracture toughness than monolithic Al_2O_3. The possibility of nanoparticles probably causes this effect due to metal particles' pinning and grain growth in Al_2O_3. The matrix is hindered so that the final composite has a grain size less than the monolithic Al_2O_3, resulting in significantly smaller grains and, consequently, better mechanical properties. Interestingly, the hardness of nano-composites with nano-Cu is lower than that of monolithic Al_2O_3. However, the nano-composites with nano-Ni-Co are higher than Al_2O_3 because it is pretty likely that Cu is softer than Al_2O_3. With the addition of nano-Cu, the nano-composites' hardness is reduced to a certain extent. Correspondingly, meta-Ni-Co is harder than Al_2O_3. Thus, a slight improvement will give nano-Ni-Co toughness to nano-composites. It states that these hybrid composites have higher Vickers hardness, impact strength, and ultimate tensile strength when compared to the singular reinforced composites due to the combined effect of SiC and B4C (Oh et al., 2006).

2.3.2 Optical Properties

The optical properties of nanomaterials—including reflection, transmission, absorption, and light emission—are entirely dependent upon their electronic structure, which varies considerably with shape, as the electronic structure of nanomaterials is closely related to the presence and excitation of surface atoms. Two factors, namely, increased energy level spacing (quantum effect) and surface plasmon resonance, profoundly contribute to the size-dependent optical properties of nanomaterials (Mittal & Banerjee, 2016). The nanometric scale of particles is reduced to 10 nm wherein significant changes would occur in their optical characteristics. Nanoparticles are so small that their electrons do not possess as much freedom from the bulk material. Electrons are therefore isolated and interact differently with light due to quantum confinement as a model that governs their movement. The reduction of dimension in the electronic structure of nanoparticles shows the most prominent influence on the energies of the highest occupied molecular orbital, which is called the valence band. In contrast, the highest empty molecular orbital is described as a conduction band. Their movements within the two bands will exhibit optical properties such as emission and absorption. This optical bandgap increases with the decrease in particle size, especially for the semiconductor nanomaterials (Karak, 2019).

2.3.3 Magnetic Properties

Nanomaterials exhibit several peculiar magnetic properties, which are generally not found in bulk materials and are primarily attributed to surface or interfacial effects caused by symmetry breaking, co-existing electromagnetic environment, charge transfer, and electromagnetic interaction. Moreover, because of the nanostructure's sizeable surface-area-to-volume ratio, the constituent atoms experience different magnetic coupling with neighboring atoms compared to bulk materials (Mohan Bhagyaraj & Oluwafemi, 2018). The electromagnetic domain distribution of ferromagnetic materials also influences the magnetization of fine particles. The conventional view is that ferromagnetic nanoparticles below the critical size are "single-domain particles" magnetized uniformly instead of larger particles, which contain several domains showing a non-uniform magnetic structure. The critical size of the single domain is affected by several factors, such as the value of the magnetic saturation, the strength of the crystal anisotropy and exchange forces, surface or domain-wall energy, and the shape of the particles (Akbarzadeh et al., 2012).

2.3.4 Electrical Properties

Properties such as conductivity or resistivity are considered electrical properties. Electrical properties are pretty much the last nanoscopic properties to be measured. The electrical properties of nanoparticles are primarily concerned with the mobility of charge carriers. When a substance is reduced to the nanometric scale, quantum-size effects and quantum confinement phenomena will probably come into play. Due to their electron energy quantization, some conductive metallic nanoparticles can become nonconductive if exposed to a specific voltage. For example, metals used as conductors, such as copper, can lose their conductivity if their size is reduced to a few nanometers. In contrast, under size reduction, insulation materials such as silicon dioxide will lose their insulating properties and become conductive (Shi et al., 2014). Increased interfacial atoms orions, sinking many defects near the grain boundaries, grain size, and doping impurities largely control nanomaterials' dielectric properties, resistivity, and electrical transports (Ramasamy & Purniah, 2001). The electrical conductivity of nanomaterials is generally lower than that of bulk materials due to the increase of the band gap energy and a decrease in particle size of the nanomaterials, especially for semiconductor nanomaterials (Singh et al., 2018).

2.4 Applications of Nanotechnology

2.4.1 Applications of Nanotechnology in Animal Science

Nanotechnology applications in animal research include the development of pharmaceuticals and nutritional release systems, the acceleration of illness detection and treatment, the development of nucleic acid releasing systems, and the manufacture of animal vaccinations. Nowadays, silver nanoparticles are employed to improve the cleanliness of livestock and poultry breeding grounds as they are effective antibacterial and antimicrobial agents (Archana et al., 2015).

2.4.2 Applications of Nanotechnology in Agricultural Science

Current nanotechnology holds the potential to transform the agriculture and food industries throughout the world. Examples of potential applications of nanotechnology in the agricultural and food industries include modern drug release systems for treating diseases, new cell and molecular biology tools, biosecurity and health assurance of agricultural and food products, the production of new systems to identify pathogenic agents, and environmental protection (Mousavi & Rezaei, 2011).

2.4.3 Nanotechnology Applications in the Nutrition Industry

Nanotechnology and allied sectors have recently expanded into a wide range of nutritional studies relating to dairy and food manufacturing, preservation, packaging, and functional food development. Furthermore, food and dairy producers, agricultural producers, and consumers are the most affected by the advancement of nanotechnology. In the food packaging industry, nanotechnology plays an important role not only in packaging materials but also in improving food safety, alerting consumers to concerns over the presence of contamination and bacteria within foods, producing stronger flavors and color quality, repairing tears in packaging, and releasing preservatives to extend packaged shelf life (Charych et al., 1996).

2.4.4 Applications of Nanotechnology in Pharmaceutical and Medicinal Sciences

Combating and stopping cancer cell development (metastasis) is one of nanotechnology's most essential applications. In reality, metastasis, defined as the spread of cancer cells from their central mass to other parts of the body via the production of secondary tumors, can be identified as the most challenging obstacle in cancer treatment. This mechanism occurs in the vast majority of malignancies. The challenges of treating metastases include their small size, high multiplicity, and dispersion to diverse organ environments (Gobin et al., 2005).

2.4.5 Nanotechnology Applications in Water Treatment

This modern technology can improve treatment capacity and facilitate the economic utilization of unconventional water sources, thus expanding the water supply. Although the application of nanotechnology opens the way for new cost-effective approaches to increasing water capacity and thus improving human life and other life cycles on the Earth, some challenges behind its application mainly refer to human health risks (Bisquert, 2003).

3 Picotechnology

3.1 Principles of Picotechnology

Picotechnology is essentially the leading edge of the precision engineering market at the Picometer scale, meaning 10–12 m, and hence is focused on controlling and manipulating atomic, subatomic, and quantum phenomena. Evolutionarily, after the nanoprocessing, Picotechnology dreams of turning the tables in material science, quantum computers, energy systems, and even fundamental physics through direct control at the levels of electron orbitals and nuclear configurations (Calegari, 2016).

3.1.1 Atomic and Subatomic Manipulation

Manipulation of atoms and subatomic structures in Picotechnology exceeds the limits of the accepted fields of nanoscience in that it deals with structures and their interactions under 100 pm. This includes the direct control of the atomic nucleus, electron configurations, and fundamental particles such as protons and neutrons. While nanotechnology usually concerns itself with manipulating matter on larger molecules (0.1–100 nm), Picotechnology addresses manipulation at the scale of electron clouds and inter-atomic forces. The Scanning Tunneling Microscope (STM) and Atomic Force Microscope (AFM) have provided some of the foundations for these capabilities, but advances in Picotechnology will most likely require more precise and non-invasive control of manipulation through quantum fields or ultrafast laser pulses. The specific aim is for deterministic structuring of matter at the most fundamental scale, leading to programmable matter, nuclear spin-based data storage, and highly high-precision sensors (Feynman, 1960).

3.1.2 Quantum Mechanical Foundations

Picotechnology operates in a regime where quantum mechanical effects are relevant and governing. Fundamental principles may include quantum tunneling, wavefunction collapse, entanglement, and the Heisenberg uncertainty principle, to name a few. At the picometer scale, one controls systems by controlling the electron wavefunctions, spin states, and probabilistic behaviors imparted by quantum mechanics. Recent quantum dots and qubits advances have used such effects in quantum computing applications. In Picotechnology, such mechanisms would extend to material structuring and data encoding much deeper than current state-of-the-art atomic-scale systems. Quantum Electrodynamics, "QED," and Quantum Chromodynamics, "QCD," may provide a theoretical background for all interactions at the subatomic level if Picotechnology aspire to manipulate forces such as the strong nuclear force in advanced energy applications or particle structuring (Eigler & Schweizer, 1990).

3.1.3 Molecular Precision Control beyond Nanoscales

Picotechnology extends the molecular precision capabilities from nanometric boundaries to enable rearrangement and bonding of atoms with accuracy well within sub-angstrom borderlines—picometer level. These include precisely locating atoms in molecules to design their electronic, optical, or mechanical properties and correcting deviations in atomic structure at femtosecond and picometer levels. One potential direction is the attosecond laser pulse, which can be coupled with that of electron movements to make this control dependent on monitoring and manipulating dynamic events at the nanometer level, paving the way to that previously unknown spatial and temporal resolution for forming and breaking bonds (Calegari, 2016). In addition, coherent control techniques can push molecular reactions along predetermined pathways via the evolution of their quantum states. This type of control has profound implications for medicine (molecular target therapy), material science (superconductors and metamaterials), and computing (memory and logic gates at the atomic scale) (Joannopoulos et al., 2008).

3.2 *Techniques and Approaches in Picotechnology*

Picotechnology concerns tools and techniques to manipulate and study matter close to the atomic level and subatomic particles. Unlike in nanotechnology, which mainly works with molecules, processes involved in Picotechnology must involve instrumentation capable of spatial resolution down to the picometer (10^{-12} m) and temporal resolution down to the picosecond (10^{-12} s). The present section considers the fundamental methodologies that practically realize these precision levels.

3.2.1 Atomic Engineering Techniques

Atomic engineering is probably the most important component in Picotechnology, which includes controlled positioning, removal, or modification of individual atoms on surfaces or in materials. One of the most promising tools from this area is the Scanning Tunneling Microscope (STM), which relies on Quantum Tunneling to image and manipulate atomic structures with Picometer scale precision. Using STM, atomic memory and logic gates have been realized, which attest to the feasibility of atomic circuitry at a large scale (Kalff, 2016). Based on cutting-edge techniques such as Atomic Layer Deposition (ALD) and Molecular Beam Epitaxy (MBE), ultra-thin films and interfaces are produced with atomic precision, which are imperatives for constructing quantum devices and next-generation semiconductors. Electron beam-induced manipulation also allows for structural reconfiguration at atomic sites by focusing high-energy electrons onto specific lattice points, thus allowing localized engineering of material properties (Krivanek, 2010).

3.2.2 High-Resolution Imaging

Developing Picotechnology without ultrahigh-resolution imaging systems capable of resolving structures at or below the picometer scale is impossible. For instance, it has reached under 50 picometers with aberration correction in Transmission Electron Microscopy (TEM), making it apt to image atomic columns and slight lattice distortion detection (Batson et al., 2002). nc-AFM, on the other hand, addresses interatomic forces, not electronic conductivity interference, generating atomic resolution images for conductive and insulating surfaces. In contrast, Cryo-EM has been conventionally known for its applications in biology but is now finding more application areas in materials science as a visualization tool for fragile molecular assemblies close to their atomic resolution, allowing applications for Picotechnology tools to be applied across disciplines. Such techniques provide the necessary capabilities for direct observation of real-time, atomic-level fabrication and development monitoring and refinement (Kühlbrandt, 2014).

3.2.3 Laser-Based Picosecond Technologies

Laser technologies working in picosecond and femtosecond regimes afford unprecedented control over ultrafast phenomena at the atomic and electronic levels. The essential method is pump-probe spectroscopy: a stimulus pulse applies an initial exciting action on the probe material, and the subsequent recording of its evolution allows for the femtosecond resolution study of transient states. This technique has illuminated dynamic chemical reactions and ultrafast phase transitions in solids. Furthermore, attosecond pulse generation makes eavesdropping electron motions around nuclei possible and temporally resolves below the scale of orbital motions for electrons. Facilities like these are instrumental in shaping molecular behaviors, starting reactions, and real-time programming energy states. Laser ablation and nanolithography complement the picometer level structuring, enabling precision cutting, etching, or altering the material surface at atomic dimensions (Zewail, 2000).

3.2.4 Quantum Tunneling and Picometer Scale Probes

The principle of Quantum Tunneling—whereby particles manage to cross energy barriers that they should not be able to cross according to classical theory—is fundamental to all measurement and manipulation of matter at the scale of Picotechnology. The Scanning Tunneling Microscope (STM) is a prime example: it resolves individual atoms using variations in Tunneling current. Moreover, it is possible through Tunneling Spectroscopy to study the local electronic density of states with subatomic resolution, which can map quantum well structures, defect states, and superconducting gaps from the perspective of quantum circuitry. Devices like the single-electron transistor and probes based on quantum dots offer unprecedented sensitivity to charge or spin flip, even at the level of single changes. Algorithms that make such probes usually coupled to artificial intelligence are used for manipulation systems in Picotechnology that are generally autonomous and adaptive under environmental fluctuation and quantum uncertainty (Tersoff & Hamann, 1985).

4 Future of Nanotechnology

The potential worldwide applications of nanotechnology promise many benefits in medicine, electronics, environmental protection, energy, and so on. With the manipulation of matter at the atomic and molecular levels creating waves of advancement in nanotechnology, a radical change is expected in drug delivery systems that achieve localized therapy to reduce side effects and increase treatment efficacy. There is much hope in detecting diseases early with the help of nanoscale devices that enhance prognosis with reduced healthcare costs. Electronics miniaturization of components would allow faster and energy-efficient computing devices for AI and quantum computing applications. Graphene and carbon nanotubes are nanomaterials incorporated into lightweight high-strength materials for the aerospace, automotive, and construction industries. Many environmental applications include the development of nanoscale filters for water purification, pollution control, and even carbon capture technologies. In the energy sector, nanotechnological solutions may improve solar panels and develop high-capacity batteries, reducing dependence on fossil fuels while enhancing global sustainability efforts. Like all emerging technologies, however, the emergence of nanotechnology brings safety, ethical, and regulatory conundrums (Roco et al., 2011). The long-term effects, largely unknown, must be analyzed and evaluated from applicable scientific principles, and international regulatory perspectives must be made with the utmost seriousness. Truly interdisciplinary actions will come into play for the responsible use of nanotechnology. As this technology matures, education and public awareness must develop faster than or commensurate with technological advancement. Hence, there is societal acceptance and informed decision-making regarding nanotechnology. Looking forward, nanotechnology is not merely a tool of the future; it is set to become the bedrock for an entirely new way of interacting with the world at the fundamentally smallest level (Drexler, 2013).

5 Future of Picotechnology

While Nanotechnology has begun to revolutionize industries, even smaller frontiers for the future could be created, called Picotechnology. This is the manipulation of matter at the picometer scale (1-trillionth of a meter), and it is an emerging theoretical thing that would afford levels of control over atomic nuclei and subatomic particles that have never previously existed, thereby ushering in an unprecedented era of scientific engineering precision. However, applications of Picotechnology will, for now, remain largely speculative: potentially in scale and impact dwarfing whatever application Nanotechnology ever achieved. Its applications may range from real-time repair of genetic mutations to direct manipulation of the atomic structure of therapeutic proteins and enzymes, revolutionizing the treatment of genetic disorders and cancers. In materials sciences, Picotechnology heralds the production of ultra-high-strength self-assembled materials with programmable properties that current nanomaterials cannot rival. Statistics on quantum computing developments are exciting: Picotechnology could fine-tune quantum states and particles for new computing power and secure communication systems (Hall & Storrs, 2005). The elegant manipulation of atomic bonds may lead to revolutionary systems in fusion energy or zero-loss energy transmission. On the road to future Picotechnology, serious stumbling blocks stand in the way—scientific and ethical considerations. Manipulating such particles demands Hiro of advancement in particle physics, quantum mechanics, and atomic-scale instrumentation, which are some decades are done. Then again, this power warrants thorough global monitoring to contain the potential threats of weaponization and irreversible environmental harm. Even so, Picotechnology, in its infancy, is poised to become the next frontier for innovation, aiming to extend the limits of what humanity can build, heal, and comprehend at the more minor scales of nature (Mishra, 2014).

6 Conclusion

Nanotechnology and Picotechnology are at the forefront of materials science, offering transformative potential across various industries, including medicine, electronics, energy, and environmental science. Both fields manipulate matter at tiny scales—nanotechnology at the atomic and molecular levels (1–100 nm) and Picotechnology at sub-nanometer levels (below 1 nanometer), involving protons, neutrons, and electrons. Nanotechnology's tangible impacts include advancements in drug delivery, nanosensors, and nanoelectronics, showcasing nanomaterials' enhanced performance and efficiency. With applications from targeted cancer therapies to strong materials and renewable energy technologies, paradigms in problem-solving are shifting. Picotechnology, though still theoretical, promises deeper manipulation of atomic particles, potentially revolutionizing our understanding of physics, chemistry, and biology, enabling atomic engineering and quantum computing. However, both fields face challenges, including technical limitations, health risks, and ethical implications. Establishing robust regulatory frameworks and promoting responsible research is essential to ensure that these technologies benefit humanity. In conclusion, exploring Nanotechnology and Picotechnology signifies a new era of scientific progress, expanding technological limits and reshaping our relationship with matter.

References

Ajayan, P. M., Stephan, O., Redlich, P., & Colliex, C. (1995). Fullerene-like tubules of carbon. *Nature, 375*, 564.

Akbarzadeh, A., Samiei, M., & Davaran, S. (2012). Magnetic nanoparticles: Preparation, physical properties, and applications in biomedicine. *Nanoscale Research Letters, 7*, 144–150.

Archana, D., Singh, B. K., Dutta, J., & Dutta, P. K. (2015). Chitosan-PVP-nano silver oxide wound dressing: In vitro and in vivo evaluation. *International Journal of Biological Macromolecules, 73*, 49–57.

Barhoum, A., Bechelany, M., & Makhlouf, A. S. H. (2019). *Handbook of nanofibers*. Springer International Publishing.

Barhoum, A., García-Betancourt, M. L., Jeevanandam, J., Hussien, E. A., Mekkawy, S. A., Mostafa, M., Omran, M. M., Abdalla, M. S., & Bechelany, M. (2022). Review on natural, incidental, bioinspired, and engineered nanomaterials: History, definitions, classifications, synthesis, properties, market, toxicities, risks, and regulations. *Nano, 12*, 177–185.

Batson, P. E., Dellby, N., & Krivanek, O. L. (2002). Sub-ångstrom resolution using aberration corrected electron optics. *Nature, 418*, 617–620.

Bisquert, J. (2003). Chemical capacitance of nanostructured semiconductors: Its origin and significance for nanocomposite solar cells. *Physical Chemistry, 5*(24), 5360–5364.

Butt, H. J., Cappella, B., & Kappl, M. (2005). *Surface Science Reports, 59*, 1–152.

Calegari, F. (2016). Advances in attosecond science. *Journal of Physics B: Atomic, Molecular and Optical Physics, 49*(6), 062001.

Charych, D., Cheng, Q., Reichert, A., Kuziemko, G., Stroh, N., Nagy, J., Spevak, W., & Stevens, R. (1996). A 'litmus test' for molecular recognition using artificial membranes. *Chemistry & Biology, 3*, 113–120.

Drexler, K. E. (1992). *Nanosystems: Molecular machinery, manufacturing, and computation*. John Wiley and Sons.

Drexler, K. E. (2013). *Radical abundance: How a revolution in nanotechnology will change civilization*. PublicAffairs.

Eigler, D. M., & Schweizer, E. K. (1990). Positioning single atoms with a scanning tunnelling microscope. *Nature, 344*, 524–526.

Feynman, R. P. (1960). There is plenty of room at the bottom. *Engineering and Applied Science at Caltech, 23*(5), 22–36.

Gaur, M., Misra, C., Yadav, A. B., Swaroop, S., Maolmhuaidh, F. Ó., Bechelany, M., & Barhoum, A. (2021). Biomedical applications of carbon nanomaterials: Fullerenes, quantum dots, nanotubes, nanofibers, and graphene. *Materials, 14*, 5978.

Gherbi, B., Laouini, S. E., Meneceur, S., Bouafia, A., Hemmami, H., Tedjani, M. L., Thiripuranathar, G., Barhoum, A., & Menaa, F. (2022). Effect of pH value on the bandgap energy and particles size for biosynthesis of ZnO nanoparticles: Efficiency for photocatalytic adsorption of methyl orange. *Sustainability, 14*, 11300.

Gobin, A. M., O'Neal, D. P., Watkins, D. M., Halas, N. J., Drezek, R. A., & West, J. L. (2005). Near infrared laser-tissue welding using nanoshells as an exogenous absorber. *Lasers in Surgery and Medicine, 37*, 123–129.

Hall, & Storrs, J. (2005). *Nanofuture: What is next for nanotechnology*. Prometheus Books.

Ijaz, I., Gilani, E., Nazir, A., & Bukhari, A. (2020). Detail review on chemical, physical and green synthesis, classification, characterizations and applications of nanoparticles. *Green Chemistry Letters and Reviews, 13*, 59–81.

Iqbal, P., Preece, J. A., & Mendes, P. M. (2012). Nanotechnology: The "top-down" and "bottom-up" approaches. In P. A. Gale & J. W. Steed (Eds.), *Supramolecular chemistry*. John Wiley and Sons.

Iravani, S., & Zolfaghari, B. (2021). Plant viruses and bacteriophages for eco-friendly synthesis of nanoparticles: Recent trends and important challenges. *Comments on Inorganic Chemistry, 42*, 226–248.

Jeevanandam, J., Chan, Y. S., & Danquah, M. K. (2016). Biosynthesis of metal and metal oxide nanoparticles. *ChemBioEng Reviews, 3*, 55–67.

Joannopoulos, J. D., Johnson, S. G., Winn, J. N., & Meade, R. D. (2008). *Photonic crystals: Molding the flow of light* (2nd ed.). Princeton University Press.

Kalff, F. E. (2016). A kilobyte rewritable atomic memory. *Nature Nanotechnology, 11*, 926–930.

Karak, N. (2019). Fundamentals of nanomaterials and polymer nanocomposites. In *Nanomaterials and polymer nanocomposites* (pp. 1–45). Elsevier.

Khan, I., Saeed, K., & Khan, I. (2019). Nanoparticles: Properties, applications and toxicities. *Arabian Journal of Chemistry, 12*, 908–931.

Krivanek, O. L. (2010). Atom-by-atom structural and chemical analysis by annular dark-field electron microscopy. *Nature, 464*, 571–574.

Kühlbrandt, W. (2014). The resolution revolution. *Science, 343*(6178), 1443–1444.
Li, J., Ma, G., Liu, H., & Liu, H. (2018). Yeast cells carrying metal nanoparticles. *Materials Chemistry and Physics, 207*, 373–379.
Marques, A. C., Vale, M., Vicente, D., Schreck, M., Tervoort, E., & Niederberger, M. (2021). Porous silica microspheres with immobilized titania nanoparticles for in-flow solar-driven purification of wastewater. *Global Challenges, 5*, 2000116.
Meftahi, A., Momeni Heravi, M. E., Barhoum, A., Samyn, P., Najarzadeh, H., & Alibakhshi, S. (2022). Cellulose nanofibers. In *Handbook of nanocelluloses* (Vol. 12, pp. 233–262). Springer.
Mishra, S. (2014). Picotechnology: Theoretical concepts and future applications. *International Journal of Scientific Research in Science and Technology, 1*(1), 45–49.
Mittal, A. K., & Banerjee, U. C. (2016). Current status and prospects of nanobiomaterials in drug delivery. In *Nanobiomaterials in drug delivery. Applications of nanobiomaterials* (pp. 147–170). Apple Academic Press.
Mittal, A. K., Chisti, Y., & Banerjee, U. C. (2013). Synthesis of metallic nanoparticles using plant extracts. *Biotechnology Advances, 31*, 346–356.
Mohan Bhagyaraj, S., & Oluwafemi, O. S. (2018). Nanotechnology: The science of the invisible. In *Synthesis of inorganic nanomaterials* (pp. 1–18). Woodhead Publishing Ltd.
Mousavi, S. R., & Rezaei, M. (2011). Nanotechnology in agriculture and food production. *Journal of Applied Environmental and Biological Sciences, 1*(10), 414–419.
Narayanan, K. B., & Sakthivel, N. (2010). Biological synthesis of metal nanoparticles by microbes. *Advances in Colloid and Interface Science, 156*, 1–13.
Nasrollahzadeh, M., Sajadi, S. M., Sajjadi, M., & Issaabadi, Z. (2019). *Interface science and technology* (Vol. 28, pp. 1–27). Elsevier.
Oh, S. T., Yoon, S. J., Choa, Y. H., Jeong, Y. K., & Niihara, K. (2006). Wear behavior of nano-sized metal particle dispersed Al_2O_3 nanocomposites. *Key Engineering Materials, 18*, 369–372.
Ovais, M., Khalil, A. T., Ayaz, M., Ahmad, I., Nethi, S. K., & Mukherjee, S. (2018). Biosynthesis of metal nanoparticles via microbial enzymes: A mechanistic approach. *International Journal of Molecular Sciences, 19*, 4100.
Paramasivam, G., Palem, V. V., Sundaram, T., Sundaram, V., Kishore, S. C., & Bellucci, S. (2021). Nanomaterials: Synthesis and applications in theranostics. *Nano, 11*, 3228.
Quan, L. N., Kang, J., Ning, C. Z., Yang, P. D., Deng, J., Su, Y., Liu, D., Liu, B., & Liu, C. (2019). Introduction: 1D nanomaterials/nanowires. *Chemical Reviews, 119*, 8955–8957.
Rai, M., & Yadav, A. (2013). Plants as potential synthesizers of precious metal nanoparticles: Progress and prospects. *IET Nanobiotechnology, 7*, 117–124.
Ramasamy, S., & Purniah, B. (2001). Electrical properties of nanostructured materials. *Proceedings of the Indian National Science Academy, 67A*, 18–26.
Ranjan, S., Dasgupta, N., Rajendran, B., Avadhani, G. S., Ramalingam, C., & Kumar, A. (2016). Microwave-irradiation-assisted hybrid chemical approach for titanium dioxide nanoparticle synthesis: Microbial and Cytotoxicological evaluation. *Environmental Science and Pollution Research, 23*, 12287–12302.
Roco, M. C., & Bainbridge, W. S. (2001). *Societal implications of nanoscience and nanotechnology*. Springer.
Roco, M. C., Mirkin, C. A., & Hersam, M. C. (2011). Nanotechnology research directions for societal needs in 2020: Summary of international study. *Journal of Nanoparticle Research, 13*(3), 897–919.
Sannino, D. (2021). Types and classification of nanomaterials. In *Nanotechnology* (pp. 15–38). Springer.
Shi, D., Guo, Z., & Bedford, N. (2014). *Nanomaterials and devices*. Elsevier.
Singh, M., Goyal, M., & Devlal, K. (2018). Size and shape effects on the band gap of semiconductor compound nanomaterials. *Journal of Taibah University for Science, 12*, 470–475.
Singh, N., & Lyon, L. A. (2007). Au nanoparticle templated synthesis of pNIPAm Nanogels. *Chemistry of Materials, 19*, 719–725.

Tersoff, J., & Hamann, D. R. (1985). Theory and application for the scanning Tunneling microscope. *Physical Review Letters, 50*(25), 1998–2001.
Thakkar, K. N., Mhatre, S. S., & Parikh, R. Y. (2010). Biological synthesis of metallic nanoparticles. *Nanomedicine, 6*, 257–262.
Zewail, A. H. (2000). Femtochemistry: Atomic-scale dynamics of the chemical bond using ultrafast lasers. *Angewandte Chemie, 39*(15), 2586–2631.
Zhang, H. (2015). Ultrathin two-dimensional nanomaterials. *ACS Nano, 9*, 9451–9469.

A Pervasive Study on the Fabrication Techniques of Nanotechnology

Jincemon Cyriac, Jisty Ann Jiji, Joice K. Joseph, Dan Mathew Sabu, Ashtami R. Nair, Tina Sebastian, V. N. Sheeja, and Tom Cherian

1 Introduction to Nanotechnology and Fabrication Methods

Nanotechnology, a revolutionary field of science and engineering, focuses on manipulating materials at the nanometer scale—one billion times smaller than a meter. It encompasses any technology that operates on the nanoscale and has diverse applications across multiple disciplines. The field involves the creation and utilization of chemical, physical, and biological systems at dimensions ranging from individual molecules or atoms to submicron scales, integrating these nanomaterials into larger functional systems. Nanotechnology has the potential to redefine technological expectations, providing innovative solutions to global challenges.

The discovery and application of carbon-based nanomaterials have paved the way for advancements in bioelectronics, biosensors, and nanomedicine. A fundamental understanding of the electrical, optical, magnetic, and mechanical properties of nanostructures is crucial for developing the next generation of functional materials. As a rapidly expanding interdisciplinary field, nanotechnology plays a key role in addressing technological and environmental challenges, particularly in solar

J. Cyriac (✉)
Department of Physics, Christ College (Autonomous) Irinjalakuda, University of Calicut, Irinjalakuda, Thrissur, Kerala, India

J. A. Jiji · J. K. Joseph · D. M. Sabu · A. R. Nair · V. N. Sheeja
Marian College Kuttikkanam (Autonomous), Kuttikkanam, Kerala, India

T. Sebastian
Department of Physics, Deva Matha College, Kuravilangad, Kerala, India

T. Cherian
Department of Chemistry, Christ College (Autonomous) Irinjalakuda University of Calicut, Irinjalakuda, Thrissur, Kerala, India

S. Kulkarni (ed.), *A Journey from Nanotechnology to Picotechnology*, Lecture Notes in Nanoscale Science and Technology 47,
https://doi.org/10.1007/978-3-032-07228-3_3

energy conversion, water treatment, medicine, and catalysis (Nasrollahzadeh et al., 2019).

At the core of nanotechnology lies quantum mechanics and statistical mechanics, which govern the molecular behavior of matter. The Heisenberg Uncertainty Principle in quantum mechanics states that an object's position and momentum cannot be precisely determined simultaneously. This principle raises intriguing questions about the feasibility of manipulating individual atoms and molecules—an essential aspect of nanotechnology (Mansoori & Soelaiman, 2005). Nanofabrication refers to the process of assembling materials to create structures with one, two, or three dimensions at the nanoscale. Understanding and analyzing nanofabrication techniques is essential, as they enable the reproducible mass production of nanoscale structures for applications in molecular adsorption, catalysis, magnetism, luminescence, nonlinear optics, and molecular sensing (Ebad-Sichani, Sharafi-Badr, et al., 2023).

Nanofabrication techniques are broadly categorized into two approaches: top-down and bottom-up. The top-down approach employs various lithographic techniques—both serial and parallel—to pattern nanoscale structures, typically in two dimensions (2D). These techniques operate on length scales that are several orders of magnitude larger than an individual nanoscale feature. In contrast, the bottom-up approach relies on molecular interactions or colloidal assembly to create nanoscale structures in both two and three dimensions. By leveraging top-down lithography and bottom-up molecular self-assembly, researchers can fabricate nanostructures with high precision, offering solutions for critical global challenges in energy, medicine, and environmental sustainability (Gates et al., 2005).

Recent advancements in nanostructured device fabrication have been driven by various imprinting techniques. The primary challenges lie in developing fabrication methods that ensure compatibility with specific materials while achieving the necessary resolution. Large-scale optical device manufacturing provides valuable insights into optimizing fabrication techniques, enhancing feature functionality, and exploring the potential of both top-down and bottom-up approaches for scalable nanotechnology applications (Tahir et al., 2021).

2 Top-Down Vs. Bottom-up Approaches

Nanomaterials can be classified based on their fabrication approach—either the top-down or bottom-up method. In the bottom-up approach, atomic or molecular components self-assemble into larger, more ordered systems, guided by natural physical principles or externally applied forces. Conversely, the top-down approach starts with a bulk material and progressively refines it using advanced tools to create smaller and more intricate structures (Arole & Munde, 2014).

A prominent industrial example of the top-down method is the semiconductor industry, where photolithography is used to manufacture metal-oxide-semiconductor field-effect transistors (MOSFETs) on silicon wafers. This lithography-based process involves a projection printing system called a stepper, which projects transistor features onto a silicon wafer coated with a light-sensitive photoresist using UV light and a photomask (Iqbal et al., 2012). Lithography remains the cornerstone of top-down nanofabrication and has been continually refined to enhance resolution and reduce feature sizes. Modern lithographic advancements now use extreme ultraviolet (EUV) photons and focused electron beams to improve precision. Although electron-beam lithography offers extremely high resolution due to its short wavelength, it is a slow process. Additionally, alternative methods, such as printing and stamping with pre-formed templates, have been explored to create nanoscale patterns. The semiconductor industry has also integrated block copolymer lithography, which combines guided self-assembly with conventional lithography. Although scanning probe lithography offers molecular-level control, it remains a low-throughput method (Khanna, 2016a).

The bottom-up approach, in contrast, aims to develop functional systems and devices capable of self-organization. This method enables the controlled assembly of atoms or molecules into multi-component nanoscale systems without material waste or extensive component removal. However, several technical challenges persist, including reactant purity, contamination control, site homogeneity, and precise surface preparation for atomic deposition (Biswas et al., 2012).

The bottom-up approach encompasses techniques such as sol-gel synthesis, vapor deposition, atomic layer deposition, molecular self-assembly, and DNA-assisted assembly, all of which play a crucial role in nanoelectronics. The sol-gel method is a straightforward approach for synthesizing nanoparticles, while vapor-phase techniques include both chemical vapor deposition (CVD)—which utilizes precursor decomposition—and physical vapor deposition (PVD)—which involves evaporating active species into a vapor phase. Atomic layer deposition (ALD) enables precise composition control and nanoscale thickness adjustment through sequential, self-limiting reaction cycles. Molecular self-assembly leverages the natural organizational tendencies of matter to form homogeneous monolayers, while self-assembly from the gaseous phase includes both physical and chemical vapor deposition. DNA-assisted assembly arranges functional nanomaterials in nanoelectronic circuits using artificial DNA nanostructures (Khanna, 2016b).

Both top-down and bottom-up techniques offer unique advantages and applications. The top-down approach is widely utilized in semiconductor fabrication, employing photolithography, extreme UV lithography, and electron beam lithography to achieve high-resolution patterning. In contrast, the bottom-up method focuses on self-assembly to create functional nanoscale systems with minimal waste, using techniques such as sol-gel synthesis, vapor deposition, and DNA-assisted assembly. While bottom-up fabrication holds great promise, challenges such as contamination management and surface preparation must be addressed to optimize its potential. The synergy between these two approaches continues to drive advancements in nanotechnology innovation.

3 Lithography Based Techniques

3.1 Photolithography: Principles and Applications

Photolithography is a complex process comprising multiple components, including light sources, exposure instruments, masks, pellicles, and photoresist materials. In advanced photolithographic processes, excimer lasers serve as the primary light source for exposure tools. When an optical system exposes mask images onto a photosensitive resist material (photoresist), latent images are formed. These images are subsequently developed in a developer solution, producing the desired resist patterns—similar to traditional photographic techniques. However, the continued development of photolithography faces challenges, particularly in achieving shorter wavelengths for enhanced resolution (Okazaki, 2021).

Micro- and nanolithography serve as the primary fabrication technologies in the semiconductor industry, enabling the production of integrated circuits and microchips. These advancements have significantly contributed to the ongoing revolution in nanotechnology (Pimpin & Srituravanich, 2012).

The photolithography process involves transferring circuit designs onto a wafer, with several key steps. The wafer is initially coated with a photoresist layer and then baked to enhance adhesion. A stepper aligns and exposes a small section of the wafer using a reticle (a template containing the design for a few chips). The exposure process is repeated in steps across the wafer to ensure full coverage using ultraviolet (UV) radiation. After exposure, the wafer undergoes development using a specialized solvent that removes the exposed photoresist, revealing the underlying pattern. A final bake is performed to ensure that the unexposed photoresist remains securely attached to the wafer (Akcalt et al., 2001).

Photolithography is a versatile and indispensable technique in microfabrication, with several key applications:

- Microcircuit Fabrication: Photolithography plays a crucial role in transferring intricate patterns onto semiconductor substrates, forming the foundation of integrated circuit manufacturing (Stevenson & Gundlach, 1986).
- Fabrication of Microlens Arrays: A photoresist layer is patterned using UV light and a mask to define lens shapes. Controlled heating (reflow) allows the photoresist to form dome-like structures due to surface tension. These patterns act as molds for replicating microlenses onto polymer substrates via injection molding or soft lithography. This technique ensures high precision and uniformity, making microlens arrays widely applicable in imaging, light collection, and beam shaping (Wu & Whitesides, 2002).
- Etching and Material Deposition: After development, the patterned areas are exposed for further processing, such as etching or material deposition. This high-resolution patterning enables the fabrication of microelectromechanical systems (MEMS), including sensors, actuators, and microfluidic components, ensuring accuracy and repeatability in their design (Miyajima & Mehregany, 1995).

Photolithography remains a fundamental and versatile technique in microfabrication, enabling the precise patterning of materials at both the microscale and nanoscale. Its high resolution, scalability, and adaptability make it indispensable in the production of advanced devices such as microcircuits, MEMS, and optical components. The continuous advancements in photolithography technology drive progress in semiconductor manufacturing and nanotechnology innovation.

3.2 *Electron Beam Lithography (EBL)*

Electron beam lithography is the process of employing a tightly focused (submicrometer diameter) electron beam to write patterns in thin coatings of electron-sensitive material. By integrating electrical scanning with mechanically monitored interferometric motion, EBL achieves exceptionally accurate patterning. For example, a pattern with one-micrometer features can extend over 100 mm while maintaining a positional accuracy of 25 μm. This process creates masks that are optically projected onto silicon wafers coated with photosensitive resists in the production of integrated circuits. This method is commonly used to create masks, which are optically projected onto silicon wafers coated with photosensitive resists in the production of integrated circuits. Although EBL can directly expose resist-coated wafers to create circuits with sub-micrometer features, the process is slow due to its point-by-point exposure nature, as well as limitations in resist sensitivity and electron beam intensity. However, emerging methods that allow simultaneous exposure of multiple points could help overcome this constraint. Alternative competing lithographic techniques include X-ray lithography and ion beam lithography (Pease, 1981). Patterns with nanoscale feature sizes can be created with electron-beam lithography due to its extremely short wavelength and appropriate energy density properties. The approach has thus been effectively applied to the fabrication of numerous nanoscale devices (Tseng et al., 2003).

In EBL, a thin layer of organic film (referred to as resist) is first scanned by a highly concentrated and carefully regulated electron beam (exposure), and the exposed or nonexposed areas of the resist are then selectively removed in a solvent (developing). This process transfers a pattern onto the substrate's surface. By either coating the chosen regions of the surface with the resist or exposing otherwise resist-covered areas, the procedure enables the patterning of extremely fine structures, frequently with sub micrometer to a few nanometer dimensions (Pala & Karabiyik, 2016). The advantages of EBL include:

- With the demonstration of dependable sub-10-nm nanofabrication processes, Electron-Beam Lithography (EBL) could be a high-quality technique that combines a respectable writing speed with the potential for high levels of integration of complicated devices (Vieu et al., 2000).
- Because of its intrinsically high spatial EBL remains the preferred technology for creating masks due to its high process margins and low resolution (less than

1 nm). Nanotechnology and nanoelectronics have developed as a result of the field's expansion from the micrometer to the nanometer scale as microelectronics continues to progress (Zhou, 2005).

- The fabrication of 3D polymer microstructures using electron beam lithography involves coating a substrate with an electron-sensitive resist, exposing it to a focused electron beam to define the desired 3D pattern, and developing it with a solvent to reveal the structure. This process enables the creation of intricate microstructures with nanoscale precision, making it invaluable for applications in micro-optics, nanoelectronics, and biomedical devices (Chen et al., 2005).

Electron beam lithography is a powerful and versatile technique for fabricating intricate micro- and nanostructures with outstanding precision and scalability. Its ability to achieve sub-10 nm resolution and maintain high spatial accuracy has made it indispensable in the advancement of nanoelectronics, photonics, and biomedical technologies. Despite its slow throughput due to the point-by-point exposure process, EBL remains the gold standard for high-resolution mask fabrication and direct nanoscale patterning. As ongoing innovations continue to improve efficiency and scalability, EBL is set to play a pivotal role in shaping the future of nanotechnology, driving breakthroughs in diverse scientific and industrial applications.

3.3 Nanoimprint Lithography (NIL)

Nanoimprint lithography (NIL) is a high-throughput, high-resolution parallel patterning technique that employs mechanical contact and three-dimensional material displacement to transfer a stamp's surface pattern onto a material (Chen et al., 2005). Currently, feature sizes as small as 5 nm have been achieved, with resolution largely dependent on the manufacturing capabilities of the stamp relief. Historically, the term "nanoimprint lithography" (NIL) referred to a hot embossing method in which a thin layer of thermoplastic material is heated until it softens, then embossed with a pattern and subsequently cooled to solidify the structure (Schift & Kristensen, 2007). Unlike conventional lithographic methods, which use photons or electrons to modify the resist's chemical and physical properties for pattern definition, NIL directly mechanically deforms the resist material. This approach allows for higher resolutions than those achievable with conventional methods, which are typically constrained by light diffraction or beam scattering (Guo, 2007).

In NIL, molds are commonly fabricated from elastomeric materials, such as poly(dimethylsiloxane) (PDMS), polyimides, and polyurethanes, with PDMS being the most frequently used. These elastomeric molds can either be attached to a rigid substrate for enhanced durability or used as standalone molds for applications on curved or uneven surfaces. However, due to the flexibility and thermal expansion of elastomeric materials like PDMS, registration errors can occur when aligning NIL patterns with pre-existing structures (Barcelo & Li, 2016).

Nanoimprint lithography is a versatile and precise nanofabrication technique that circumvents the resolution limitations of conventional photon-based lithography by directly patterning materials through mechanical deformation. Its ability to produce nanoscale features with high fidelity and efficiency makes it an invaluable tool for advanced nanofabrication.

4 Vapor Phase Deposition Methods

4.1 Chemical Vapor Deposition (CVD): Types and Applications

In chemical vapor deposition (CVD), a solid film is deposited through the reaction of vapor-phase substances, often diluted with an inert carrier gas. CVD is a crucial technique due to its ability to deposit a wide range of elements and compounds under atmospheric pressure and at relatively low temperatures. It enables the deposition of high-purity films, including amorphous, polycrystalline, epitaxial, and uniaxially oriented polycrystalline films. The process involves material and energy transfer to and from the reaction site, alongside the thermodynamics, kinetics, and chemical reactions within the reactor (Frey, 2015).

A CVD process can be explained through mass transport mechanisms (including forced convection and gaseous diffusion), gaseous flow behavior (such as flow types, flow patterns, and velocities), and systems chemistry (encompassing chemical species, thermodynamics, kinetics, and reaction processes) (Spear, 1982). For surface modification, CVD polymerization techniques often adapt solution-phase organic synthesis into heterogeneous vapor-phase processes. Initiated CVD (iCVD) provides a flexible approach to forming chain growth polymers, while oxidative chemical vapor deposition (oCVD) facilitates the stepwise growth of conjugated polymers, which often exhibit semiconducting or conductive properties. Additionally, molecular layer deposition (MLD) and vapor deposition polymerization (VDP) support the synthesis of step-growth polymers via condensation processes (Gleason, 2015).

The electrical properties of CVD-grown graphene offer promising applications. For instance, flexible transparent conductive electrodes for organic solar cells can be fabricated using few-layer graphene grown on Ni. Similarly, large-grain graphene synthesized on Cu foil exhibits electrical characteristics suitable for field-effect transistors (FETs) (Zhang et al., 2013).

The increasing research interest in two-dimensional (2D) materials stems from their adjustable electrical, optical, and chemical properties. Heterostructures based on 2D materials, which combine multiple distinct 2D layers, create unique platforms for the development of new materials, exotic physical phenomena, and novel applications. The ability to produce high-quality, cost-effective 2D materials and heterostructures at scale remains a major challenge. Despite these obstacles, CVD has emerged as a powerful technique for the synthesis of 2D materials and their heterostructures, offering scalability and material versatility (Cai et al., 2018).

Recent advancements in plasma CVD, metal-organic CVD, and other emerging technologies further illustrate the adaptability and continuous evolution of CVD. The expanding range of applications reflects the extensive research conducted by scientists in government laboratories, universities, and the semiconductor industry. As new designs, products, and materials continue to emerge, CVD remains at the forefront of dynamic technological change, driving innovations across multiple disciplines (Pierson, 1999).

4.2 Physical Vapor Deposition (PVD)

Physical vapor deposition (PVD) is a process in which thin films are formed by the condensation of a vaporized material onto a substrate. The PVD process takes place under vacuum conditions, ensuring precise deposition and minimal contamination. Various PVD techniques include ion plating, enhanced sputtering, cathode arc deposition, electron beam physical vapor deposition, evaporative deposition, and sputtering. The deposition process involves the vaporization of a solid coating material through heat or ion bombardment (sputtering) (Shahidi et al., 2015).

Since the early 1980s, hard coatings such as titanium nitride (TiN) have been industrially produced using PVD techniques. The initial commercial success of PVD coatings on high-speed steel tooling was primarily driven by cathodic arc deposition and low-voltage electron beam methods. Over time, additional PVD techniques, including unbalanced magnetron sputtering and high-voltage triode electron beam deposition, have gained prominence in industrial applications (Sproul, 1996).

4.3 Sputtering and Evaporation

When it comes to the sputtering process, the magnetron sputtering method forms thin layers of many materials. A target is the raw material used in this vacuum coating technique. Sputtering processes include the placement of a magnetron close to the target. The release of atomic-sized particles from the target occurs when an inert gas is introduced into the vacuum chamber and is accelerated by a high voltage supplied between the target and the substrate in the magnetron's direction. As a result of the gas ions' kinetic energy reaching the target and traveling to the substrate to form a solid thin film, these particles are projected. By flipping the voltage polarity between the substrate and the target, a process known as cathodic cleaning, the technology enables the removal of prior contaminants from the substrate's surface (Baptista et al., 2018). In principle, the same stress mechanisms apply to films that are sputter-deposited. Nevertheless, an added compressive stress contribution is produced by the atomic peening mechanism at low sputter pressures. Chemical reactions between the film and sputter gas impurities become more prominent at higher sputter pressures, which affect the behavior of stress (Koch, 2010).

With arc evaporation, high-quality hard films with a very low macroparticle content may now be produced. There is also talk about the reactive deposition of hard coatings. On large and complexly shaped substrates, particular emphasis is paid to producing uniform films and forming films with desirable qualities (Vyskočil & Musil, 1990). The benefit of the evaporation approach is that it permits direct synthesis using pure metals as starting materials, eliminating interactions with costly and possibly hazardous intermediates (Förster et al., 2012). Many efforts have been made over the past 10 years to enhance the technical designs of arc deposition systems in order to minimize their primary disadvantage (high roughness due to microdroplets) and take advantage of the advantages this technique exhibits over other PVD methods (high ionization rate, good adhesion). Almost any tool can be coated using the current arc-PVD systems, which offer a wide variety of commercial coatings to choose from based on the application. Arc-PVD coatings offer a useful method of improving the treated tools and component's resistance to wear as well as other characteristics like corrosion and oxidation (Rodriguez et al., 2002).

Thus, magnetron sputtering and arc evaporation are two widely used PVD techniques for producing thin films with distinct advantages. Magnetron sputtering excels in creating uniform films and controlling stress mechanisms, while arc evaporation offers high-quality coatings with excellent adhesion and ionization rates. Despite challenges like microdroplets in arc deposition, advancements in design have significantly improved its performance. Both methods play a crucial role in enhancing the wear, corrosion, and oxidation resistance of tools and components, making them indispensable in various industrial applications.

4.4 Atomic Layer Deposition (ALD)

Atomic layer deposition (ALD) is a vapor-phase process used to create thin films of a wide range of materials. It is based on sequential, self-limiting surface reactions, allowing precise control over film composition and thickness at the angstrom level, along with exceptional conformality on high-aspect-ratio structures. Due to these advantages, ALD has become a powerful tool for numerous industrial and scientific applications. This section provides a brief introduction to ALD and highlights some of its applications, including solid oxide fuel cells, high-k transistors, and solar cell devices (Johnson et al., 2014).

The demand for atomic-level control over thin film deposition has intensified with the miniaturization of semiconductor devices. As miniaturization leads to highly intricate structures, conformal coatings become essential. The conformality achieved by ALD on high-aspect-ratio structures is unparalleled compared to other thin-film deposition techniques. The development of ALD has been driven by the need for continuous, pinhole-free films in semiconductor devices. Additionally,

ALD finds applications beyond the semiconductor industry, including diffusion barrier coatings with low gas permeability and low-electron-leakage dielectrics for magnetic read/write heads (George, 2010).

Unlike conventional chemical vapor deposition methods, ALD introduces precursor vapors into the reactor one at a time, with alternating purging and evacuation steps. After each exposure stage, a monomolecular layer of the precursor adheres to the surface, enabling a self-limiting film growth process. This unique mechanism provides several advantages, such as superior conformality, homogeneity, and precise control over film thickness (Ritala & Leskelä, 2002).

ALD has witnessed rapid advancements, largely driven by the continuous downscaling of electronic devices. Moreover, an increasing number of emerging technologies are benefiting from its ability to deposit superior thin films (Knoops et al., 2015). Over the past three decades, significant progress has been made in ALD, leading to the extensive study of crystalline and polycrystalline compound and elemental semiconductors for applications in electronics, optoelectronics, and light-emitting diodes. In particular, ALD enables nanoscale and sub-nanometer thickness control, facilitating the fabrication of high-speed electronic logic circuits. This capability contributed to the surge in ALD research in the late 1990s and early 2000s (Parsons et al., 2011).

As one of the most promising techniques for high-quality thin film deposition, ALD continues to evolve to meet diverse application requirements. The continuous expansion of material choices and the demand for conformal, ultra-thin, high-quality films present both opportunities and challenges for ALD process development (Oviroh et al., 2019).

ALD operates through sequential alternating pulses of precursors and co-reactants, ensuring self-limiting surface reactions. The general ALD process involves cyclic surface reactions: (1) adsorption of precursor A, (2) purging of excess precursor and byproducts using an inert gas, (3) adsorption of precursor B, and (4) repeated purging followed by cyclic repetition to achieve layer-by-layer growth until the desired film thickness is obtained. The adsorption process includes both physisorption and chemisorption, while excess physisorbed species are removed through inert gas purging. Each cycle is constrained by the number of available active surface sites, resulting in a self-limiting chemisorption process.

The ALD process is highly temperature-dependent, with an "ALD temperature window" defining the optimal range for film growth. Below this window, insufficient reactivity or precursor condensation may occur, whereas higher temperatures can lead to precursor decomposition or desorption. Only within the ALD temperature window can self-limiting ALD reactions take place successfully (Zhang, Li, et al., 2022).

Despite recent advancements, several challenges remain before ALD can be fully integrated into microelectronics applications. Fundamental studies on precursor chemistry and film deposition remain crucial, particularly as ALD expands to new materials and device architectures, such as ferroelectric oxides for non-volatile memory applications. The seamless integration of ALD processes and materials into overall device fabrication workflows is a key research area that must be addressed early in process development (Leskelä & Ritala, 2003).

5 Wet Chemical Synthesis

Wet chemical synthesis is a widely utilized bottom-up approach for fabricating nanomaterials by assembling structures atom by atom or molecule by molecule in a liquid-phase environment. This method offers precise control over particle size, shape, and composition, making it highly effective for synthesizing nanoparticles, quantum dots, thin films, and nanostructures with tailored properties. Unlike top-down approaches, which involve the physical reduction of bulk materials to nanoscale dimensions, wet chemical synthesis relies on solution-based chemical reactions to drive nucleation and growth.

This approach is particularly advantageous for producing high-purity materials with uniform morphology, often at low temperatures and ambient pressure. As a bottom-up strategy, wet chemical synthesis provides a modular and scalable means of assembling atomic or molecular species into functional nanostructures. Its versatility has enabled precise tuning of nanomaterials' optical, electronic, and surface properties by manipulating size, shape, and composition. This flexibility, both kinetic and thermodynamic, makes wet chemical synthesis a preferred choice for nanomaterial fabrication (Nikam et al., 2018).

5.1 Sol-Gel Method: Process and Applications

The sol-gel process has been extensively studied and widely applied in recent years due to its simplicity compared to other coating techniques and its ability to significantly enhance metallic biomaterial coatings on substrates (Zhang et al., 2011). The sol-gel method involves forming networks by structuring a colloidal suspension (sol), which then undergoes gelation to create a system with a continuous liquid phase (gel) (Pierre, 1998; Zhang et al., 2004). A sol is a dispersion in which colloidal particles, typically solid structures ranging from 1 to 100 nm in diameter, are suspended in a liquid medium (Davis & Rideal, 1963). In contrast, a gel is a rigid, interconnected network containing sub micrometer-sized pores and polymeric chains that extend beyond a micrometer in length.

The term "gel" encompasses a wide variety of material combinations, classified into four categories as defined by Flory: (1) highly organized lamellar structures, (2) completely disordered covalent polymer networks, (3) polymer networks formed through physical aggregation with predominantly disordered arrangements, and (4) unique disordered structures (Flory, 1953). Depending on the type of precursor used, the sol-gel process can be categorized into two main types: those utilizing inorganic precursors (such as chlorides, nitrates, and sulfides) and those based on alkoxide precursors.

5.1.1 Sol-Gel Process

There are three primary methods for producing sol-gel nanoparticles:

- Transformation of a colloidal suspension into a gel.
- Hydrolysis and polycondensation of alkoxide or nitrate precursors, followed by hypercritical drying of the resulting gel.
- Hydrolysis and polycondensation of alkoxide precursors, followed by aging and drying under normal atmospheric conditions.

The sol-gel process follows a structured sequence:

Step 1: Mixing: A sol or colloidal suspension is prepared by dispersing colloidal particles in water while maintaining a pH level that prevents precipitation (Iler, 1955; Iler, 1979).

Step 2: Casting: The low-viscosity sol is poured into a mold, which is selected to prevent adhesion of the gel to its surface.

Step 3: Gelation: Over time, colloidal particles and condensed silica species form a three-dimensional network. The physical properties of the gel network depend on the particle size and degree of cross-linking prior to gelation. As gelation progresses, viscosity increases sharply, leading to the formation of a solid structure shaped by the mold. By carefully controlling the sol's viscosity, fibers can also be pulled or spun during gelation.

Step 4: Aging: Also known as syneresis, aging involves immersing the gel in a liquid medium for several hours or days. During this stage, polycondensation continues, accompanied by localized dissolution and reprecipitation within the gel network. This process strengthens the interparticle necks, reduces porosity, and enhances the gel structure. Sufficient strength must be developed during aging to prevent cracking in the subsequent drying phase.

Step 5: Drying: During drying, liquid is removed from the gel's interconnected pore network. If the pores are smaller than 20 nm, significant capillary stresses can develop, potentially leading to gel cracking if the drying process is not carefully managed. Several strategies can mitigate this issue, including reducing the liquid's surface energy by adding surfactants, eliminating very small pores, employing hypercritical evaporation to avoid a solid–liquid interface, and carefully controlling hydrolysis and condensation rates to ensure uniform pore sizes.

Step 6: Densification: The final step involves heating the porous gel at high temperatures to achieve densification, wherein pores are eliminated. The densification temperature depends on factors such as pore size, pore connectivity, and surface area. Alkoxide-based gels can be densified at approximately 100 °C, whereas gels synthesized through commercial processes require temperatures in the range of 1500–1720 °C (Shoup, 1988; Vasconcelos et al., 1989).

The sol-gel process offers several advantages, making it highly versatile for various applications:

1. Synthesis can be achieved at relatively low temperatures.
2. High purity products can be obtained.
3. The production process is highly efficient.
4. The method offers flexibility in designing chemical compositions and achieving material homogeneity.
5. The synthesized materials can be used in specialized forms, such as fibers, aerogels, and surface coatings.
6. The process allows the synthesis of amorphous materials that can be applied as thin films.
7. Porous materials can be fabricated and subsequently modified with organic or polymeric compounds.
8. The solution-phase synthesis leads to high chemical reactivity of precursors, enabling precise control over material properties (Li et al., 2021; Yang et al., 2020).

5.2 *Hydrothermal and Solvothermal Techniques*

Hydrothermal and solvothermal methods are significant areas of inorganic synthesis. Hydrothermal synthesis refers to chemical reactions occurring in an aqueous solution at temperatures higher than water's boiling point, whereas solvothermal synthesis happens in a nonaqueous solution at relatively high temperatures. These methods are extensively used for synthesizing both traditional and advanced materials, managing waste, and replicating geothermal and biohydrothermal processes.

Today, hydrothermal or solvothermal synthesis is generally defined as the chemical reaction-based production of materials in a sealed and heated aqueous solution or organic solvent at controlled temperatures (100–1000 °C) and pressures (1–100 MPa). These reactions are typically carried out in specially designed sealed vessels or high-pressure autoclaves under subcritical or supercritical conditions of the solvent. The main difference between hydrothermal and solid-state reactions is their "reactivity," which is reflected in their distinct reaction mechanisms. Solid-state reactions depend on the diffusion of raw materials at the interface, while hydrothermal and solvothermal reactions involve the interaction of reactant ions or molecules within a solution. Additionally, hydrothermal or solvothermal reactions often serve as alternative, milder synthetic methods by enabling reactions at lower temperatures compared to solid-state processes (Xu et al., 2007).

5.2.1 Reaction Medium

Among solvents, water stands out as an environmental-friendly and cost-effective option, making it the most commonly used solvent in hydrothermal and solvothermal syntheses compared to organic alternatives. Under high-temperature and high-pressure hydrothermal conditions, the properties of water as a solvent undergo significant changes. A hydrothermal system has three notable features: (1) it accelerates reaction rates among complex ions, (2) enhances hydrolysis reactions, and (3) significantly alters the redox potential of the reactants. Generally, two fundamental types of chemical reactions are observed. The first is ionic reactions, such as the metathesis of inorganic materials, which can often occur rapidly at ambient temperatures. The second type involves free radical reactions, such as the explosive reactions of organic compounds. Based on electronic theory, organic compounds with polar bonds often display behaviors similar to ionic reactions.

Additionally, the increased ionization constant of water at elevated temperatures promotes ionic and hydrolysis reactions, enabling insoluble inorganic materials and organic compounds to react effectively under high-temperature and high-pressure hydrothermal conditions (Rabenau, 1985). In solvothermal synthesis, the organic solvent serves as both the reaction medium and a means to dissolve or partially dissolve the reactants, forming solvent-reactant complexes that influence the chemical reaction rate. In addition, the organic solvent affects the concentration and state of the reactants' active species, thereby influencing the reaction pathway.

5.3 Electrochemical Deposition

Electrochemical deposition is a versatile technique for fabricating high-aspect-ratio microtools from various materials by concentrating metal deposition to directly form small, shaped electrodes. This method is widely recognized as a simple and cost-effective approach for producing microelectrodes used in micromachining applications. Additionally, it enables the direct fabrication of complex three-dimensional (3D) microstructures on high-strength metals (Madden & Hunter, 1996).

Electrochemical deposition in a specific, controlled area is achieved using nonconductive masking. This process involves the reduction of metallic ions into solid metal, which then deposits onto the cathode surface when an appropriate electric current passes through an electrolyte solution. The electrolyte, formed by dissolving a metallic salt in water, contains charged ions essential for deposition. An acidic metallic salt solution serves as the electrolyte, with an anode submerged in it, while the cathode—where deposition occurs—is positioned above the anode. A nonconductive mask is placed between the electrodes to control the deposition pattern.

During the deposition process, a small, constant gap is maintained between the anode and the mask. As current flows, metal ions migrate and deposit onto the cathode surface through the mask, allowing the formation of intricate 3D microstructures by designing the mask pattern accordingly. The mask, typically made of nonconductive materials such as polymethyl methacrylate (PMMA), can be precisely machined into different cross-sectional shapes using micromachining techniques. A pulsed voltage is applied to ensure localized deposition with high precision.

The porosity and accuracy of the electrochemically deposited structure can be controlled by adjusting process parameters, including applied voltage, pulse frequency, duty cycle, and the distance between the electrodes and the mask. This method facilitates the fabrication of microtools and microfeatures from materials such as metals, metal alloys, conductive polymers, and semiconductors, making it suitable for a wide range of applications (Bhattacharyya, 2015).

6 Molecular Beam Epitaxy (MBE)

Molecular beam epitaxy (MBE) is a widely used technique for growing thin, epitaxial films of various materials, including oxides, semiconductors, and metals. Initially developed for the growth of compound semiconductors, this remains its most common application due to the critical role these materials play in the electronics industry (Herman & Sitter, 1996).

6.1 Fundamentals of MBE

In MBE, thin films are formed through the interaction of molecular or atomic beams of constituent elements, which are thermally heated and directed onto a substrate surface maintained at high temperatures under ultrahigh vacuum (UHV) conditions. The key distinction between MBE and earlier vacuum deposition methods is its precise control over beam fluxes and growth conditions. Due to the UHV environment, MBE operates under nonequilibrium conditions, where growth is predominantly governed by the kinetics of surface interactions between the incoming beams and the atomic layers of the substrate crystal. In contrast to other epitaxial growth techniques, such as liquid phase epitaxy or vapor phase epitaxy—both of which function near thermodynamic equilibrium and are largely dictated by diffusion processes—MBE relies on surface kinetics for precise film deposition (Herman & Sitter, 1996).

As the name suggests, MBE involves directing focused streams of atoms or molecules in a UHV chamber toward a substrate crystal. These streams impact the crystal, which is maintained at a temperature that allows the incoming atoms to diffuse across its surface and integrate into the lattice at appropriate sites. The UHV environment minimizes contamination of the growing surface. In this setup, the atoms and molecules travel freely without colliding, either reaching the substrate or condensing onto the cooler chamber walls, where they are effectively removed from the system. A shutter in the path of each beam allows near-instantaneous control of material deposition. These characteristics enable slow, contamination-free film growth while allowing rapid compositional changes—much faster than the time required to grow a single atomic layer.

6.2 Advantages of MBE in Semiconductor Fabrication

Compared to other deposition methods, MBE requires an exceptionally high vacuum to maintain minimal impurity levels. The ultrahigh vacuum conditions allow for the deposition of extremely pure films. The source temperature is a crucial factor, as it controls both the material flux reaching the substrate and the rate of surface diffusion or desorption. The term "beam" refers to the fact that evaporated atoms travel without interacting with each other or residual gases in the vacuum chamber, due to their extended mean free paths, until they reach the wafer.

Throughout the process, reflection high-energy electron diffraction (RHEED) is used to monitor the formation of crystal layers in real time. A computer-controlled system regulates the shutters in front of each furnace, allowing precise control over layer thickness down to a single atomic layer. This level of precision enables the fabrication of complex multilayer structures from different materials, which are now fundamental to modern semiconductor devices, including semiconductor lasers and light-emitting diodes (Kumar et al., 2020).

In addition to inorganic semiconductors, MBE is also employed for depositing organic semiconductors. The method features a low growth rate of approximately 1 μm per hour, allowing for highly controlled thermal evaporation and the separate evaporation of each component. Moreover, MBE operates at relatively low substrate temperatures and produces ultrasharp interface profiles (Morresi, 2013).

6.3 Recent Advances in MBE Technology

The primary challenges associated with MBE production include achieving large-area uniformity, minimizing defect densities, and ensuring high throughput. Significant advancements have been made in all three areas in recent years. In terms of uniformity, commercial MBE systems have achieved thickness variations of less than 1% across a 3-inch GaAs wafer and defect densities as low as 100 cm^{-2} (Chaix, 1986; Fronius et al., 1986).

A new multi-wafer MBE system has been introduced, incorporating a 7.5-inch-diameter rotating substrate holder. This system enables the simultaneous deposition of seven 2-inch wafers or three 3-inch wafers using solid Ga and AB sources. The thickness and carrier concentration uniformity for GaAs and $Al_xGa_{1-x}As$ layers are maintained within ±1% across a 7.5-inch diameter. Additionally, oval defects originating from the Ga source have been reduced to a density below 40 cm^{-2} (Saito et al., 1986).

Further refinements in interface smoothness down to a single atomic layer have been achieved through interrupted growth cycles or periodic high-temperature pulses applied to the substrate between heterostructures (Sakaki et al., 1985). Innovations such as the avalanche multiplication effect in quantum well superlattices, highly sensitive photodetectors using effective mass filters, and novel transistors based on resonant tunneling have broadened the scope of device fabrication.

The ability of MBE to precisely synthesize and engineer the microstructures of heterojunction thin films has facilitated groundbreaking developments in high-speed semiconductor devices (Capasso et al., 1985; Capasso et al., 1986; Chang et al., 1974).

7 Self-Assembly and Template-Assisted Methods

In the last 20 years, significant advancements have been made in creating synthetic methods for nanomaterials (NMs) that allow precise control over their size, shape, composition, and spatial arrangement. In template-assisted synthesis, the "template" serves as the key component. It is a nanostructured entity whose size, morphology, and charge distribution significantly influence the properties of the synthesized NMs. This method generally involves three steps: preparation of the template, synthesis of NMs using the template, and, if necessary, the removal of the template (Scarpelli et al., 2018; Vazquez, 2015).

Self-assembly, a simple yet effective bottom-up approach, enables the creation of well-ordered structures with novel functionalities. Inorganic and organic nanoparticles, nanostructures, polymers, and natural biomolecules such as DNA, proteins, and enzymes play a crucial role in forming hierarchical composites. These composites range from simple to complex structures, achieved through self-assembly processes. Biocompatible composite nanostructures find applications in diverse fields, including biomedicine, pharmaceuticals, agriculture, the food industry, and other bio-related areas. Their wide usage is attributed to their well-ordered structures, unique molecular properties, customizable functionalities, and a variety of physical, chemical, and biological characteristics (Hou, 2016).

7.1 Molecular and Block Copolymer Self-Assembly

In biomaterials, the self-assembly mechanism between nanostructures, as well as between nanostructures and biomolecules, is governed by noncovalent internal interactions and external stimuli. The complexity of these interactions determines the formation of the desired structures. Electrostatic interactions between materials and between materials and biomolecules play a critical role in biomaterial applications. These interactions enable layer-by-layer assembly, resulting in the creation of self-assembled monolayers, multilayers, synthetic polyelectrolytes, and highly organized hybrid bio-nanostructures, such as biomaterials derived from DNA, proteins, peptides, and enzymes. Wang and his team produced peptide nanofibers, PNF-based silver nanowires, and graphene nanosheets using electrostatic interactions (Wang et al., 2019). Similarly, Liu et al. developed micelle-induced protein nanowires through electrostatic bonding (Sun et al., 2016). In the creation of biomaterials, hydrogen bonds refer to the interactions between electronegative atoms and hydrogen atoms in materials or biomolecules. These bonds play a crucial role in the formation of self-assembled long-chain one-dimensional biomolecular biomaterials and hybrid biomaterials that combine biomolecules with materials. Li and colleagues proposed that the self-assembly process responsible for the formation of Diphenylalanine micro-rods is driven by hydrogen bonds (Li et al., 2015). Additionally, Yang and colleagues synthesized Diphenylalanine peptide nanowires on graphene surfaces using hydrogen bonds (Li et al., 2013).

The self-assembly (SA) of block copolymers (BCPs) into organized nanostructures on solid surfaces offers a promising method for future lithography, with its applications gradually moving toward practical implementation. Block copolymer (BCP) lithography relies on the microphase separation of block copolymers composed of two (or more) incompatible polymer blocks. The separation occurs due to factors such as the block length ratio, the Flory–Huggins parameter, and the degree of polymerization. This process results in the formation of arrays of distinct polymer blocks, producing versatile structures with nanoscale feature sizes and periodicity. A critical factor in block copolymer (BCP) lithography, requiring insights from various research disciplines, is the interaction between block copolymer thin films and solid substrates. Two main mechanisms drive pattern formation. The first involves the (de-)wetting behavior of (block co-) polymer films on surfaces, which greatly influences the film's morphology and topography at the microscale. The second mechanism is the microphase separation of block copolymers into organized nanoscale patterns, driven by their internal interfacial energies (Brassat et al., 2019; Farrell et al., 2011; Lee et al., 2004).

Dewetting of extremely thin films typically occurs via two mechanisms: spinodal dewetting or nucleation and growth. Spinodal dewetting is caused by fluctuations in film thickness, which creates capillary waves that can spontaneously intensify (Cheng & Perahia, 2016). This mechanism usually takes place in polymer films with very thin layers. In contrast, thicker films are more likely to develop holes at defect sites, which then nucleate, leading to dewetting through the nucleation and growth process.

In block copolymer (BCP) films, dewetting has been observed to follow the nucleation and growth mechanism when microphase separation is induced by solvent annealing. On the other hand, spinodal dewetting occurs in thin BCP films during thermal annealing (Yang & Zhao, 2005).

7.2 Hard and Soft Templates for Nanostructure Formation

The hard template method, a nanocasting technique, utilizes solid porous materials as molds to fabricate nanomaterials. These solid molds possess appropriate size, shape, and internal porosity, allowing liquid precursors to fill the mold and form a desired structure. After the nanomaterial is formed, the mold is removed without altering the material's properties (Xie et al., 2016). Commonly used templates include inorganic and silica-based materials, as well as polymer microspheres like polystyrene-acrylate (PSA), porous membranes, mesoporous carbon, porous anodic aluminum oxide (AAO), and ion exchange resins, due to their controlled particle sizes and distinct structures (Kawamura et al., 2014; Nugraha et al., 2022; Platschek et al., 2011; Savic et al., 2018).

When synthesizing nanomaterials via the hard template method, several factors must be considered. The precursor should be chemically inert with respect to the template, necessitating careful selection of both materials. Strong interactions between the precursor and the template must be avoided to prevent pore blockage. Additionally, the solvent must facilitate the easy diffusion of the precursor into the mesoporous template. To maintain the structural integrity of the desired product after template removal, an adequate amount of precursor is essential (Meng et al., 2005).

Unlike the hard template method, soft template approaches primarily utilize organic materials such as surfactants, flexible organic molecules, block copolymers, and interfaces as templates (Li et al., 2014; Meng et al., 2005; Poolakkandy & Menamparambath, 2020; Savic et al., 2018). The synthesis of nanomaterials using soft templates typically follows one of three mechanisms: cooperative self-assembly (CSA), true liquid crystal templating (TLCT), or evaporation-induced self-assembly (EISA) (Soler-Illia & Azzaroni, 2011).

In the CSA mechanism, small amounts of surfactants are used, resulting in the formation of micelles through the concurrent assembly of precursors and templates. This process, known as "cooperative assembly," culminates in the formation of a liquid crystal phase, which precipitates to yield the desired nanomaterials (Wang et al., 2011).

The TLCT mechanism relies on high surfactant concentrations. In this case, a liquid crystal phase forms in the absence of precursors, which are later introduced and combined with the pre-existing liquid crystal phase (Savic et al., 2018; Wang et al., 2011).

The EISA mechanism involves three steps. First, a dilute solution of precursors, templates, and a volatile solvent is prepared with the appropriate stoichiometry. Next, this solution is spread over a large area to achieve the critical micelle concentration (CMC) through solvent evaporation. Finally, the precursors condense near the liquid crystal phase, forming the final nanomaterial (Grosso et al., 2004; Hu et al., 2023; Petkovich & Stein, 2013).

Following the synthesis of nanomaterials, the soft template is removed using various methods such as combustion, depolymerization through thermal treatment or calcination, or low-temperature chemical treatment. These processes are designed to preserve the morphology and properties of the nanomaterials (Olivier et al., 2013).

7.3 Applications in Biomedicine

Self-assembly and template-assisted methods have led to remarkable advancements in biomedical applications, particularly in biosensing, drug delivery, and tissue engineering.

Biosensors have demonstrated immense potential for detecting antibodies, proteins, DNA, and glucose with high specificity. The strong binding affinity between antibodies and analytes allows for efficient molecular recognition in chemical and biological sensors. For example, Olivier and colleagues developed antibody-mimetic materials using functionalized peptoid nanosheets, which provided both chemical and biological stability while serving as templates for the growth of gold nanoparticles (Liu et al., 2014).

Similarly, Liu and colleagues modified graphene oxide (GO) with gelatin to mimic charged protein behavior during bone formation (Tibbitt et al., 2015). The biomimetic mineralization of hydroxyapatite (HA) on the GO-gelatin (GO-Gel) surface demonstrated enhanced cell adhesion, proliferation, and alkaline phosphatase (ALP) activity compared to GO or glass surfaces, highlighting its potential in bone tissue engineering.

Self-assembling two-dimensional organic biomaterials (2DOBMs) have revolutionized drug and gene delivery by providing bioinert carriers that prolong circulation times in the body (Davis et al., 2006). These biomaterials have significantly improved the parenteral administration of small-molecule drugs and have enabled the development of sophisticated carriers for gene therapy (Shao et al., 2012). The combination of self-assembly principles with nanotechnology continues to open new possibilities for the controlled delivery of therapeutics and the engineering of functional tissues.

8 Mechanical Fabrication Methods

Mechanical fabrication methods use a variety of processes to shape and modify materials. Ball milling, and nanoimprint lithography, are two key technologies. Roll-to-roll processing is the continuous handling of materials for flexible electronics. Milling and drilling are the common examples of traditional machining methods that cut and shape materials precisely.

8.1 Ball Milling: Overview and Applications

Ball milling is an important technique in grinding other than the roller mills, stirred media mills and high-pressure grinding rolls. Ball milling uses high energy to repeatedly weld and fracture sample particles. This is a unique synthesis approach for preparing nanometer-scale materials under non-equilibrium conditions. Researchers have frequently used it to develop Mg-based hydrogen storage materials. Ball milling is a one of the synthesis processes for Mg-based compounds utilized in hydrogen storage studies. Ball milling are of three types based on its functions: mechanical grinding, mechanical alloying, and reactive ball milling (Gorrasi & Sorrentino, 2015). Ball milling at room temperature on dry powder mixes is advantageous since it eliminates the need for high temperatures, toxic solvents, and complex polymerization procedures. Ball milling is a viable option for mass-producing hybrid organic-inorganic materials, as well as creating nanocomposites with temperature-sensitive molecules that are both environmentally and economically sustainable (Delogu et al., 2017). Ball milling modifies filler morphology, improving compatibility with polymer mixes and enabling successful grafting reactions. Ball milling has been used to combine filler nanoparticles with polymer matrix in solid form before melt compounding (Suryanarayana, 2001).

Ball milling involves collisions between milling instruments in a batch reactor with powdered materials. The collisions are the result of the reactor, or a portion of it, moving. During a collision, a portion of the powder is trapped and subjected to mechanical load (Baláž & Baláž, 2008). This typically leads to cold-welding, fracture, and plastic deformation caused by lattice flaws. Powders can undergo physical and chemical alterations during microstructural refinement, depending on their composition and chemical environment (Madou, 2018).

Ball milling is an effective method used in a variety of industries to fabricate and change nanoscale materials. In polymer nanocomposites, polymers are mixed with scattered heterophases to produce uniform distribution and improved material characteristics. This technique also alters filler morphology, improves compatibility with polymer blends, and makes grafting easier. One of the most well-known uses of ball milling is mechanical alloying of elemental metals, which allows for the creation of unique alloys and composite materials. Aside from metals, ball milling is used to produce ceramics, polymers, and composite materials with fine control over their microstructure. Mechanical activation via ball milling is important in biomass process engineering because it increases reactivity and efficiency in bioconversion processes (Suryanarayana, 2001).

8.2 Nanopatterning via Mechanical Deformation

The demand for sub-wavelength structures in optoelectronics and photonics is rising. While traditional optical lithography is reliable, diffraction restricts the feature sizes and limitation in scalability for large-area, low-cost manufacturing (Ok et al.,

2013). Alternative technologies, such as nanoimprint lithography (NIL), micro/nanoscale patterning with block copolymer arrays, controlled wetting, pattern-induced polymer self-organization, and plasmonic nano-ruling, have emerged to overcome these constraints. NIL provides sub-10 nm resolution with great precision and reproducibility, allowing mechanically defined sub-micron patterns beyond the optical diffraction limit. Traditional NIL has restrictions in throughput and replica area because to its stamping method (Guo, 2007).

Roll-to-roll nanoimprint lithography, dynamic mold sweeping, and NanoChannel-guided Lithography (NCL) are some of the most common techniques. In Roll-to-Roll NIL, a flexible master mold wrapped around a roll is used to constantly imprint patterns on a flexible or rigid substrate. This approach dramatically increases throughput and replica area while preserving nanoscale resolution. Fluoropolymers, thermally curable polydimethylsiloxane (PDMS), and UV-cured low surface-energy compounds are all suitable mold materials (Ahn & Guo, 2009). Liquid-state UV-curable polymers, such as epoxysilicone and epoxy-silsesquioxane, are effective resist materials for quick and continuous patterning at room temperature. This system can incorporate continuous resist coating techniques like as roll-dipping, continuous drop-casting, and on-demand droplet dispensing (Guo, 2007).

Dynamic Nano-Inscribing (DNI) uses a stiff master mold's split edge to inscribe flawless nanograting patterns on a moving substrate. DNI can pattern a variety of materials at fast rates, with feature sizes as small as 50 nm. It is carried out at room temperature or with brief heating to minimize damage to sensitive materials. DNI provides dynamic patterning on both flat and flexible/curved surfaces (Ok et al., 2011). NanoChannel-guided Lithography (NCL) employs a horizontal sliding cleaved edge of a master mold to constantly produce nanogratings with higher aspect ratios than DNI. NCL uses UV-curable liquid resist that is free of elastic recovery, resulting in a well-defined cross-sectional profile. The liquid resist is extruded smoothly from nanochannels on the master mold and cured with UV light. NCL is a residual layer-free process, eliminating the need for post-etching (Davisson & Germer, 1928).

Roll to Roll NIL is applied to large-area metal wire-grid polarizers, organic photovoltaic cells, and flexible plasmonic devices. DNI/NCL techniques enable the construction of nonlinear (circular) grating patterns, conformal patterns on curved surfaces, and metal grating structures for optical metamaterials.

9 Plasma-Assisted Techniques

Plasma technology is frequently used in modern processing, mainly for non-thermal plasmas at low temperatures. For the fabrication of nanomaterials plasma technology is one of most important green synthesis method, due to its distinguishing qualities as compared to solid, liquid, and gas-phase synthesis methodologies. For describing the ionized gas, Irving Langmuir introduced the word plasma for the first time. When applying energy or heat to a substance, it typically changes phase.

Melting and evaporation are mainly two types of shifting which occurs at constant temperatures. In melting, the solid substance is shift to liquid and in evaporation the liquid is changed to gaseous. Adding energy to a gas causes neutral particles to lose electrons and form ions. Plasma is also an ionized gas, but not a gas-phase transition (Conrads & Schmidt, 2000).

Plasmas, often considered to as a fourth state of matter, can be classified based on various criteria, including their origin, temperature, degree of ionization, and pressure. Natural plasmas and artificial plasmas are the plasmas based on its origin (Pimpin & Srituravanich, 2012). Depending on temperature and ionization, plasmas can be thermal, weakly ionized and highly ionized plasmas. Final classification is based on pressure: low-pressure plasmas, atmospheric pressure plasmas and high-pressure plasmas. Pressure plays a crucial role in plasma behavior, affecting the density and mean free path of particles, while temperature and the degree of ionization influence the energy states and properties of the plasma, making each category uniquely suited to specific applications (Chiappim et al., 2022). Low-pressure plasma applications fall into three categories: material deposition, surface etching/modification, and gas/liquid treatment. The operations in the three subdivisions are carried out at low temperatures, which is favorable when compared to traditional thermal processes (Oehrlein, 1986).

9.1 Reactive Ion Etching (RIE)

Reactive-ion etching is a dry etching process that uses a plasma to transfer patterns onto microelectronic devices. It is favored over wet etching because it offers more control over critical lateral dimensions, which is essential in the manufacture of micron and submicron-scale devices (Chapman & Vossen, 1981).

A glow-discharge plasma creates the etching environment with a feed gas such as CF_4 for silicon and silicon dioxide. This plasma is composed of positive and negative ions, electrons, radicals, and neutrals (Huff, 2021). The material to bc etched is placed on a high-frequency, DC-biased electrode. The electrode gains a negative charge as electron mobility increases, exposing the material to energetic positive ion bombardment. Chemical interactions between radicals and neutrals (e.g., fluorine, CF_3, CF_2) and the material being etched (e.g., silicon) create volatile species such as SiF4, as well as their predecessors SiF, SiF_2, or SiF_3. Positive ions, such as CF_3^+, speed across the plasma sheath and remove material by sputtering. The combination of chemical activity and physical sputtering produces higher material degradation rates in the vertical direction, making it crucial for duplicating device patterns (Chapman & Vossen, 1981). When considering reactive-ion etching, it's important to include both gas-phase processes and plasma–surface interactions. In gas-phase processes, we aim to measure the densities of electrons, ions, radicals, parent molecules, and product species, as well as their energy-distribution functions and temporal and spatial variations. We need to understand the creation and loss rates of reactive species through electron–impact interactions, collisions, and heterogeneous processes (e.g., walls). The effective electric field, including its spatial and temporal dependencies, must be understood.

Plasma–surface interactions involve the chemistry and physics of gas-surface processes, particularly those involving energetic ions, electrons, or photons. Surface impingement rates of plasma species, energy content, sticking coefficients, species concentrations on the surface, and the types and rates of surface chemical reactions are all important aspects to understand. These parameters are influenced by ion, electron, and photon bombardment, which affect the generation and ejection of volatile etch products as well as the chemically changed near-surface region. Future techniques, like magnetron and hollow-cathode discharges, aim to increase etch rates by increasing the concentration of reactive intermediates. These techniques use magnetic and electric fields or cavities inside the cathode to trap electrons and increase plasma ionization efficiency. A fundamental understanding of glow-discharge gas-phase phenomena and plasma-surface interactions is crucial for the development of reactive-ion etching (Chapman & Vossen, 1981).

Reactive Ion Etching (RIE) is a simple and powerful process commonly used in semiconductor manufacturing, microelectromechanical systems (MEMS), and nanotechnology. One significant application of RIE is the creation of high-aspect-ratio microstructures, which are necessary for modern integrated circuits and MEMS devices. For example, inductively coupled plasma (ICP) deep reactive ion etching (DRIE) is a widely adopted process for etching deep silicon structures with high aspect ratios, allowing the construction of sophisticated micro- and nano-machined devices (Yi et al., 2021).

9.2 Plasma–Enhanced CVD (PECVD)

Plasma-enhanced CVD (PECVD) is an industrial-grade microelectronics technology. This technique is used to deposit thin-film within the industry. It employs plasma to decompose precursors into highly reactive species, and growth can occur directly on noncatalytic substrates such as SiO_2/Si or sapphire at significantly lower temperatures than standard CVD (Neubert & Vergöhl, 2018). We mainly use PECVD technique for growing SiO_2, ZnO and carbon nanotubes, but also to pattern or modify these materials (Neubert & Vergöhl, 2018). The fundamental advantage of PECVD is that the temperature can potentially be decreased from 400 to 2000 °C to room temperature, allowing temperature-sensitive substrates like polymers to be coated. Plasma can be created in a variety of ways, including electric fields ranging from low frequency (kHz) to high frequency (MHz), as well as microwaves in the GHz range. To create plasma in a practical manner, a wide range of procedure variants have been devised. Plasma sources include hollow cathodes and sputter magnetrons. Magnetic fields are often used to improve plasma density (Wei et al., 2013). Energy is transferred from charged particles to precursors as potential and kinetic energy, even at ambient temperature. Precursors can collide with charged particles. Reactive substances, like as free radicals and charged ions, contribute significantly to growth. High breakdown efficiency allows for easy deposition under optimum conditions without the need for catalysts. PECVD's noncatalytic nature enables for in-situ growth of target materials on a variety of substrate forms, avoiding complex transfer methods (Boyd et al., 2015).

To create 2D materials using PECVD, insert a substrate in a quartz tube and heat to the desired temperature. The substrates used in the experiment include noncatalytic SiO_2/Si, Si, quartz, and sapphire, which yield similar findings, as well as catalytic substrates like Cu, which result in rapid deposition but lack control. Plasma was created by introducing precursors upstream (Chen et al., 2015). PECVD can be classified as microwave (MW-PECVD), radio frequency (RF-PECVD), or direct current (DC-PECVD) based on plasma source power frequencies. Microwave PECVD operates in the frequency range of 0.5–10 GHz. Radio Frequency PECVD functions within the frequency range of 1–500 MHz and Direct Current PECVD does not operate within a specific frequency range like other methods. Instead, it uses a constant direct current to generate plasma (Jiang et al., 2003).

The PECVD technique produces pinhole-free films with great control over thickness and substrate morphology. PECVD deposits thin films that are chemically inert, insoluble, mechanically tough, and thermally durable due to their highly branched and cross-linked structures. Plasma-polymerized thin films have strong substrate adhesion and can withstand solvent treatments, multiple exposures to heat, humidity, environment, and mechanical wear. Depositing a monomer as a stable, thin film on a substrate can be difficult due to wet synthesis techniques requiring complex chemistries, harsh solvents, and surface modifications for molecular attachment. Many of these difficulties are resolved with PECVD because the plasma causes radicalisation in the monomer and activates the substrate surface. PECVD allows for thin film deposition and surface modification, as well as the creation of nanostructures. Such a method has been developed for usage with soft matter and is suitable for low temperatures and power (Vandenabeele & Lucas, 2020).

9.3 Role of Plasma in Nanomaterial Modification

Plasma is an important tool for modifying nanomaterials, as it provides a diverse way for surface treatment and functionalisation while also improving material characteristics. Plasma modification of nanomaterials has attracted significant attention due to its effectiveness in manipulating surface properties without affecting the material's bulk characteristics. Also, it is critical for increasing the affinity between fillers and the matrix and creating repulsive interactions between particles. There are various ways, including wet technique and plasma treatments. Plasma treatments offer technical, environmental, and economic advantages over wet procedures by requiring fewer chemicals and stages. Plasma treatments can be used to graft a range of chemical functions, which increases the dispersion of nanomaterials and their attachment to the matrix (He et al., 2022). Nanomaterials, which have found uses in sensors, adsorption, separation, and photocatalysis, can be synthesized and processed quickly using Dielectric Barrier Discharge (DBD) plasma. DBD plasma alters the surface of nanomaterials by revealing active sites, refining electronic structures, or changing surface characteristics like adsorption capacity, luminescence, and catalytic performance. When the plasma comes into contact with the material's surface, active groups may be chemically reduced, oxidized, etched, doped, or grafted (Peppas & Langer, 1994).

10 Biological and Biomimetic Approaches

Biomaterials are made up of all or a portion of a living structure that performs, enhances, or replaces a natural function. These biomaterials are typically designed for medical biotechnology and pharmaceutical purposes. A very significant aspect of a biomaterial is that it is biocompatible, which indicates that it remains inactive in the host or does not react with the host material (Gebeshuber, 2022). Biomimetic nanotechnology focusses on the fundamental characteristics of living systems and how their properties might be applied to humans. The majority of biological materials, structures, and processes rely on nanoscale capabilities. Biomimetics of nanoscale features in living systems is a demanding, exciting, and rewarding field. However, because most biological activities are naturally multifunctional, it might be difficult to separate specific traits that are attractive for potential revolutionary technological applications (Hadjipanayis & Siegel, 1994).

10.1 Bio-Templates for Nanostructures

Various methods have been developed for manufacturing nanomaterials, including milling, chemical, and lithographic processes. The main limitation is the lack of control over the final morphology of nanostructures. The template synthesis process involves preparing micro- and nanomaterials with a desired shape to enhance nanostructure order (Kivelson & Heeger, 1988). A template is a core structure that forms a network and creates a hollow with similar morphological and stereochemical properties when removed. Nanostructures are synthesized within porous templates using several approaches. Templates with uniformly sized cylindrical pores produce monodisperse nanocylinders of the required material within their voids. Nanocylinders can be a nanorod or a nanotube based on operating characteristics. Nanostructures can either remain inside the template pores or be gathered as free nanoparticles. Alternatively, they can extend from the surface like brush bristles. The template approach allows for the creation of monodisperse nanorods and nanotubules with a variety of shapes. The process has been used to create nanotubules and nanofibrils from conductive polymers, metals, semiconductors, carbon, and other materials (Ozin, 1992).

Template-based synthesis can be categorized into three modes: negative, positive, and surface step edge methods. Negative template synthesis is the most widely used of the three categories listed. Template pores are filled and deposited with appropriate materials using electrochemical processes. After dissolving or removing the pattern, free-standing photocopies of wires, whiskers, and cylinders are created. Nanomaterials replicate the geometry and morphology of the template pores. Template materials include polymeric foils, mica, and glass (Fullam et al., 2000). The Positive Template Method involves depositing desired material on the outer surface of wire-like template structures. It produces wire/tube-like structures after removing the template (Zach et al., 2000). Surface step-edge templates generate conductive metal oxide by selectively depositing materials on defect spots in the substrate (Singh & Chakarvarti, 2016).

Bio-templates are organic and biological in nature. These bio-templates have the advantage of having highly ordered morphologies, well-defined structures, and organizations, which results in considerably superior desirable nanomaterial characteristics. There are several types of bio-templates, such as bacteria, textiles/paper, hair cells, insect wings, spider silk, wool, and wood. However, just a few structures have been documented that have showed technological interest and are briefly examined. Butterfly wings can be utilized to create nanostructures with better optical properties, making them more suitable for photonic sensor applications than artificial ones. Butterfly wings are made up of several small scales that are divided into more than two layers separated by air. Now, these thousand-scale layers are responsible for various light reflections. The reflectance spectrum encompasses the entire visible range but decreases in the UV area. Wings with highly precise morphology in the form of aligned lamella make their perfect replica, coupled with superior optical qualities (Wang et al., 2014). Butterfly wings are employed as quasi-periodic resonance cavities in optoelectronic systems, functioning as lasers with random lasing (Zhang et al., 2009). Butterfly wing scales were employed to synthesize Titania photoanodes, resulting in dye-sensitized solar cells (Herrera-Beurnio et al., 2021).

Bio-templating can be divided into three molecules, biological structures, plant and animal products. Different molecular templates are native or modified proteins, DNA, microtubules, and cellulose. Viral envelopes, bacterial cell surface layer proteins, microalgae, eukaryotic and prokaryotic cells, and biofilms are some of the biological structures. Similarly, Wood, cotton, aquatic plants, green leaves, butterfly wings, and eggshells are Plant and animal products (Wang et al., 2016).

Bio-templating find applications in the field of adsorbents, cancer therapy, bone tissue engineering, light emitting materials, sensors, energy capture and storage and catalysis. Bio-templating is a versatile technique used to create adsorbents with higher adsorption capacities. Examples include eggshell membranes and cotton. Bio-template materials are also employed in targeted drug delivery and cancer treatment, utilizing gold nanoshells produced from adenovirus vectors and tobacco mosaic virus components. Graphene and chitosan-based hydrogels serve as efficient drug delivery systems. Additionally, gold nanoclusters coated with a green-emitting dye are created as imaging probes. Yttrium disilicate, derived from the sponge Luffa cylindrica, is used for light emission applications. Cellulose crystals are utilized as templates to create SnO_2 crystal films for carbon monoxide sensors. Rose petals and lotus leaves are employed to make pressure and strain sensors. Furthermore, rape bee pollen serves as a prototype for food sector sensors that detect the freshness of seafood (Wang et al., 2016).

10.2 Biomimetic Synthesis of Nanostructures

Biomolecules and microorganisms have enormous potential for the development of functional nanomaterials and nanodevices. In one method, biomolecules or microbes can serve as templates and building blocks for nanoparticles assembly to produce complex nanostructures. In another manner, certain biomolecules can self-assemble into organized suprastructures, which can then be exploited to create functional nanomaterials (Zan & Wu, 2016). Otto Schmitt developed the term of biomimetics in 1957. It generally refers to the replication of the structures, traits, models, and compositions of natural items to solve various issues.

The fabrication of biomimetic nanostructures is a bottom-up approach. Biomolecules such as DNA, proteins, peptides, and viruses, as well as microorganisms like bacteria and plant leaves, can be used to create these nanostructures. Biomimetic synthesis of nanomaterials is the most rapidly expanding and promising field. Biomimetic synthesis may be classified into two types: functional biomimetic synthesis (FBS) and process biomimetic synthesis (PBS). Functional biomimetic synthesis mimics certain qualities of natural materials, structures, and systems using different chemicals and processes. Artificial bones made from bio-ceramics, polymers, and composites have been synthesized using diverse ways to achieve high strength, fracture toughness, and biocompatibility. Process biomimetic synthesis is another type of synthesis method, which tries to create various useful nanomaterials/structures by following the synthesis methods, processes, or procedures of natural chemicals and materials/structures. For example, by emulating the protein production process, various unique nanostructures, dendrimer-like structures, pyramids, cubes, and 2D nanoparticle arrays (Kumar et al., 2024).

The discovery of new vaccines is a top priority, with cationic nanostructures improving immune responses. Lipid nanoparticles in mRNA vaccines are also investigated. Biomimetic nanocarriers are being investigated for drug delivery across the blood–brain barrier to treat neuroinflammation, neurodegeneration, and Alzheimer's disease. Bioinspired chlorin-based photosensitisers immobilized on carbon materials are being studied for use in photodynamic and photothermal therapy. Nanoporous polylactic acid microtubes are being developed to simulate capillary vessels, which are necessary for vascularization in engineered tissues. Researchers are working to improve renal vasculature, a major obstacle in generating modified kidneys for donation. Oil-in-water nanocapsules coated with chitosan and containing quorum-sensing inhibitors are being utilized to reduce biofilm growth in the oral cavity. Synthetic nanomembranes are being investigated for use in separation science, chemical engineering, and medicine. Hierarchical materials, ranging from the nano to the macroscale, are being employed to boost photocatalytic effectiveness for water splitting and hydrogen generation (Hadjipanayis & Siegel, 1994).

10.3 Advantages in Sustainable Nanotechnology

Nanotechnology is the manipulation of materials at the nanoscale, which ranges between 1 and 100 nm, and it has emerged as a revolutionary force in sustainable development. Nanotechnology frequently demonstrates enhanced qualities like as strength, catalytic activity, and conductivity. This enables the creation of more efficient and effective solutions for a wide range of applications, including energy generation and environmental cleanup. Nanotechnology can be used to optimize operations in order to reduce raw material and energy consumption, hence minimizing waste and environmental effect. Nanocatalysts, for example, can speed up chemical reactions at lower temperatures and pressures, saving considerable amounts of energy. Nanotechnology advancements make these solutions more scalable and economically viable. As technology develop and production costs fall, the benefits become available to a greater range of companies and communities around the world (Zhao et al., 2019).

Nanotechnology provides innovations that go beyond incremental improvements, addressing social, environmental, and global issues. Nanotechnology breakthroughs are making an impact in fields such as energy, environmental protection, resource management, and healthcare by enabling the development of smart materials and connected devices. Nanoscience and nanotechnology have improved communication skills by bringing together scientific, technical, medical, and other groups, influencing a wide range of professions (Smale et al., 2021). Nanotechnology enables the creation of improved filtration systems that can remove impurities and pathogens from water more effectively than traditional approaches. Nanomaterials are utilized to improve the performance of solar cells, energy storage systems like as batteries and supercapacitors, and to develop new triggers for producing cleaner energy. Nanotechnology drives innovation throughout industries, from healthcare to agriculture, by enabling the creation of novel materials, sensors, and technologies that improve production and efficiency while consuming fewer resources. Nanotechnology provides solutions for monitoring and reducing environmental contaminants, improving carbon capture and storage systems, and increasing the efficiency of renewable energy sources. Nanotechnology helps to monitor and preserve biodiversity, increase agricultural output through precision farming techniques, and provide sustainable packaging materials with little environmental impact (Nagar & Pradeep, 2020). Nanotechnology and materials are also predicted to help achieve a sustainable society by improving water purification (Zhou et al., 2012), lowering CO_2 emissions, and encouraging material circulation through recycling methods. Nanotechnology and materials also benefit human health and well-being by providing wearable biosensors and biomaterials for regenerative medicine.

11 Laser-Assisted Nanofabrication

The use of lasers provides advantages in nanofabrication. Laser irradiation on material surfaces produces a number of effects, including localized heating, melting, ablation, breakdown, photochemical reaction, and the development of different nanostructures. Both photothermal and photochemical processes may be involved. A broad range of laser wavelengths enables deliberate energy delivery to a target material via resonant excitation by delivering photons with matching energy to interact with the target material, resulting in laser-induced bond breakage and bond-selective chemical reactions. The ease with which laser beams can be manipulated encourages the development of mask-free direct laser nanostructure writing techniques (Kabashin & Meunier, 2004).

11.1 Laser Ablation Techniques

A pure laser ablation occurs when the plasma is transparent to laser energy. Radiation ablates the target, resulting in nanoclusters that cool down as they interact with the environment. Nanoclusters created in a gaseous atmosphere can be placed on a substrate to form a thin, nanostructured layer. Ablation in liquids can release material and generate colloidal nanoparticle solutions (Chichkov et al., 1996). The properties of these NPs can be controlled by adjusting the parameters of the laser, the target material, and the surrounding environment.

To understand the mechanisms of NP generation by laser ablation, we must first examine the interactions between a material and an intense laser beam. The process begins when a strong laser beam strikes a material's surface, causing ablation, or material removal (Semaltianos, 2010). The laser's energy is initially absorbed by the material's carriers, such as electrons. The energy is subsequently transmitted to the lattice by electron–phonon interaction. The ablation mechanism is heavily influenced by the laser pulse width in proportion to the electron-phonon coupling period. If when the pulse width exceeds the electron-phonon coupling time constant, the energy is wasted as heat, resulting in melting and vaporization of the material. The creation of a prominent heat affected zone, recast layers, surface debris, mechanical cracks, and other flaws in the area where the laser beam strikes the material surface. This process causes flaws around the irradiated area. Otherwise, the energy is generally contained inside the irradiated area. Because of the ultrashort pulse width, this results in direct sublimation (solid-to-vapor transition); nevertheless, with greater energies, melting and vaporization occur.

Nanoparticle synthesis (NPs) is a simple method for producing nanoparticles via laser ablation. This process involves ablating a solid target in a gaseous or liquid environment and collecting the NPs as nanopowder or colloidal solution. It is environmentally friendly and does not necessitate lengthy reaction times, high temperatures, or complex chemical synthesis techniques. Metallic, semiconducting, polymeric, and complex multi-element alloys are all possible forms of nanoparticles. The properties of the NPs formed are determined by the laser parameters used for ablation, such as wavelength, pulse repetition rate, pulse width, pulse energy, and fluence, as well as the environment in which they are generated (vacuum, controlled gas atmosphere, or liquid). This is an environmentally sustainable and laboratory-safe alternative to other procedures (Serra & Piqué, 2019).

11.2 Laser-Induced Forward Transfer

Laser-induced forward transfer (LIFT) is a digital printing technology that uses a pulsed laser beam to project material from a donor thin film onto a receiving substrate, where it is eventually deposited as a voxel. This working concept enables LIFT to operate with both solid and liquid donor films, giving the technology with a remarkable broad range of printable materials and making it very competitive against other digital technologies such as inkjet printing. Not only does LIFT have access to a far wider variety of ink viscosities and loading particle sizes, but the ability to print from solid films enables the single-step printing of multilayers and complete devices, as well as 3D printing (Morales et al., 2017).

The first process related to LIFT was in 1969, using a CW Argon laser to transfer inks from a Mylar tape. This was initially called laser writing (LW). In 1986, Bohandy et al. used a pulsed excimer laser to transfer copper, which is considered the origin of LIFT processes. The term LIFT was first used in 1988, when Bohandy used a Nd:YAG laser to transfer copper and silver (Arnold et al., 2007).

LIFT is a non-contact method where a laser is used to transfer material from a donor to an acceptor substrate (Banks et al., 2006). The main components include a donor, an acceptor, a gap, and a laser. Donor consists of several layers, which includes a substrate transparent to the laser, a thin film of the material to be transferred, and sometimes an intermediate layer. Glass is used as the major transparent substrate. Acceptor is the substrate where the donor material is deposited. Its wettability affects the final shape of the transferred material. The distance between the donor film and the acceptor surface, usually ranging from 0.1 to 1000 μm. Typically uses short pulses or ultrashort pulses to reduce thermal effects and obtain smaller voxels (Colina et al., 2006). The wavelength is chosen based on the interaction material, with UV being common. The laser fluence is a key parameter. The laser pulse interacts with the donor, and the induced physical change pushes material toward the acceptor. There is a threshold fluence for the transfer; using values just above this threshold results in the smallest voxels (Banks et al., 2008).

The transfer mechanism is mainly controlled by the solid or the fluid nature of the donor material. The donor film is solid, and material is transferred in the solid phase. This involves thermal stresses or ablation, leading to the mechanical breakage of the donor film. It is used when the donor material needs to remain unchanged during transfer (Patrascioiu et al., 2014). In the fluid LIFT, the transfer is dominated by fluid motion. The laser evaporates the material, creating a cavitation bubble that pushes the fluid toward the acceptor (Cotell & Grabowski, 1992).

LIFT can be used for both additive and subtractive processes. It allows for printing of various active and passive elements, including capacitors, inductors, resistors, and organic polymers for OLEDs and OTFTs. LIFT can also be used to create chemical sensors. LIFT is used for metallization in solar cells, offering personalized free-form metallization patterns. It allows for printing of metallic contacts, fingers, and busbars. LIFT can create 2D and 3D microstructures. It can be used to print vertical structures like bridges, pyramids, and pillars. It can also be used for surface modification, including creating plasmonic or superhydrophobic structures. LIFT is used for transferring biomaterials such as cells, DNA, and proteins. This includes applications like tissue engineering and regenerative medicine. LIFT is used for printing cells in 3D structures, with a focus on maintaining cell viability. Hydrogels are often used as a matrix for cells (Arnold et al., 2007).

11.3 Pulsed Laser Deposition

PLD is a physical vapor deposition technique that uses laser energy to eject material from a target, forming a plasma. The ejected material then deposits onto a substrate to create a thin film. PLD is useful for producing crystalline films due to its flexibility in controlling atmospheric conditions and target chemical complexity. However, this flexibility leads to complexity, often requiring rigorous optimization of growth parameters. The key learning points for PLD include performing initial experiments, understanding laser ablation and growth dynamics, tailoring growth parameters, recognizing the utility of in situ monitoring, and correlating film properties to ablated species (Shepelin et al., 2023).

PLD offers advantages such as high film purity, a relatively simple and inexpensive deposition apparatus, and a high instantaneous rate of deposition. The process has seen significant growth since 1987, with an increasing number of materials being deposited and more laboratories involved in its development (Shepelin et al., 2023).

Thin films produced by PLD have a wide range of applications, including magnetic, electrical, anti-corrosive, reflective, and anti-reflective coatings, as well as uses in electronic materials, sensors, and actuators. The PLD process involves three critical stages: laser ablation, plasma formation, and film nucleation and growth. Instrumentation includes a vacuum chamber, a target, a substrate holder, and a laser. The laser, typically an excimer or solid-state laser, is used to ablate the target material. The laser's pulse length affects the ablation process with nanosecond lasers being the most commonly used for PLD. The vacuum chamber's base pressure is usually between 10^{-6} mbar and 10^{-9} mbar, which influences the rate of contaminant formation and the mean free path of species. Target holders can be rod-shaped or disk-shaped, with each having advantages and disadvantages. Substrate holders include heaters to enable film growth at temperatures up to 1000 °C or even 1500 °C with laser-based heating (Hubler, 1992).

PLD is widely used for the fabrication of high-temperature superconducting thin films. PLD is used for creating passive microwave devices for communications technology and high-speed digital technology. PLD can deposit biocompatible coatings such as hydroxylapatite for biomedical applications. PLD is a useful method for the growth of epitaxial ferroelectric films, which have applications in non-volatile memories and electro-optics. PLD is also used for growing ferrite films for applications in microwave circuits. PLD opens possibilities for integrating different functions, such as superconductivity, ferroelectricity, and magnetism, into compact, planar electronic circuitry (Miskovic & Kaufhold, 2022).

12 Advanced 3D Printing in Nanotechnology

Printing is an additive process. Meaning materials are placed only where they are needed which results in the reduction of waste production where it also happens to be more efficient than subtractive method[1] (Muldoon et al., 2022). Thin films, coatings, bulk 3D constituents of varied properties can all be created by 3D printing which constructs objects layer by layer. Nanomaterials are embedded for the sole purpose of enhancing strength, reduction of weight and tunable photonic emissions (Shah et al., 2019).

12.1 Nano-Resolution Additive Manufacturing

Minuscule technology is a fad plaguing the globe. Along with that, additive manufacturing has also found their way into microscale 3D fabrication. Some technologies have achieved to sub-micron resolutions where the defined nanocriterion is ≤500 nm such as laser-induced photoreduction, meniscus-confined electroplating, SICM configurations, focused ion beam-induced deposition (FIBID) in addition to electrohydrodynamic printing (EHD). A broad array of materials can be processed where structures with resolutions below 100 nm is obtained via optimizing the parameters and ink. The EHD technique stands out compared to others for additive manufacturing of sub-micrometer structure because of allowing straightforward integration into constructing process lines. In EHD, controlled environments or

[1] Miskovic and Kaufhold, "Additive Manufacturing for Nano-Feature Applications."

pre-fabricated carrier substrates are antiquated. These characteristics separate this particular method from others. All these traits conclude the dominance of EHD printing for additive manufacturing in Nano-scale (Chen et al., 2010).

12.2 Applications in Micro-Nano Devices

The usage of micro- and nano-scale structures in 3D printed drug delivery systems are recently progressing which enabled drug release in a controlled way according to the patient. Also, blood vessels have been 3D printed to customize the vasculature structure, comprising porous features that facilitate the flow of cells and nutrients. Furthermore, their role in 3D printed scaffolds for tissue engineering fosters bio-compatibility and generation of tissue.

12.2.1 Physical Sensor

The successful fabrication of many sensor devices implies the effectiveness of 3D printing techniques in sensor fabrication including acoustic, tactile, microcantilever, transistors, strain, pressure photosensors, etc. For electronic skin and human-machine interface, touch sensor could be the contender. For the purpose of navigation of drones, strain and pressure sensors, MEMS devices, an acoustic sensor is used. An extensive range of applications are accessible such as Antibody detection, electrochemical sensors, wearable electronics, etc. with the usage of materials namely MXene quantum dot, CNT-PDMS, PNIPAM (Kang et al., 2020).

12.3 Challenges in Scaling Down

Even though there are advancements and improvements in the available materials and print quality, when creating larger objects, challenges are faced by material extrusion 3D printers. These challenges are mostly confined to the design aspects as object being printed gets bigger; it is difficult to scale up the size of prints. Therefore, there are well known challenges faced in upscaling for industrial purposes (Mu et al., 2022).

13 In Situ Monitoring and Characterization

Variable temperature assembly, humidity chamber, variable pressure device are all accessories which can be paired to In situ XRD which could investigate the phase transformations and stability in the programmed heating, high humidity, hydrothermal synthesis in photocatalyst materials.

Zhang, Shen, et al. (2022) used in situ time-resolved XRD to analyze the creation of Zn-rich (Ga1-xZnx) (N1-xOx) solid solution photocatalysts under high temperature and pressure conditions. AbouelNour & Gupta (2022) created a hexagonal/monoclinic WO_3 phase junction that split photogenerated carriers efficiently. The phase transition of WO_3 from hexagonal to monoclinic took place at 400 °C which was revealed by In situ XRD. Instead of focusing on the photocatalytic reaction, most in situ XRD techniques applied in photocatalysis were onto the process of manufacturing semiconductor materials.

In Situ X-ray Absorption Spectroscopy (XAS)

In order to study the geometric and electronic structure of active sites of photocatalysts, a powerful tool like X-ray spectroscopy is essential. In situ XAS methods monitor the dynamic behavior of geometric structure and electronic environment of catalytic sites which offer insights into molecular structure activity for photocatalytic systems. In particular, XANES provide high signal to noise ratio, relative insensitivity to thermal disorder and short acquisition times turning it suitable for in situ studies at high temperatures. As it is challenging to achieve efficient illumination in the reaction setup for in situ XAS, there use in exploring photocatalysts remains scarce.

Metal additive manufacturing (AM) provides benefits like unique designs, improved material properties, and faster production, but difficulties in sustaining part quality stays on. Monitoring key process factors such as molten pool behavior and surface texture is a must. In situ and real-time monitoring methods are presently the most popular (Ebad-Sichani et al., 2023).

13.1 Real-Time Monitoring during Fabrication

In situ process monitoring methods allows a closed-loop AM process that can identify and correct defects which ensures the consistency and stability in production. Including machine learning and monitoring techniques into these processes ensured for continuous evaluation of material deposition and developing in situ intervention methods to rectify defects. It has been utilized across all AM classes to gain deep understanding. These methods give an astounding way to monitor and control print quality by identifying defects, reducing material waste and enabling defect correction (Betancourt & Brannon-Peppas, 2006).

13.2 Characterization Techniques for Nanostructures

There are various advanced techniques used to interact with nanoscale features, classified into three main categories: topology, internal structure, and composition. Topological techniques include field emission scanning electron microscopy

(FESEM)—an advanced technology for capturing the microstructure image of the materials. Scanning probe microscopy (SPM)—a type of microscopy that images a surface using a physical probe that scans the specimen. And optical microscopy (confocal and NSOM). Techniques for investigating internal structures contain transmission electron microscopy (TEM), magnetic resonance force microscopy (MRFM), and X-ray diffraction (XRD) (Seniutinas et al., 2017).

14 Industrial Applications of Nanofabrication Techniques

Nanotechnology has been widely applied across various fields, including electronics, environmental remediation, medicine, and energy. In the electronics industry, nanoscale materials are essential for manufacturing devices such as transistors and solar cells. In the medical field, nanoparticles play a crucial role in drug delivery and imaging. Within the energy sector, nanotechnology is being explored to enhance the efficiency of batteries and solar cells. Additionally, nanoparticles contribute to environmental applications by aiding in the removal of contaminants and pollutants.

14.1 Applications in Electronics and Photonics

Nanofabrication is essential for producing components smaller than 1 μm, enabling precision at the atomic scale. This technology plays a crucial role in manufacturing highly efficient transistors, sensors, and memory components. Carbon nanotubes (CNTs) and graphene are particularly well-suited for use in these electronic devices due to their superior electrical, mechanical, and thermal properties. Moreover, nanotechnology contributes to the development of stronger, lighter, and more durable materials, broadening its industrial applications (Nandipati et al., 2024).

14.2 Energy Storage and Conversion

Nanostructured arrays offer innovative platforms for developing efficient energy storage and conversion devices. Traditional planar architectures in energy storage systems have limitations that are difficult to overcome without restructuring their design. The integration of ordered nanostructure arrays has emerged as an effective strategy for improving energy storage efficiency.

Electrochemical methods have demonstrated high efficiency in portable high-power energy storage applications. These include ion batteries, fuel cells, and supercapacitors. Ion batteries and fuel cells store and generate energy through redox reactions, whereas supercapacitors rely on the alignment of electrolyte ions at the electrode interface to form and release electrical double layers. Additionally, electrochemically active electrodes enhance energy storage capacity during charge-discharge cycles. To surpass the performance limitations of conventional planar designs, nanostructured arrays are integrated, providing high surface-to-volume ratios that enable faster and more efficient energy conversion and storage.

14.3 Drug Delivery and Biomedical Devices

Micro- and nanofabrication techniques allow precise control over size, morphology, topology, and functionality, significantly impacting cell biology, molecular biotechnology, and medicine. These advancements enable researchers to study biomaterial interactions at the cellular and molecular levels, leading to the development of synthetic systems capable of modifying physiological responses.

Nanofabrication plays a key role in developing advanced drug, peptide, protein, and gene delivery systems. By precisely controlling size, architecture, and surface properties, these techniques enhance the predictability and performance of both in vitro and in vivo applications, surpassing conventional drug delivery methods.

One recent advancement in controlled drug delivery involves injectable self-assembled micro-reservoirs composed of biocompatible metallic cylinders (such as titanium or gold) loaded with drugs and sealed with temperature-sensitive polymeric membranes. By applying electromagnetic radiation, the drug release can be externally controlled, as the membranes degrade or increase in permeability in response to external stimuli (Liu et al., 2019).

15 Future Trends and Challenges

Optical microscopy has evolved beyond imaging and is now widely used in materials modification and fabrication, as seen in scanning electron microscopy (SEM), focused ion beam (FIB) milling/imaging, and atomic force microscopy (AFM). The fabrication of future nanodevices requires precise three-dimensional (3D) structuring within the 1–100 nm range, allowing integration across platforms such as electronics, nano-fluidics, and photonics. This level of precision is crucial for interfacing with biological and molecular systems.

Nanofabrication techniques for micro-optical, magnetic, fluidic, and microelectromechanical systems (MEMS) devices are rapidly advancing, particularly in FIB applications. FIB is also being explored for resist exposure, partially replacing electron beam lithography (EBL). The introduction of multi-species ion sources, enabling the selection of ions such as silicon (Si) or gold (Au), has further expanded fabrication possibilities.

15.1 Emerging Fabrication Techniques

Along with FIB advancements, micro-optical element fabrication is progressing, leveraging its long depth-of-focus for high-tolerance 3D structuring. This technique is instrumental in creating micro-lenses and high-aspect-ratio holes.

In theragnostic applications, therapeutic compounds are combined with diagnostic agents within a single nanodevice, enabling simultaneous therapy and real-time disease monitoring. However, developing such nanoscale devices presents several challenges. One major issue is ensuring stability between bioactive and diagnostic agents without compromising their individual functionalities. Scaling up production is another significant challenge, as transitioning from laboratory-scale to large-scale manufacturing requires uniformity, reproducibility, and high-quality standards. Additionally, commercialization demands substantial investment in research and large-scale production (Farrell et al., 2011).

15.2 Challenges in Commercialization and Scale-Up

Cost-effective production using a top-down approach for complex nanopatterns remains a challenge. Overcoming these difficulties requires advancements in real-time and post-process technologies, data analysis, and process control. The integration of machine learning (ML) approaches can significantly improve process efficiency by optimizing high-speed metrology, modeling relationships between yield and process parameters, and employing multimodal data fusion for process monitoring.

15.3 Role of AI and Machine Learning in Nanofabrication

Artificial intelligence (AI) is revolutionizing nanofabrication by automating processes that traditionally required human expertise. AI-driven systems can optimize the fabrication process by enhancing precision, efficiency, and scalability.

For instance, AI-assisted capsule design for sustained drug delivery has shown promise in cancer treatment. AI enhances mechanical, electrochemical, and spectral properties of nanosensors, improving their diagnostic capabilities for cancer patients. These advancements allow for early detection, personalized treatment planning, and continuous disease monitoring. AI algorithms can also be integrated into data analysis to identify optimal combinatory treatments. For example, convolutional neural networks (CNNs) can be employed for image analysis during disease monitoring, significantly improving diagnostic accuracy and treatment outcomes (Farrell et al., 2011).

References

AbouelNour, Y., & Gupta, N. (2022). In-situ monitoring of sub-surface and internal defects in additive manufacturing: A review. *Materials and Design, 222*, 111063. https://doi.org/10.1016/j.matdes.2022.111063

Ahn, S. H., & Guo, L. J. (2009). Dynamic nanoinscribing for continuous and seamless metal and polymer nanogratings. *Nano Letters, 9*(12), 4392–4397.

Akcalt, E., Nemoto, K., & Uzsoy, R. (2001). Cycle-time improvements for photolithography process in semiconductor manufacturing. *IEEE Transactions on Semiconductor Manufacturing, 14*(1), 48–56. https://doi.org/10.1109/66.909654

Arnold, C. B., Serra, P., & Piqué, A. (2007). Laser direct-write techniques for printing of complex materials. *MRS Bulletin, 32*(1), 23–31.

Arole, V. M., & Munde, S. V. (2014). Fabrication of nanomaterials by top-down and bottom-up approaches-an overview. *Journal of Materials Science, 1*, 89–93.

Baláž, P., & Baláž, P. (2008). Applied mechanochemistry. In *Mechanochemistry in nanoscience and minerals engineering* (pp. 297–405). Springer.

Banks, D. P., Grivas, C., Mills, J. D., Eason, R. W., & Zergioti, I. (2006). Nanodroplets deposited in microarrays by femtosecond Ti: Sapphire laser-induced forward transfer. *Applied Physics Letters, 89*(19), 193107.

Banks, D. P., Grivas, C., Zergioti, I., & Robert, W. E. (2008). Ballistic laser-assisted solid transfer (BLAST) from a thin film precursor. *Optics Express, 16*(5), 3249–3254.

Baptista, A., Silva, F., Porteiro, J., Míguez, J., & Pinto, G. (2018). Sputtering physical vapour deposition (PVD) coatings: A critical review on process improvement and market trend demands. *Coatings, 8*(11), 402. https://doi.org/10.3390/coatings8110402

Barcelo, S., & Li, Z. (2016). Nanoimprint lithography for Nanodevice fabrication. *Nano Convergence, 3*(1), 21. https://doi.org/10.1186/s40580-016-0081-y

Betancourt, T., & Brannon-Peppas, L. (2006). Micro- and nanofabrication methods in nanotechnological medical and pharmaceutical devices. *International Journal of Nanomedicine, 1*(4), 483–495. https://doi.org/10.2147/nano.2006.1.4.483

Bhattacharyya, B. (2015). Design and developments of microtools. In *Electrochemical micromachining for nanofabrication, MEMS and nanotechnology* (pp. 101–122). William Andrew Publishing.

Biswas, A., Bayer, I. S., Biris, A. S., Wang, T., Dervishi, E., & Faupel, F. (2012). Advances in top–down and bottom–up surface nanofabrication: Techniques, applications and future prospects. *Advances in Colloid and Interface Science, 170*(1–2), 2–27. https://doi.org/10.1016/j.cis.2011.11.001

Boyd, D. A., Lin, W.-H., Hsu, C.-C., Teague, M. L., Chen, C.-C., Lo, Y.-Y., Chan, W.-Y., Su, W.-B., Cheng, T.-C., & Chang, C.-S. (2015). Single-step deposition of high-mobility graphene at reduced temperatures. *Nature Communications, 6*(1), 6620.

Brassat, K., Kool, D., Nallet, C. G. A., & Lindner, J. K. N. (2019). *Advanced Materials Interfaces, 7*, 1901565.

Cai, Z., Liu, B., Zou, X., & Cheng, H.-M. (2018). Chemical vapor deposition growth and applications of two-dimensional materials and their heterostructures. *Chemical Reviews, 118*(13), 6091–6133. https://doi.org/10.1021/acs.chemrev.7b00536

Capasso, F., Allam, J., Cho, A. Y., Mohammed, K., Malik, R. J., Hutchinson, A. L., & Sivco, D. L. (1986). *Applied Physics Letters, 48*, 1294.

Capasso, F., Mohammed, K., Cho, A. Y., Hull, R., & Hutchinson, A. L. (1985). *Applied Physics Letters, 47*, 420.

Chaix, C. (1986). *Presented at Riber MBE user meeting*, York, September 11.

Chang, L. L., Esaki, L., & Tsui, R. (1974). *Applied Physics Letters, 24*, 593.

Chapman, B., & Vossen, J. L. (1981). *Glow discharge processes: Sputtering and plasma etching*. American Institute of Physics.

Chen, H., Wang, L., Bai, J., Hanson, J. C., Warren, J. B., Muckerman, J. T., Fujita, E., & Rodriguez, J. A. (2010). *Journal of Physical Chemistry C, 114*, 1809.

Chen, J., Bo, Z., & Lu, G. (2015). *Vertically-oriented graphene* (Vol. 10, pp. 973–978). Springer.

Chen, K.-S., Lin, I.-K., & Ko, F.-H. (2005). Fabrication of 3D polymer microstructures using electron beam lithography and nanoimprinting technologies. *Journal of Micromechanics and Microengineering, 15*(10), 1894–1903. https://doi.org/10.1088/0960-1317/15/10/015

Cheng, G., & Perahia, D. (2016). *Polymer International, 65*, 39.

Chiappim, W., Neto, B. B., Shiotani, M., Karnopp, J., Gonçalves, L., Chaves, J. P., Sobrinho, A. S., Leitão, J. P., Fraga, M., & Pessoa, R. (2022). Plasma-assisted nanofabrication: The potential and challenges in atomic layer deposition and etching. *Nanomaterials, 12*, 3497. https://doi.org/10.3390/nano12193497

Chichkov, B. N., Momma, C., Nolte, S., Von Alvensleben, F., & Tünnermann, A. (1996). Femtosecond, picosecond and nanosecond laser ablation of solids. *Applied Physics A, 63*, 109–115.

Colina, M., Duocastella, M., Fernández-Pradas, J. M., Serra, P., & Morenza, J. L. (2006). Laser-induced forward transfer of liquids: Study of the droplet ejection process. *Journal of Applied Physics, 99*(8), 084909.

Conrads, H., & Schmidt, M. (2000). Plasma generation and plasma sources. *Plasma Sources Science and Technology, 9*, 441.

Cotell, C. M., & Grabowski, K. S. (1992). *Novel materials applications of pulsed laser deposition*. Cambridge University Press.

Davis, J. T., & Rideal, E. K. (1963). *In interfacial phenomena*. Academic Press.

Davis, M. E., Hsieh, P. C., Takahashi, T., Song, Q., Zhang, S., Kamm, R. D., Grodzinsky, A. J., Anversa, P., & Lee, R. T. (2006). *Proceedings of the National Academy of Sciences of the United States of America, 103*, 8155–8160.

Davisson, & Germer. (1928). *Proceedings of the National Academy of Sciences of the United States of America, 14*, 317.

Delogu, F., Gorrasi, G., & Sorrentino, A. (2017). Fabrication of polymer nanocomposites via ball milling: Present status and future perspectives. *Progress in Materials Science, 1*, 75–126. https://doi.org/10.1016/j.pmatsci.2017.01.003

Ebad-Sichani, S., Sharafi-Badr, P., Hayati, P., & Jaafar Soltanian-Fard, M. (2023). *Nano-fabrication methods*. IntechOpen. https://doi.org/10.5772/intechopen.112429

Ebad-Sichani, S., et al. (2023). Nano-fabrication methods. In *Nanofabrication techniques-principles, processes and applications*. IntechOpen.

Farrell, R. A., Kehagias, N., Shaw, M. T., Reboud, V., Zelsmann, M., Holmes, J. D., Torres, C. M. S., & Morris, M. A. (2011). *ACS Nano, 5*, 1073.

Flory, P. J. (1953). *Principles of polymer chemistry*. Cornell University Press.

Förster, H., Wolfrum, C., & Peukert, W. (2012). Experimental study of metal nanoparticle synthesis by an arc evaporation/condensation process. *Journal of Nanoparticle Research, 14*(7), 926. https://doi.org/10.1007/s11051-012-0926-1

Frey, H. (2015). Chemical vapor deposition (CVD). In *Handbook of thin-film technology* (pp. 225–252). Springer. https://doi.org/10.1007/978-3-642-05430-3_9

Fronius, H., Fischer, A., & Ploog, K. (1986). *Applied Physics Letters, 48*, 1392.

Fullam, S., Cottell, D., Rensmo, H., & Fitzmaurice, D. (2000). Carbon nanotube templated self-assembly and thermal processing of gold nanowires. *Advanced Materials, 12*(19), 1430–1432.

Gates, B. D., Xu, Q., Stewart, M., Ryan, D., Willson, C. G., & Whitesides, G. M. (2005). New approaches to nanofabrication: Molding, printing, and other techniques. *Chemical Reviews, 105*(4), 1171–1196. https://doi.org/10.1021/cr030076o

Gebeshuber, I. C. (2022). *Biomimetic nanotechnology, vol. 2. Biomimetics*. MDPI. https://doi.org/10.3390/biomimetics7010016

George, S. M. (2010). Atomic layer deposition: An overview. *Chemical Reviews, 110*(1), 111–131. https://doi.org/10.1021/cr900056b

Gleason, K. K. (2015). Overview of chemically vapor deposited (CVD) polymers. In *CVD polymers* (pp. 1–11). Wiley. https://doi.org/10.1002/9783527690275.ch1

Gorrasi, G., & Sorrentino, A. (2015). Mechanical milling as a technology to produce structural and functional bio-nanocomposites. *Green Chemistry, 17*(5), 2610–2625.

Grosso, D., Cagnol, F., Soler-Illia, G. J. D. A., Crepaldi, E. L., Amenitsch, H., Brunet-Bruneau, A., Bourgeois, A., & Sanchez, C. (2004). *Advanced Functional Materials, 14*, 309.

Guo, L. J. (2007). Nanoimprint lithography: Methods and material requirements. *Advanced Materials, 19*(4), 495–513. https://doi.org/10.1002/adma.200600882

Hadjipanayis, G. C., & Siegel, R. R. W. (1994). *Nanophase Materials - Synthesis, properties, applications: Proceedings of the NATO Advanced Study Institute*. Kluwer.

He, J., Wen, X., Wu, L., Chen, H., Hu, J., & Hou, X. (2022). Dielectric barrier discharge plasma for nanomaterials: Fabrication, modification and analytical applications. *TrAC - Trends in Analytical Chemistry.* https://doi.org/10.1016/j.trac.2022.116715

Herman, M. A., & Sitter, H. (1996). *Molecular beam epitaxy*. Springer.

Herrera-Beurnio, M. C., Hidalgo-Carrillo, J., López-Tenllado, F. J., Martin-Gómez, J., Estévez, R. C., Urbano, F. J., & Marinas, A. (2021). Bio-templating: An emerging synthetic technique for catalysts. A Review. *Catalysts, 11*(11), 1364. https://doi.org/10.3390/catal11111364

Hou, X. (2016). *Design, fabrication, properties and applications of smart and advanced materials*. CRC Press.

Hu, D., Wang, E., Wang, A., Li, H., & Duan, A. (2023). Synthesis of wrinkle silica nanospheres by solvent evaporation-induced self-assembly method for hydrogenation reaction. *Materials Today Chemistry, 27*, 101280.

Hubler, G. K. (1992). Pulsed laser deposition. *MRS Bulletin, 17*(2), 26–29. https://doi.org/10.1557/s0883769400040586

Huff, M. (2021). Recent advances in reactive ion etching and applications of high-aspect-ratio microfabrication. *Micromachines, 12*(8), 991. https://doi.org/10.3390/mi12080991

Iler, R. K. (1955). *The chemistry of silica*. Wiley.

Iler, R. K. (1979). *The chemistry of silica*. Wiley.

Iqbal, P., Preece, J. A., & Mendes, P. M. (2012). Nanotechnology: The "top-down" and "bottom-up" approaches. In *Supramolecular chemistry*. Wiley. https://doi.org/10.1002/9780470661345.smc195

Jiang, H., Johnson, W. E., Grant, J. T., Eyink, K., Johnson, E. M., Tomlin, D. W., & Bunning, T. J. (2003). Plasma polymerized multi-layered photonic films. *Chemistry of Materials, 15*(1), 340–347.

Johnson, R. W., Hultqvist, A., & Bent, S. F. (2014). A brief review of atomic layer deposition: From fundamentals to applications. *Materials Today, 17*(5), 236–246. https://doi.org/10.1016/j.mattod.2014.04.026

Kabashin, A. V., & Meunier, M. (2004). Laser-assisted methods for nanofabrication. In *Photon processing in microelectronics and photonics III* (Vol. 5339, p. 631). SPIE. https://doi.org/10.1117/12.532281

Kang, M., Liang, J., Wang, F., Chen, X., Lu, Y., & Zhang, J. (2020). *Materials Research Bulletin, 121*, 110614.

Kawamura, G., Muto, H., & Matsuda, A. (2014). Hard template synthesis of metal nanowires. *Frontiers in Chemistry, 2*, 104.

Khanna, V. K. (2016a). *Top-down nanofabrication* (pp. 381–396). Frontiers Media. https://doi.org/10.1007/978-81-322-3625-2_23

Khanna, V. K. (2016b). *Bottom-up nanofabrication* (pp. 397–417). American Scientific Publishers. https://doi.org/10.1007/978-81-322-3625-2_24

Kivelson, S., & Heeger, A. J. (1988). Intrinsic conductivity of conducting polymers. *Synthetic Metals, 22*(4), 371–384.

Knoops, H. C. M., Potts, S. E., Bol, A. A., & Kessels, W. M. M. (2015). Atomic layer deposition. In *Handbook of crystal growth* (pp. 1101–1134). Elsevier. https://doi.org/10.1016/B978-0-444-63304-0.00027-5

Koch, R. (2010). Stress in evaporated and sputtered thin films—A comparison. *Surface and Coating Technology, 204*(12-13), 1973–1982. https://doi.org/10.1016/j.surfcoat.2009.09.047

Kumar, A., Parveen, Z., & Sharma, A. (2024). Nanotechnology for sustainable development: Innovations and challenges. *International Journal of Research Publication and Reviews, 5*, 368–375.

Kumar, S., Saralch, S., Jabeen, U., & Pathak, D. (2020). Metal oxides for energy applications. In *Colloidal metal oxide nanoparticles* (pp. 471–504). Elsevier.

Lee, H., Yoo, P. J., Kwon, S. J., & Lee, H. H. (2004). *The Journal of Chemical Physics, 121*, 4346.

Leskelä, M., & Ritala, M. (2003). Atomic layer deposition chemistry: Recent developments and future challenges. *Angewandte Chemie International Edition, 42*(45), 5548–5554. https://doi.org/10.1002/anie.200301652

Li, P., Chen, X., & Yang, W. (2013). Graphene-induced self-assembly of peptides into macroscopic-scale organized nanowire arrays for electrochemical NADH sensing. *Langmuir, 29*, 8629–8635.

Li, Q., Jia, Y., Dai, L., Yang, Y., & Li, J. (2015). Controlled rod nanostructured assembly of diphenylalanine and their optical waveguide properties. *ACS Nano, 9*, 2689–2695.

Li, W., Wu, Z., Wang, J., Elzatahry, A. A., & Zhao, D. (2014). A perspective on mesoporous TiO_2 materials. *Chemistry of Materials, 26*(1), 287.

Li, Z., Shi, Y., Zhu, A., et al. (2021). Light-responsive, reversible emulsification and demulsification of oil-in-water Pickering emulsions for catalysis. *Angewandte Chemie International Edition, 60*(8), 3928–3933.

Liu, D., Shadike, Z., Lin, R., Qian, K., Li, H., Li, K., Wang, S., Yu, Q., Liu, M., Ganapathy, S., Qin, X., Yang, Q.-H., Wagemaker, M., Kang, F., Yang, X.-Q., & Li, B. (2019). *Review of recent development of in situ/operando characterization techniques for lithium battery research*. Wiley-VCH.

Liu, H., Cheng, J., Chen, F., Bai, D., Shao, C., Wang, J., Xi, P., & Zeng, Z. (2014). *Nanoscale, 6*, 5315–5322.

Madden, J. D., & Hunter, I. W. (1996). Three-dimensional micro-fabrication by localized electrochemical deposition. *Journal of Microelectromechanical Systems, 5*, 24–32.

Madou, M. J. (2018). *Fundamentals of microfabrication and nanotechnology three-volume set*. CRC Press.

Mansoori, G. A., & Soelaiman, T. A. F. (2005). *Nanotechnology–An introduction for the standards community*. ASTM International Chicago.

Meng, Y., Gu, D., Zhang, F., Shi, Y., Yang, H., Li, Z., Yu, C., Tu, B., & Zhao, D. (2005). Ordered mesoporous polymers and homologous carbon frameworks: Amphiphilic surfactant templating and direct transformation. *Angewandte Chemie, International Edition, 44*(43), 7053.

Miskovic, G., & Kaufhold, R. (2022). *Additive manufacturing for nano-feature applications: Electrohydrodynamic printing as a next-generation enabling technology*. Silicon Austria Labs.

Miyajima, H., & Mehregany, M. (1995). High-aspect-ratio photolithography for MEMS applications. *Journal of Microelectromechanical Systems, 4*(4), 220–229. https://doi.org/10.1109/84.475549

Morales, M., Munoz-Martin, D., Marquez, A., Lauzurica, S., & Molpeceres, C. (2017). Laser-induced forward transfer techniques and applications. In *Advances in laser materials processing: Technology, research and applications* (pp. 339–379). Elsevier. https://doi.org/10.1016/B978-0-08-101252-9.00013-3

Morresi, L. (2013). *Basics of molecular beam epitaxy (MBE) technique*. Bentham Science Publishers.

Mu, C., Lv, C., Meng, X., Sun, J., Tong, Z., & Huang, K. (2022). In situ characterization techniques applied in Photocatalysis: A review. *Advanced Materials Interfaces, 22*, 1842. https://doi.org/10.1002/admi.202201842

Muldoon, K., Song, Y., Ahmad, Z., Chen, X., & Chang, M.-W. (2022). High precision 3D printing for micro to nano scale biomedical and electronic devices. *Micromachines, 13*(4), 642.

Nagar, A., & Pradeep, T. (2020). Clean water through nanotechnology: Needs, gaps, and fulfillment. *ACS Nano, 14*(6), 6420–6435.

Nandipati, M., Fatoki, O., & Desai, S. (2024). Bridging nanomanufacturing and artificial intelligence—A comprehensive review. *Materials, 17*(7), 1621. https://doi.org/10.3390/ma17071621

Nasrollahzadeh, M., Sajadi, S. M., Sajjadi, M., & Issaabadi, Z. (2019). An introduction to nanotechnology. In *Interface science and technology* (Vol. 28, pp. 1–27). Elsevier. https://doi.org/10.1016/B978-0-12-813586-0.00001-8

Neubert, T., & Vergöhl, M. (2018). Organic optical coatings. In *Optical thin films and coatings: From materials to applications* (pp. 425–447). Elsevier. https://doi.org/10.1016/B978-0-08-102073-9.00010-2

Nikam, A., Bhagavatula, P., & Kulkarni, A. (2018). Wet chemical synthesis of metal oxide nanoparticles: A review. *CrystEngComm, 20*, 5091–5107. https://doi.org/10.1039/C8CE00487K

Nugraha, R. E., Fauziyah, N. A., & Wira, G. R. (2022). Templated growth formation of mesoporous silica materials: A soft-hard template approach. *Nusantara Science and Technology Proceedings, 318*, 2747.

Oehrlein, G. S. (1986). Reactive-ion etching. *Physics Today, 39*(10), 26–33. https://doi.org/10.1063/1.881066

Ok, J. G., Ahn, S. H., Kwak, M. K., & Guo, L. J. (2013). Continuous and high-throughput Nanopatterning methodologies based on mechanical deformation. *Journal of Materials Chemistry C, 1*(46), 7681–7691. https://doi.org/10.1039/c3tc30908h

Ok, J. G., Park, H. J., Kwak, M. K., Pina-Hernandez, C. A., Ahn, S. H., & Guo, L. J. (2011). Continuous patterning of nanogratings by Nanochannel-guided lithography on liquid resists. *Advanced Materials, 23*, 4444–4448.

Okazaki, S. (2021). Photolithography. In *Handbook of laser technology and applications* (pp. 105–110). CRC Press.

Olivier, G. K., Cho, A., Sanii, B., Connolly, M. D., Tran, H., & Zuckermann, R. N. (2013). *ACS Nano, 7*, 9276–9286.

Oviroh, P. O., Akbarzadeh, R., Pan, D., Coetzee, R. A. M., & Jen, T.-C. (2019). New development of atomic layer deposition: Processes, methods and applications. *Science and Technology of Advanced Materials, 20*(1), 465–496. https://doi.org/10.1080/14686996.2019.1599694

Ozin, G. A. (1992). Nanochemistry: Synthesis in diminishing dimensions. *Advanced Materials, 4*(10), 612–649.

Pala, N., & Karabiyik, M. (2016). Electron beam lithography (EBL). In *Encyclopedia of nanotechnology* (pp. 1033–1057). Springer. https://doi.org/10.1007/978-94-017-9780-1_344

Parsons, G. N., George, S. M., & Knez, M. (2011). Progress and future directions for atomic layer deposition and ALD-based chemistry. *MRS Bulletin, 36*(11), 865–871.

Patrascioiu, A., Fernández-Pradas, J. M., Palla-Papavlu, A., Morenza, J. L., & Serra, P. (2014). Laser-generated liquid microjets: Correlation between bubble dynamics and liquid ejection. *Microfluidics and Nanofluidics, 16*, 55–63.

Pease, R. F. W. (1981). Electron beam lithography. *Contemporary Physics, 22*(3), 265–290. https://doi.org/10.1080/00107518108231531

Peppas, N. A., & Langer, R. (1994). New challenges in biomaterials. *Science, 263*(5154), 1715–1720.

Petkovich, N. D., & Stein, A. (2013). Controlling macro-and mesostructures with hierarchical porosity through combined hard and soft templating. *Chemical Society Reviews, 42*(9), 3721.

Pierre, A. C. (1998). *Introduction to sol-gel processing*. Springer.

Pierson, H. O. (1999). *Handbook of chemical vapor deposition: Principles, technology, and applications*. Noyes Publications.

Pimpin, A., & Srituravanich, W. (2012). Review on micro-and nanolithography techniques and their applications. *Engineering Journal, 16*(1), 37–56.

Platschek, B., Keilbach, A., & Bein, T. (2011). Mesoporous structures confined in anodic alumina membranes. *Advanced Materials, 23*(21), 2395.

Poolakkandy, R. R., & Menamparambath, M. M. (2020). Soft-template-assisted synthesis: A promising approach for the fabrication of transition metal oxides. *Nanoscale Advances, 2*(11), 5015.

Rabenau, A. (1985). *Angewandte Chemie International Edition, 24*, 1026.

Ritala, M., & Leskelä, M. (2002). Atomic layer deposition. In *Handbook of thin films* (pp. 103–159). Elsevier. https://doi.org/10.1016/B978-012512908-4/50005-9

Rodriguez, R. J., Garcia, J. A., Medrano, A., Rico, M., Sánchez, R., Martinez, R., Labrugère, C., Lahaye, M., & Guette, A. (2002). Tribological behaviour of hard coatings deposited by arc-evaporation PVD. *Vacuum, 67*(3-4), 559–566. https://doi.org/10.1016/S0042-207X(02)00248-8

Saito, J., Igarashi, T., Nakamura, T., Kondo, K., & Shibatomi, A. (1986). Fourth international conference on MBE, York, England, 7-10 September.

Sakaki, H., Tanaka, M., & Yoshino, J. (1985). *Japanese Journal of Applied Physics, 24*, L417.

Savic, S., Vojisavljevic, K., Zivojevic, K., Mladenovic, M., & Knezevic, N. (2018). Hard template synthesis of nanomaterials based on mesoporous silica. *Metallurgical and Materials Engineering, 24*(4), 225.

Scarpelli, F., Mastropietro, T. F., Poerio, T., & Godbert, N. (2018). Mesoporous TiO_2 thin films: State of the art. *Journal of Sustainable Materials and Environmental Times, 508*(1), 135.

Schift, H., & Kristensen, A. (2007). Nanoimprint lithography. In *Springer handbook of nanotechnology* (p. 239). Springer.

Semaltianos, N. G. (2010). Nanoparticles by laser ablation. *Critical Reviews in Solid State and Materials Sciences, 35*(2), 105–124. https://doi.org/10.1080/10408431003788233

Seniutinas, G., Balčytis, A., Reklaitis, I., Chen, F., Davis, J., David, C., & Juodkazis, S. (2017). Tipping solutions: Emerging 3D nano-fabrication/-imaging technologies. *Nano, 6*, 8. https://doi.org/10.1515/nanoph-2017-0008

Serra, P., & Piqué, A. (2019). Laser-induced forward transfer: Fundamentals and applications. *Advanced Materials Technologies*. https://doi.org/10.1002/admt.201800099

Shah, J., Snider, B., Clarke, T., Kozutsky, S., Lacki, M., & Hosseini, A. (2019). Large-scale 3D printers for additive manufacturing: Design considerations and challenges. *The International Journal of Advanced Manufacturing Technology, 104*, 3679–3693.

Shahidi, S., Moazzenchi, B., & Ghoranneviss, M. (2015). A review-application of physical vapor deposition (PVD) and related methods in the textile industry. *The European Physical Journal Applied Physics, 71*(3), 31302. https://doi.org/10.1051/epjap/2015140439

Shao, H., Xin, G., Zheng, J., Li, X., & Akiba, E. (2012). Nanotechnology in Mg-based materials for hydrogen storage. *Nano Energy, 1*, 590–601. https://doi.org/10.1016/j.nanoen.2012.05.005

Shepelin, N. A., Tehrani, Z. P., Ohannessian, N., Schneider, C. W., Pergolesi, D., & Lippert, T. (2023). A practical guide to pulsed laser deposition. *Chemical Society Reviews, 52*, 2294–2321. https://doi.org/10.1039/d2cs00938b

Shoup, R. D. (1988). *Ultrasonic processing of advanced ceramics* (p. 347). Wiley.

Singh, V., & Chakarvarti, S. K. (2016). Biotemplates and their uses in nanomaterials synthesis: A review. *American Journal of Bioengineering and Biotechnology*. https://doi.org/10.7726/ajbebt.2016.1001

Smale, E. M., Egberts, T. C. G., Heerdink, E. R., van den Bemt, B. J. F., & Bekker, C. L. (2021). Waste-minimising measures to achieve sustainable supply and use of medication. *Sustainable Chemistry and Pharmacy, 20*, 100400.

Soler-Illia, G. J. A. A., & Azzaroni, O. (2011). Multifunctional hybrids by combining ordered mesoporous materials and macromolecular building blocks. *Chemical Society Reviews, 40*(2), 1107.

Spear, K. E. (1982). Principles and applications of chemical vapor deposition (CVD). *Pure and Applied Chemistry, 54*(7), 1297–1311. https://doi.org/10.1351/pac198254071297

Sproul, W. D. (1996). Physical vapor deposition tool coatings. *Surface and Coating Technology, 81*(1), 1–7. https://doi.org/10.1016/0257-8972(95)02616-9

Stevenson, J. T. M., & Gundlach, A. M. (1986). The application of photolithography to the fabrication of microcircuits. *Journal of Physics E, 19*(9), 654–667. https://doi.org/10.1088/0022-3735/19/9/001

Sun, H., Zhang, X., Miao, L., Zhao, L., Luo, Q., Xu, J., et al. (2016). Micelle-induced self-assembling protein nanowires: Versatile supramolecular scaffolds for designing the light-harvesting system. *ACS Nano, 10*, 421–428.

Suryanarayana, C. (2001). Mechanical alloying and milling. *Progress in Materials Science, 46*(1–2), 1–184.

Tahir, U., Shim, Y. B., Kamran, M. A., Kim, D.-I., & Jeong, M. Y. (2021). Nanofabrication techniques: Challenges and future prospects. *Journal of Nanoscience and Nanotechnology, 21*(10), 4981–5013. https://doi.org/10.1166/jnn.2021.19327

Tibbitt, M. W., Rodell, C. B., Burdick, J. A., & Anseth, K. S. (2015). *Proceedings of the National Academy of Sciences of the United States of America, 112*, 14444.

Tseng, A. A., Chen, K., Chen, C. D., & Ma, K. J. (2003). Electron beam lithography in nanoscale fabrication: Recent development. *IEEE Transactions on Electronics Packaging Manufacturing, 26*(2), 141–149. https://doi.org/10.1109/TEPM.2003.817714

Vandenabeele, C. R., & Lucas, S. (2020). Technological challenges and Progress in nanomaterials plasma surface modification—A review. *Materials Science and Engineering R: Reports, 1*, 100521. https://doi.org/10.1016/j.mser.2019.100521

Vasconcelos, W., DeHoff, R. T., & Hench, L. L. (1989). *Proceedings of the 4th international conference on ultrastructure processing of ceramics, glasses, and composites*. Defense Technical Information Center.

Vazquez, M. (2015). *Magnetic nano-and microwires: Design, synthesis, properties and applications*. Woodhead Publishing.

Vieu, C., Carcenac, F., Pépin, A., Chen, Y., Mejias, M., Lebib, A., Manin-Ferlazzo, L., Couraud, L., & Launois, H. (2000). Electron beam lithography: Resolution limits and applications. *Applied Surface Science, 164*(1–4), 111–117. https://doi.org/10.1016/S0169-4332(00)00352-4

Vyskočil, J., & Musil, J. (1990). Arc evaporation of hard coatings: Process and film properties. *Surface and Coating Technology, 43–44*, 299–311. https://doi.org/10.1016/0257-8972(90)90083-O

Wang, C.-S., Chang, T.-Y., Lin, T.-Y., & Chen, Y.-F. (2014). Biologically inspired flexible quasi-single-mode random laser: An integration of Pieris Canidia butterfly wing and semiconductors. *Scientific Reports, 4*(1), 6736.

Wang, L., Gong, C., Yuan, X., & Wei, G. (2019). Controlling the self-assembly of biomolecules into functional nanomaterials through internal interactions and external stimulations: A review. *Nanomaterials, 9*, 285–312.

Wang, L., Sun, Y., Li, Z., Wu, A., & Wei, G. (2016). Bottom-up synthesis and sensor applications of biomimetic nanostructures. *Materials, 9*(1), 53. https://doi.org/10.3390/ma9010053

Wang, W., Faulkner, D., Moir, J., & Ozin, G. A. (2011). The effect of solvent in evaporation-induced self-assembly: A case study of benzene periodic mesoporous organosilica. *Science China. Chemistry, 54*, 1920.

Wei, D., Lu, Y., Han, C., Niu, T., Chen, W., & Wee, A. T. S. (2013). Critical crystal growth of graphene on dielectric substrates at low temperature for electronic devices. *Angewandte Chemie, 52*, 14121–14126.

Wu, M.-H., & Whitesides, G. M. (2002). Fabrication of two-dimensional arrays of microlenses and their applications in photolithography. *Journal of Micromechanics and Microengineering, 12*(6), 747–758. https://doi.org/10.1088/0960-1317/12/6/305

Xie, Y., Kocaefe, D., Chen, C., & Kocaefe, Y. (2016). Review of research on template methods in preparation of nanomaterials. *Journal of Nanomaterials*. https://doi.org/10.1155/2016/2302595

Xu, R. R., Pang, W. Q., Yu, J. H., Huo, Q. S., & Chen, J. S. (2007). *Chemistry of zeolites and related porous materials*. John Wiley.

Yang, H., & Zhao, D. (2005). Synthesis of replica mesostructures by the nanocasting strategy. *Journal of Materials Chemistry, 15*(12), 1217.

Yang, L., Dai, Q., Liu, L., et al. (2020). Rapid sintering method for highly conductive $Li_7La_3Zr_2O_{12}$ ceramic electrolyte. *Ceramics International, 46*(8), 10924.

Yi, K., Liu, D., Chen, X., Yang, J., Wei, D., Liu, Y., & Wei, D. (2021). Plasma-enhanced chemical vapor deposition of two-dimensional materials for applications. *Accounts of Chemical Research, 54*(4), 1011–1022. https://doi.org/10.1021/acs.accounts.0c00757

Zach, M. P., Ng, K. H., & Penner, R. M. (2000). Molybdenum nanowires by electrodeposition. *Science, 290*(5499), 2120–2123.

Zan, G., & Wu, Q. (2016). Biomimetic and bioinspired synthesis of nanomaterials/nanostructures. *Advanced Materials, 16*, 2099–2147. https://doi.org/10.1002/adma.201503215

Zhang, H., Yang, D., Ji, Y., Ma, X., Xu, J., & Que, D. (2004). Low temperature synthesis of flower-like ZnO nanostructures by cetyltrimethylammonium bromide-assisted hydrothermal process. *The Journal of Physical Chemistry. B, 108*, 3955–3958. https://doi.org/10.1021/jp036826f

Zhang, J., Li, Y., Cao, K., & Chen, R. (2022). Advances in atomic layer deposition. *Nanomanufacturing and Metrology, 5*(3), 191–208. https://doi.org/10.1007/s41871-022-00136-8

Zhang, J. X., Guan, R. F., & Zhang, X. P. (2011). Synthesis and characterization of sol–gel hydroxyapatite coatings deposited on porous NiTi alloys. *Journal of Alloys and Compounds, 509*(13), 4643–4648.

Zhang, W., Zhang, D., Fan, T., Gu, J., Ding, J., Wang, H., Guo, Q., & Ogawa, H. (2009). Novel Photoanode structure templated from butterfly wing scales. *Chemistry of Materials, 21*(1), 33–40.

Zhang, Y., Shen, S., Li, H., & Hu, Y. (2022). Review of in situ and real-time monitoring of metal additive manufacturing based on image processing. *Critical Review, 123*, 1–20.

Zhang, Y., Zhang, L., & Zhou, C. (2013). Review of chemical vapor deposition of graphene and related applications. *Accounts of Chemical Research, 46*(10), 2329–2339. https://doi.org/10.1021/ar300203n

Zhao, Y., Bai, C., Brinker, C. J., Chi, L., Dawson, K. A., Gogotsi, Y., Halas, N. J., Lee, S.-T., Lee, T., & Liz-Marzán, L. (2019). Nano as a rosetta stone: The global roles and opportunities for nanoscience and nanotechnology. *ACS Nano, 13*, 10853–10855.

Zhou, Y. S., Xiong, W., Park, J., Qian, M., Mahjouri-Samani, M., Gao, Y., Jiang, L., & Lu, Y. (2012). Laser-assisted nanofabrication of carbon nanostructures. *Journal of Laser Applications, 24*(4), 6046. https://doi.org/10.2351/1.4716046

Zhou, Z. J. (2005). Electron beam lithography. In *Handbook of microscopy for nanotechnology* (pp. 287–321). Springer.

Exploring the Realm of Nanotechnology

Advanced Techniques for Material Characterization and Analysis in Nano-Technology

Raja Kaliyaperumal and Karuppiah Nagaraj

1 Introduction

Nanotechnology has emerged as a transformative field, with advancements that depend heavily on the precise characterization of nanomaterials. Understanding the mechanical, chemical, and physical properties of these materials is crucial for their effective application across diverse industries, including electronics, biotechnology, and medicine. Traditional characterization methods, such as light microscopy and standard X-ray diffraction, face significant limitations when applied to nanoscale materials, as they lack the resolution and depth needed to examine atomic structures and surface characteristics at the nanoscale. In order to comprehend and maximize the mechanical, chemical and physical characteristics of nanoparticles, material characterization is a crucial field in nanotechnology. For more uses of nanomaterials in electronics, biotechnology, and medicine, precise characterization techniques are becoming essential (Liu et al., 2012; Shnoudeh et al., 2019). In order to properly realize the planned applications with better efficiency and innovation across several sectors, appropriate characterization aids in the manipulation of nanomaterials. The most common methods used for traditional characterization are light microscopy or conventional X-ray diffraction. However, the inefficiency of these techniques in handling nanomaterials and barely depicting atomic structures makes them very

R. Kaliyaperumal (✉)
Department of Biological and Chemical Sciences, Liberal Arts and Sciences, Mohan Babu University (Erstwhile Sree Vidyanikethan Engineering College), Tirupathi, Andhra Pradesh, India

K. Nagaraj
Surfactant Biomedical Research Laboratory, Center for Global Health Research (CGHR), Saveetha Medical College and Hospitals Saveetha Institute of Medical and Technical Sciences (SIMATS), Chennai, Tamilnadu, India

S. Kulkarni (ed.), *A Journey from Nanotechnology to Picotechnology*, Lecture Notes in Nanoscale Science and Technology 47,
https://doi.org/10.1007/978-3-032-07228-3_4

limited in their ability to describe molecular interactions at the nanoscale. Moreover, these techniques typically yield substantially less information regarding the mechanical properties of nanomaterials, their chemical makeup, and their surface shape.

To address these challenges, this study explores the use of advanced characterization techniques such as scanning electron microscopy (SEM), transmission electron microscopy (TEM), atomic force microscopy (AFM), X-ray photoelectron spectroscopy (XPS), and Raman spectroscopy. Each of these methods offers unique advantages: SEM and TEM provide high-resolution imaging, AFM delivers precise surface topography, XPS enables chemical composition analysis, and Raman spectroscopy offers insights into molecular interactions and vibrational modes. By combining these complementary approaches, the study aims to enhance our ability to analyze and manipulate nanomaterials, enabling the development of innovative applications and overcoming the limitations of traditional techniques.

Its goal is to investigate and unearth novel approaches to the high resolution, sensitivity, and accuracy of nanomaterial characterisation that beyond those of conventional techniques. The study discusses a few novel techniques for imaging and spectroscopy, with a focus on unique benefits in nanotechnology. This investigation highlights the importance of advanced characterization techniques in pushing the boundaries of nanotechnology, ultimately contributing to the design and fabrication of materials for cutting-edge applications in science and industry.

2 New Methods for Material Characterization

2.1 Scanning Electron Microscopy (SEM)

SEM uses a concentrated electron beam to attack the samples surface. The signals derived from the electron interactions within the sample are sufficiently detailed to provide information on the surface topography and composition of the materials. SEM resolution capabilities for in-depth morphological examination make it a popular tool in nanotechnology. The methodology in this way directs the synthesis of nanomaterials for in-depth morphological investigation. High spatial resolution, enabling the provision of nanoscale features. To analyze elements, use EDX. Method that is non-destructive the sample is not harmed and is preserved for additional examination. Seeing surface features in materials at high magnifications and obtaining comprehensive three-dimensional images are two of the many powerful imaging capabilities of scanning electron microscopy.

SEM works by directing a concentrated electron beam onto the sample surface. In order to produce high-resolution photographs that can scan surface topography, texture, and composition, these electrons interact with the material atoms and produce multiple secondary signals. These signals are typically secondary electrons, backscattered electrons and X-rays. The ability to create highly resolved images at magnifications ranging from 10× to over 500,000× is one of the SEM's main advantages. This makes SEM a vital tool for seeing minute details of materials at the nanoscale that are otherwise invisible to the unaided eye or a standard optical microscope. SEM is used extensively in nanotechnology to investigate the morphology of thin films, nanowires, nanofibers, and many other nanostructures (Long et al., 2011; Patil et al., 2017). Apart from that, compositional analysis can be obtained by using SEM with analytical instruments such as EDX. Determining and quantitatively analyzing a sample's elemental composition is another benefit of the EDX. Surface imaging gains the chemical dimension from EDX.

SEM offers a way to closely examine minute surface details. This is crucial for understanding the structure and morphology of nanomaterials. SEM is able to capture images that resemble three dimensions while maintaining a large depth of field, which enhances its capacity to reveal surface textures and contours. The most varied materials, including metals, ceramics, polymers, biological tissues, etc., can be analyzed with SEM. It offers compositional and morphological information when paired with EDX. It serves as a versatile tool for material research as a result. The fields of materials science, semiconductor technology, biotechnology, and many more are responsible for the creation and characterization of the great majority of contemporary sophisticated materials. Its significance stems from its effect on failure analysis, quality control, and detail surface analysis at the nanoscale for nanotechnology research and innovation.

2.2 *Electron Microscopy by Transmission*

TEM relies upon the transmission principle, which makes a thin sample transmit the beam of electrons. By creating an image on the detector, these transmitted electrons make it possible to see interior structures at atomic resolutions. The highest resolution currently possible for microscopy methods, atomic-level resolution of TEM allows for the observation of individual atoms in materials, hence providing structural information (Hayat, 2012). TEM is crucial to the study of nanomaterial atomic organization and crystallography because it helps scientists better understand the characteristics and hence direct the various nanoparticle synthesis (Almarri et al., 2023; Hosseinkhani et al., 2023; Kaleem et al., 2024; Liu et al., 2023) and development of innovative nanotechnologies (Figs. 1, 2, 3, and 4). Using a very thin sample as a medium, extremely intense electrons are transmitted through it to interact with the material to create an internal, high-resolution image known as transmission electron microscopy. An image that may be enlarged to atomic-scale detail is created when the electrons interact with the atoms as they go through the sample. When examining the morphology, chemical content, and crystallography of a material at the nanoscale, TEM is crucial. TEMs use electron beams rather than microscopes, and because electrons have shorter wavelengths than light, they offer significantly higher resolution than optical microscopes. For the characterization of nanomaterials, including thin films, nanowires, and nanoparticles (Reimer, 2013;

Fig. 1 Synthesis of iron oxide nanoparticles (Stadtlander, 2007)

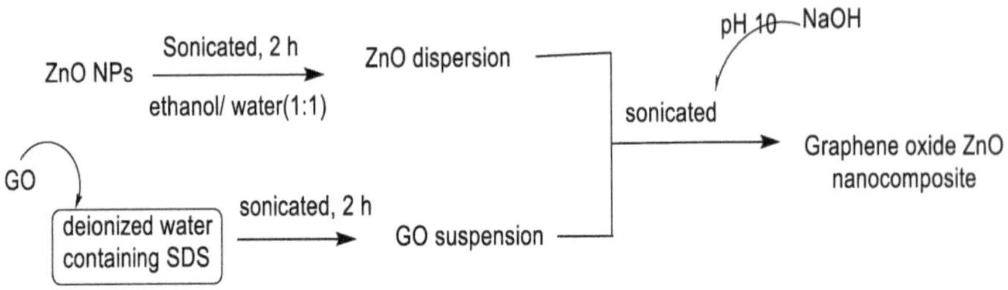

Fig. 2 Synthesis of GO-ZnO nanocomposite (Stevie & Donley, 2020)

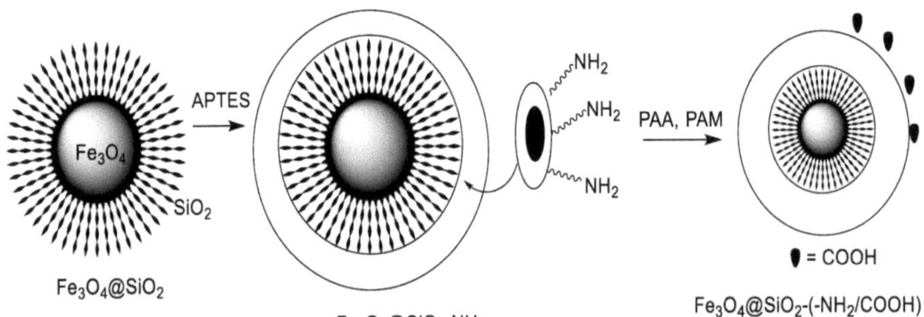

Fig. 3 Synthesis of $Fe_3O_4@SiO_2$-(-NH_2/-COOH) NPs (Van der Heide, 2011)

Stadtlander, 2007), TEM is therefore a crucial instrument and one of the most effective instruments for atomic and molecular imaging resolution, TEM resolution can reach the size of individual atoms.

TEM allows one to see the interior structure of nanomaterials, including crystal orientation and grain boundaries. Crystallography and chemistry composition detail data can be obtained by hyphenating TEM with methods like electron diffraction and EDX. Thus, to investigate the size, shape, and internal structure of the

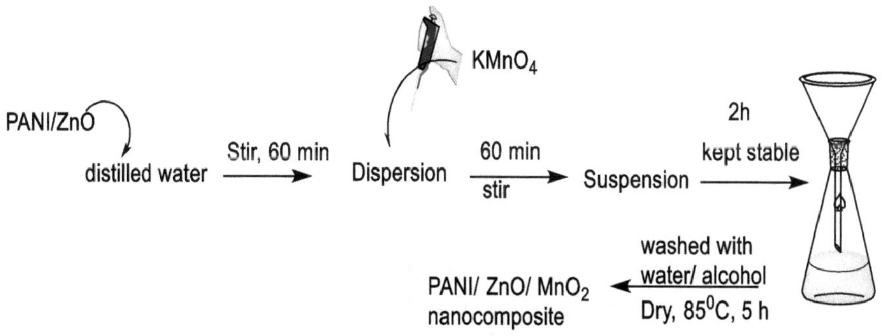

Fig. 4 Synthesis of $CuO/La_2O_3/GO$-based nanocomposite

nanoparticles, TEM is employed. The characteristics of the nanoparticles can be modified for use in electronic, medicinal, and energy-related applications. When it comes to comprehending the crystal structures of nanomaterials and how different atom arrangements impact the material's characteristics. The direct observation of flaws, dislocations, and grain boundaries in a nanomaterial's TEM can provide an image of the structure, which frequently has a substantial impact on both its mechanical and electrical properties. In this sense, the resolution and application of TEM in nanotechnology research are good and diverse. These characteristics make it a powerful instrument in the hands of a scientist to examine materials at the atomic level and direct their research toward gaining some understanding of structure-property correlations. AFM creates a three-dimensional picture of the surface topology by measuring the forces that interact with a material's surface as the tip moves across it (Jandt, 2001). For nano mechanical research, AFM is extremely valuable because it can detect mechanical properties at the nanoscale, such as stiffness, hardness, and adhesion. It also produces high-resolution images. AFM is also used in the characterization of thin films, friction, wear, mechanical reaction on surface phenomena at the nanoscale level, and the surface properties of nanomaterials like nanotubes and nanorods. One high-resolution method of scanning probe microscopy used to examine the surface characteristics of materials at the nanoscale is atomic force microscopy.

Through the use of a cantilever's flexible end with a sharp tip affixed on it, it scans the surface of a sample. The laser beam senses forces between the tip and substance, which create displacements in the cantilever as a result of interactions with the surface. The surface topography was then represented in a very high-resolution 3D picture (James & Robson, 2012). An atomic-level resolution that is superior to that of traditional optical microscopes makes AFM a suitable tool for research on surface structures and nanomaterials. In addition to imaging, force measurements between the tip and the surface enabled by AFM can provide valuable insights on mechanical attributes like stiffness, hardness, and stickiness. When describing a nanomaterial's surface, AFM provides atomic or molecular precision. The mechanical characteristics of friction, elasticity, and adhesion can also be measured with the AFM in addition to imaging. Consequently, it turns into a very useful instrument for comprehending the behavior of materials at the nanoscale. In its various modes, AFM can operate in liquid, vacuum, or ambient environments. It can be used to scan delicate biological samples, biomaterials, polymers, and some kinds of biomolecules in a non-destructive manner. Since AFM offers a wealth of data on surface shape, roughness, and even mechanical properties, it is essentially one of the most widely used techniques in nanoparticle, nanowire, nanotube, and thin-film research. Biomaterials and Biological Samples: AFM is a useful tool for resolving the structure of soft biological material. Biomaterials and biological samples can be investigated at very high resolution, particularly at the cellular, protein, and DNA levels, which are important components of biophysical and nanomedical research (Cheong et al., 2019; Marrese et al., 2017). Surface Engineering and Nanofabrication In order to create innovative materials and technologies, AFM is utilized for surface patterning, nanolithography, and nanoscale studies on tribology features including wear and friction. AFM is an extremely valuable tool in the field of nanotechnology because it allows scientists to gain deeper insights into surface phenomena and material interactions that are critical to the development of next-generation nanomaterials. It can also provide atomic-level visualization and, beyond dimensional characterization, mechanistic insight into materials.

3 Spectral Analysis Applications

3.1 X-ray Photoelectron Spectroscopy (XPS)

The underlying idea behind XPS is that electrons can be emitted from a substance when X-rays pass through it. The main reason for this is that these released electrons have kinetic energy that can be measured (Stevie & Donley, 2020). Elemental composition and chemical states of materials are the main information that XPS offers. as it comes to identifying alterations that occur as materials transition from their bulk structure to nanoscale frameworks, XPS is incredibly helpful in the research of thin-film chemical characteristics. For this reason, understanding surface chemistry, oxidation states, and electrical environments is crucial and can only be accomplished with the use of XPS. Consequently, it turns into an excellent tool for describing thin films and nanomaterials. X-ray photoelectron spectroscopy that uses surface sensitivity to determine a material's electronic structure, chemical states, and elemental makeup. Working principle: X-ray radiation is applied to a sample, causing some of the material's surface to expel electrons. The kinetic energy measurement of expelled (photoemitted) electrons is used to provide details on the elements present and their corresponding oxidation states. Surface analysis, which is crucial for nanotechnology, thin films, coatings, and catalytic research, can benefit greatly from XPS because it usually examines a material's initial 1–10 nm.

Insight into surface contaminants, layer thicknesses, and surface reactions can be obtained, along with knowledge of chemical interactions. In addition to detecting almost all of the elements in the periodic table, with the exception of hydrogen and helium, XPS can also quantitatively ascertain the relative concentrations of each element that makes up a material. Beyond determining an element's elemental composition, XPS can differentiate between an element's many chemical states, such as its oxidation states, and can be a useful tool for learning about the element's chemical environment. Applications requiring thin films, such as corrosion studies and surface alterations, benefit greatly from XPS's strong sensitivity to surface and limited investigation of the material's topmost layers. XPS has been used to characterize the composition and bonding states of nanomaterials and nanocoating's for use in energy storage, electronics, and optical applications. It is an extremely sensitive instrument when it comes to evaluating the oxidation states and corrosion behavior of nanomaterials, which can lead to improvements in material stability and performance. An essential platform for cutting-edge nanotechnology research and development (Van der Heide, 2011), the XPS is a reasonably versatile and non-destructive tool that offers highly accurate surface analysis as well as fundamental insights into the chemical and electronic properties of materials.

3.2 *Electrophoretic Mobility (EPM) and Nanoparticle Tracking Analysis (NTA)*

Electrophoretic mobility (EPM) is used to measure surface charge, affecting nanoparticle (NP) stability and aggregation. Low EPM values indicate larger NP aggregates, while higher values suggest stable, evenly dispersed particles. EPM tracks NP composition changes, such as in silica–polyelectrolyte–iron oxide systems, in conjunction with methods like dynamic light scattering (DLS). Studies comparing electrophoretic light scattering (ELS) have demonstrated that EPM also aids in the calculation of ζ-potential. When combined, these techniques provide thorough understandings of NP stability and behavior in diverse settings.

By monitoring the Brownian motion of individual particles in liquid dispersions, nanoparticle tracking analysis (NTA) is a potent method for examining the size distribution of nanoparticles (NPs). Because it eliminates bias against larger particles and achieves higher sensitivity, it offers advantages over dynamic light scattering (DLS), particularly for bimodal or polydisperse samples. Based on light scattering characteristics, NTA's reliable, real-time tracking of NPs provides precise measurements of particle size, concentration, and occasionally material composition. NTA allows for thorough characterization across a variety of media and improved statistical confidence, albeit being a little slower than DLS.

3.3 Raman Spectroscopy

Raman spectroscopy plays a crucial role in characterizing nanomaterials both during synthesis and post-synthesis, offering invaluable insights into their molecular composition, structural properties, and phase transitions. This technique is nondestructive and highly sensitive, making it essential for comprehensively understanding nanoscale systems (Junior et al., 2015). Raman spectroscopy identifies the vibrational modes of molecules, reflecting their bonding and structural configurations. During synthesis, it provides real-time feedback on the formation of specific chemical bonds, enabling researchers to verify the successful synthesis of desired nanomaterials. Furthermore, various nanomaterials such as metal oxides, semiconductors, and carbon-based structures like graphene and carbon nanotubes exhibit distinct Raman signatures that can elucidate phase transitions or polymorphs occurring during synthesis. The technique assesses the crystallinity of nanomaterials, distinguishing between amorphous and crystalline structures. Sharp, well-defined peaks in the Raman spectra indicate higher crystallinity, while broad or shifted peaks suggest defects or disordered regions (Dresselhaus et al., 2010). Raman spectroscopy can detect defects, strain, and disorder, particularly in carbon-based materials, through the intensity and position of the D-band, G-band, and 2D-band peaks.

Size and Morphology Information for Raman spectra can shift based on nanoparticle size, allowing for the estimation of size and distribution in semiconductor nanoparticles (quantum dots). The sensitivity of Raman to lattice strain and stress is useful for tuning material properties, as shifts in peak positions can indicate these factors arising from particle size effects or substrate interactions. The technique is adept at detecting changes in vibrational modes due to doping with different elements, impacting the electronic, optical, and mechanical properties of nanomaterials. It can confirm the presence of surface modifications through specific vibrational signatures, which is crucial for enhancing performance or stabilization. Raman spectroscopy allows for in situ monitoring of reaction progress during nanomaterial synthesis. This capability to detect intermediate products and track reactant conversions provides feedback essential for optimizing synthesis conditions like temperature and pressure. The technique is sensitive to phase transitions that nanomaterials may undergo during synthesis, such as changes from amorphous to crystalline states. Raman can identify different polymorphic forms, which is vital for applications where specific material properties are required, such as in catalysis or electronic devices.

Evaluation of Chemical Reactions and Interactions: Raman spectroscopy elucidates the chemical mechanisms involved in nanomaterial synthesis by identifying intermediate species and reaction byproducts. It also detects interactions between nanomaterials and their environments, such as molecule adsorption, which is significant for catalyst development and environmental sensors. The technique can monitor temperature-dependent behavior, essential for understanding thermal stability during synthesis. Changes in peak intensities or positions can reveal responses to varying thermal conditions. Similarly, Raman spectroscopy can observe how

Table 1 Summary of nanoparticle characterization techniques

Instruments	Acquired details
XRD (group: X-ray based techniques)	Crystal structure, composition, and grain size reveal nanoparticle arrangement, elements, and crystallite size
XAS (EXAFS, XANES)	X-ray absorption coefficients provide insights into the chemical state of species, interatomic distances, and Debye-Waller factors, even for non-crystalline nanoparticles
SAXS	Particle size, size distribution, and growth kinetics are crucial for understanding nanoparticle properties
XPS	Electronic structure, elemental composition, oxidation states, and ligand binding are critical for understanding surface-sensitive properties in materials
FTIR	Surface composition, ligand binding
Photoluminescence	Structural features like defects, size, and composition affect optical properties, influencing light absorption, reflection, and transmission

nanomaterial properties change under different pressures, particularly during high-pressure synthesis methods like hydrothermal or solvothermal processes. For carbon nanomaterials, such as graphene and carbon nanotubes, Raman spectroscopy is an invaluable characterization tool. It provides insights into the number of graphene layers, defect density, doping levels, and the degree of graphitization. By analyzing the intensities and shifts of the D-band and G-band, researchers can distinguish between ordered and disordered carbon materials, helping to evaluate their quality and functionalization (Lee et al., 2021). Therefore, Raman spectroscopy offers a multifaceted approach to characterize nanomaterials throughout the synthesis process and beyond, enabling researchers to optimize and innovate in nanotechnology applications effectively shown in Table 1.

4 Nanotechnology Applications and Advantages

Nanomaterials can now be more easily characterized thanks to Raman spectroscopy. It gives a material access to high-resolution chemical data. It is very helpful in: Raman spectroscopy is a key instrument used to determine the structural characteristics of CNTs, such as diameter, chirality, and flaws (Georgakilas et al., 2015). It is employed in the assessment of carbon nanomaterials' quality and purity. In Raman spectroscopy, it is extremely crucial for studying graphene, specifically its number of layers, flaws, and strain within layers. The distinctive Raman fingerprints of graphene enable comprehensive study of its structural and electrical characteristics. Raman spectroscopy can be used to investigate surface plasmon resonances in nanomaterials, such as metallic nanoparticles, and to augment signals detected by trace amounts of molecules adsorbed onto the nanoparticles by surface-enhanced Raman scattering. Composites of carbon nanotubes and many other nanostructures are used extensively in biosensors; Raman spectroscopy can be used to examine chemical interactions at the nanoscale. As biological targets, proteins, DNA, or viruses may be detected (Bonhommeau et al., 2022).

SEM offers high-resolution surface imaging with an excellent depth of field, making it perfect for examining surface morphology (Dresselhaus et al., 2010; Junior et al., 2015; Lee et al., 2021; Van der Heide, 2011). TEM provides even better resolution, allowing observation of interior structures at the atomic level, making it vital for deep structural study. AFM excels in providing 3D surface topography at the nanoscale and can function in various conditions, including liquids, which is useful for biological samples. Raman spectroscopy is a non-destructive technique that offers molecular and chemical information, making it useful for identifying and studying materials in biological and aquatic environments. When Raman spectroscopy is combined with a microscope, a powerful technique known as Raman microscopy is achieved, enabling high-resolution measurements on individual nanostructures with spatial precision down to the nanoscale (Bonhommeau et al., 2022; Dresselhaus et al., 2010; Georgakilas et al., 2015; Lee et al., 2021). This technique is versatile and can be employed in a variety of environments, including biological tissues, vacuums, and across all three states of matter solid, liquid, and gas. A key advantage of Raman spectroscopy is its reduced interference from water, making it ideal for analyzing aqueous samples, which pose challenges for infrared spectroscopy due to water's strong absorption in the infrared region. XRD is particularly useful for determining crystalline formations, phase identification, and measuring crystallite sizes. When comparing these techniques, TEM has the highest spatial resolution, followed by AFM and SEM, while Raman and XRD are complementing for structural and chemical research. SEM, AFM, and Raman are non-destructive techniques, however TEM requires thin sample preparation. SEM and AFM are more surface-sensitive, whereas TEM and XRD investigate interior structures, with XRD concentrating on crystallinity (Bonhommeau et al., 2022; Dresselhaus et al., 2010; Georgakilas et al., 2015; Junior et al., 2015; Lee et al., 2021; Van der Heide, 2011). Each technique is appropriate for distinct areas of material investigation, and combining them frequently offers comprehensive insights into structural, chemical, and morphological features shown in Table 2.

Table 2 Key parameters and characterization techniques

Characterization	Entity
Size (structural properties)	Characterization techniques such as TEM, XRD, DLS, and NTA assess size, shape, and composition. SAXS, HRTEM, and SEM provide structural details, while AFM and EXAFS focus on surface features and atomic environments. Techniques like FMR, DCS, and ICP-MS analyze magnetic properties and elemental composition. Optical properties are evaluated using UV-vis and MALDI, while NMR and TRPS aid in understanding molecular interactions. EPLS and magnetic susceptibility measurements further enhance characterization, offering a comprehensive insight into nanoparticle properties
Shape	Effectively determine nanoparticle shape. EPLS and FMR provide insights into electronic and magnetic properties, while 3D tomography offers comprehensive spatial information

5 Conclusion

Characterization of nanoparticles is critical to realizing their full potential in various domains. Advanced techniques including as SEM, TEM, AFM, XPS, and Raman Spectroscopy provide more resolution, sensitivity, and accuracy than conventional approaches. These techniques provide for a more in-depth study of nanoscale structures, surface phenomena, and molecular interactions, as well as key insights into nanomaterial mechanical, chemical, and physical properties. Using these techniques, researchers may better design, modify, and optimize nanomaterials for cutting-edge applications in electronics, biology, energy storage, and medicine. The continued development of these techniques will move nanotechnology forward, fostering ideas with broad industrial applications.

References

Almarri, M. N., Khalaf, M. M., Gouda, M., Heakal, F. E., Elmushyakhi, A., Taleb, M. F. A., & El-Lateef, H. M. (2023). Chemical, surface, and thermal studies of mixed oxides cupric oxide(CuO). Lan thanum oxide(La_2O_3), and graphene oxide for dye degradation from aqueous solution. *JMR&T, 23*, 2263–2274.

Bonhommeau, S., Cooney, G. S., & Huang, Y. (2022). Nanoscale chemical characterization of molecules using tip-enhanced Raman spectroscopy. *Chemical Society Reviews, 51*(7), 2416–2430.

Cheong, L. Z., Zhao, W., Song, S., & Shen, C. (2019). Lab on a tip: Applications of functional atomic force microscopy for the study of electrical properties in biology. *Acta Biomaterialia, 99*, 33–52.

Dresselhaus, M. S., Jorio, A., Souza Filho, A. G., & Saito, R. (2010). Defect characterization in graphene and carbon nanotubes using Raman spectroscopy. *Philosophical transactions of the Royal Society A, 368*, 5355–5377.

Georgakilas, V., Perman, J. A., Tucek, J., & Zboril, R. (2015). Broad family of carbon nanoallotropes: Classification, chemistry, and applications of fullerenes, carbon dots, nanotubes, graphene, nanodiamonds, and combined superstructures. *Chemical Reviews, 115*(11), 4744–4822.

Hayat, M. E. (2012). *Basic techniques for transmission electron microscopy*. Elsevier.

Hosseinkhani, O., Hamzehlouy, A., Dan, S., Sanchouli, N., Tavakkolo, M., & Hashemipour, H. (2023). Graphene oxide/ZnO nanocomposites for efficient removal of heavy metal and organic contaminants from water. *Arabian Journal of Chemistry, 16*, 105176.

James, M. R., & Robson, S. (2012). Straightforward reconstruction of 3D surfaces and topography with a camera: Accuracy and geoscience application. *Journal of Geophysical Research: Earth Surface, 117*(F3), 3017.

Jandt, K. D. (2001). Atomic force microscopy of biomaterials surfaces and interfaces. *Surface Science, 491*(3), 303–332.

Junior, E. S., La Porta, F. A., Liu, M. S., Andres, J., Varela, J. A., & Longo, E. (2015). Atomic force microscopy of biomaterials surfaces and interfaces. *Dalton Transactions, 44*(7), 3159–3175.

Kaleem, M., Minhas, L. A., Hashmi, M. Z., Farooqi, H. M. U., Waqar, R., Kamal, K., Aljaluod, R. S., Alarjani, K. M., & Mumtaz, A. S. (2024). Biogenic synthesis of iron oxide nanoparticles and experimental modeling studies on the removal of heavy metals from wastewater. *Journal of Saudi Chemical Society, 28*, 101777.

Lee, A. Y., Yang, K., Anh, N. D., Park, C., Lee, S. M., Lee, T. G., & Jeong, M. S. (2021). Raman study of D*band in graphene oxide and its correlation with reduction. *Applied Surface Science, 536*, 147990.

Liu, J. W., Liang, H. W., & Yu, S. H. (2012). Macroscopic-scale assembled nanowire thin films and their functionalities. *Chemical Reviews, 112*(8), 4770–4799.

Liu, Z., Lei, M., Zeng, W., Li, B., Liu, D., & Liu, C. (2023). Synthesis of magnetic Fe_3O_4@SiO_2-(-NH_2/-COOH) nanoparticles and their application for the removal of heavy metals fromwastewater. *Ceramics International, 49*, 20470–20479.

Long, Y. Z., Li, M. M., Gu, C., Wan, M., Duvail, J. L., Liu, Z., & Fan, Z. (2011). Recent advances in synthesis, physical properties and applications of conducting polymer nanotubes and nanofibers. *Progress in Polymer Science, 36*(10), 1415–1442.

Marrese, M., Guarino, V., & Ambrosio, L. (2017). Atomic force microscopy: A powerful tool to address scaffold design in tissue engineering. *Journal of functional biomaterials, 8*(1), 7.

Patil, J. V., Mali, S. S., Kamble, A. S., Hong, C. K., Kim, J. H., & Patil, P. S. (2017). Electrospinning: A versatile technique for making of 1D growth of nanostructured nanofibers and its applications: An experimental approach. *Applied Surface Science, 423*, 641–674.

Reimer, L. (2013). *Transmission electron microscopy: Physics of image formation and microanalysis* (Vol. 36). Springer.

Shnoudeh, A. J., Hamad, I., Abdo, R. W., Qadumii, L., Jaber, A. Y., Surchi, H. S., & Alkelany, S. Z. (2019). *Biomaterials and bionanotechnology* (pp. 527–612). Academic Press.

Stadtlander, C. T. K. H. (2007). Scanning electron microscopy and transmission electron microscopy of mollicutes: Challenges and opportunities. *Modern Research and Educational Topic in Microscopy, 1*, 122–131.

Stevie, F. A., & Donley, C. L. (2020). Introduction to X-ray photoelectron spectroscopy. *Journal of Vacuum Science & Technology A, 38*(6), 063204.

Van der Heide, P. (2011). *X-ray photoelectron spectroscopy: An introduction to principles and practices*. John Wiley & Sons.

Nanotechnology in Medicines and Healthcare

Priyanka S. Deorankar, Amol B. Kumbhar, Nilakshi V. Nerkar, and Shakila S. Shaikh

1 Introduction

The term nanotechnology is used to refer the modification and engineering of matter at the nanoscale—generally considered to be between 1 and 100 nm. It encompasses nanotechnology, which is the design, manufacture, and usage of materials and technologies having special qualities that arise because of their nanoscale size, as well as nanoscience, which is concerned with comprehending phenomena at this scale (Karnwal et al., 2024).

At the nanoscale, matter can exhibit peculiar physical, chemical, and biological characteristics in contrast to larger materials, individual atoms, and molecules. Some nanostructured materials have properties that are unique to that size or form and do not exist in larger forms of the same material. Some are better conductors of heat or electricity. With changes in size or structure, they may become more chemically reactive, alter color, or reflects light more effectively (Bhushan, 2017).

It entails developing and utilizing technologies and nanomaterials with special qualities, then incorporating them into more extensive systems for real-world uses. This field holds significant potential to address many contemporary challenges. Its applications are diverse, encompassing the modification of physical properties in existing devices, the development of entirely novel approaches through the development of sophisticated materials with nanoscale properties, molecular self-assembly, and the exploration of direct atomic-level matter manipulation. Through these

P. S. Deorankar (✉) · A. B. Kumbhar · N. V. Nerkar
Department of Pharmaceutics, Rajmata Jijau Shikshan Prasarak Mandal, College of Pharmacy, Pune, India

S. S. Shaikh
Department of Pharmacology, Rajmata Jijau Shikshan Prasarak Mandal, College of Pharmacy, Pune, India

S. Kulkarni (ed.), *A Journey from Nanotechnology to Picotechnology*, Lecture Notes in Nanoscale Science and Technology 47,
https://doi.org/10.1007/978-3-032-07228-3_5

capabilities, nanotechnology promises to revolutionize multiple domains of science and industry (Suhag et al., 2023). Nanotechnology is a hybrid of concepts from chemistry, physics, material science and engineering. As an emerging field the influence of nanotechnology (the application of nanomaterials) is far reaching. Its importance in the twenty-first century is comparable to the revolutionary nature information and approach for electronic components in the previous century. It addresses problems of utmost necessity in sectors like electronics, industry, medicine, and the environment. As a "megatrend" that affects many different industries, nanotechnology propels revolutionary innovation. For many, it is a type of general-purpose technology that will have an impact on science, the economy, and society in the future by offering novel solutions (Bhushan, 2015; Bhushan, 2016).

Methods of producing nanoparticles and nanostructures are typically performed either by the "top-down" or the "bottom-up" technique, whereby the top-down technique begins with bigger particles that cab be reduced through grinding or pulverization, while the bottom-up approach increases them on a molecule-by-molecule or atom-by-atom scale. While the production of nanostructures has only recently begun, the limit to getting things smaller appears to have been achieved. In the latter scenario, process control and preventing random reactions are essential for the creation of nanostructures. The application of the bottom-up procedure has been made easier by the advancement of enhanced microscopy methods such as scanning tunneling microscopy and atomic force microscopy. Nanomaterials utilized in active surfaces, coatings, electronics, and nanostructured materials are a few applications of nanotechnology (Vishwakarma et al., 2013).

2 Rational of Nanotechnology

Materials at the nanoscale exhibit a significantly higher surface area-to-volume ratio, enhancing their mechanical strength and reactivity. In view of this characteristic, nanomaterials are very useful for processes like delivering medications and catalysis. Additionally, quantum phenomena become dominant at this scale, leading to unique behaviors such as improved electrical conductivity, color changes, and altered magnetic properties compared to their bulk counterparts. As per to projections, the world nanotechnology market is expected to expand considerably in years to come, exceeding $125 billion. Investments in this field have surged as governments and private sectors recognize its potential as a driver of economic growth and technological advancement (Nicholas et al., 2020).

Nanotechnology is increasingly recognized as a transformative force in medicine and healthcare, which could bring new innovative solutions that enhance diagnosis, treatment, and disease prevention. Nanotechnology enables the creation of sophisticated medical applications, like innovations for targeted drug administration that can efficiently deliver medicinal substances to particular regions inside the body, therefore mitigating side effects and increasing mode efficacy. This is peculiarly useful in cancer treatment, since nanoparticles provide drugs directly to tumor sites while waiving normal cells. Moreover, nanotechnology improves the accuracy of diagnostics using nanosensors and imaging agents that recognize diseases at their earliest stages by locating cells and molecular biomarkers with high sensitivity and specificity. Nanotechnology works on a scale smaller than most other forms of manipulation of matter, allowing for innovations that can change the face of an industry, or even improve someone's quality of life (Karahmet et al., 2024).

3 Types of Nanotechnology

A wide variety of technologies and applications are included in nanotechnology as mentioned in Fig. 1, which is mainly divided into a number of groups according to its properties and manufacturing processes. Nanomaterials are one of the main categories; they are designed at the nanoscale to have special qualities that distinguish them from the others in bulk. Examples include quantum dots, carbon nanotubes, and nanoparticles; each has a variety of uses in environmental technology, electronics, and medicine. Nanoelectronics is another important field that focuses on creating more compact and effective electronic parts, like memory devices and nanoscale transistors, to improve the functionality of contemporary technologies.

By modifying light at the nanoscale, nano-optics produces sophisticated optical devices that enhance medical imaging and communication. In nanomedicine, nanoparticles employed for high-resolution imaging hardware systems for diagnostic cause and also for tailored drug administration way of systems that reduce unwanted side effects as well as increase the effectiveness of treatment. In addition, nano-energy is the application of nanotechnology to production energy, also nano-batteries and nano-solar cells that offer an improved level of performance and

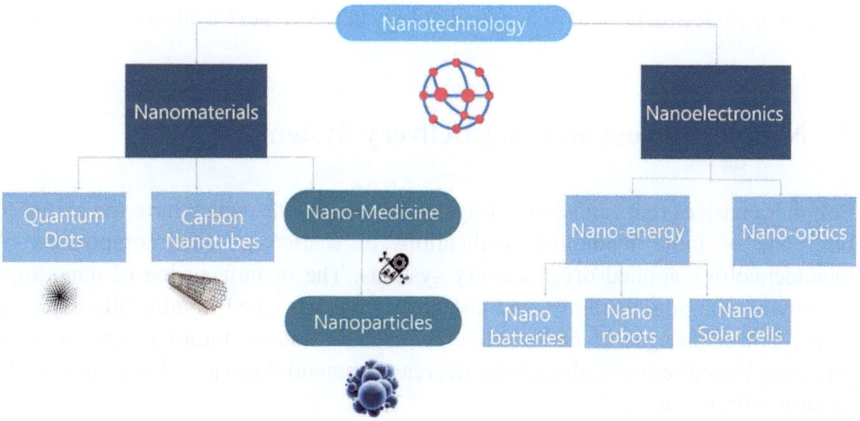

Fig. 1 Different types of nanotechnology

efficiency compared to traditional technologies. Last but not least, nanorobotics is the study of creating minuscule robots that can carry out molecular level operations. This field may find value in medicine for targeted treatments and bodily repairs. These diverse forms of nanotechnology demonstrate how it can revolutionize a number of industries and spur technological innovation. There are various types of nanotechnology, and each has its own characteristics and methods that will contribute to some common (e.g., health, energy, electronics, and environmental science) (Bhatia, 2016).

4 The Significance of Nanotechnology in Medicine

Nanomedicine and bionanotechnology are two of the most promising applications of nanotechnology. Applications will probably be in drug delivery, prosthetic materials, imaging methods, and diagnostics. Additional uses could be as scaffolds for tissue engineering or as structural supports in medical equipment and biological materials. Material surfaces may have micro- and nano- grooves that can direct cell development, but chemical coating also affects cell behavior. Three aspects of nanotechnology—drugs, prosthetics and implants, and diagnostic methods—are especially well-suited to biomedicine. Outside-the-body biomedical devices, such as diagnostic sensors and "labon-a-chip" techniques that can test blood and other samples and be integrated into drug research analytical instruments and development, are gaining a lot of interest. Numerous businesses are creating nanotechnology uses for gene therapy, implanted insulin pumps, and anticancer medications inside the body (Shyam et al., 2012).

Nanostructured materials are being used in implants and prostheses by other researchers. While some techniques are in different parts of testing or are used now, others are just envisaged. Applications in currently developing nanoparticles and longer-term studies utilizing manufactured nanorobots to perform cellular repairs are both included in the field of nanotechnology in medicine (Cavalcanti et al., 2008).

5 Nanotechnology in Drug Delivery Systems

Polymer matrices that can adsorb drugs and nanoparticles which have the ability to bind one or more medicinal medications or scatter are first components of nanotechnology-applied drug delivery systems. The manufacturing of nanodrugs employing imaging, therapies, and diagnostics has advanced significantly in recent years. Lower dosages of nanomedicines result in notable improvements in their pharmacological effects, along with decreased potential of side effects and health hazards (Abid et al., 2023).

5.1 Advancements of Nanoparticle in Drug Delivery

For enhancing pharmacological and therapeutic qualities of medications, lipid- or polymer-containing nanoparticles and drug transportation systems can be created. Drug delivery systems' power lies in its capacity to modify the drug's pharmacokinetics and bio distribution. Nanoparticles offer advantageous qualities that can be utilized to enhance medicine delivery when they are made to evade the body's defense processes. By virtue of their nanoscale size and high surface-area-to-volume ratio, nanoparticles can deliver large amounts of medication into the bloodstream with ease. Drugs that can pass across cell membranes and enter the cytoplasm of cells are among the sophisticated drug delivery systems under development. Efficiency is crucial since a lot of illnesses rely on cellular functions that can only be disrupted by medications that enter the cell. One method that pharmacological molecules can be employed more effectively is through triggered response. They are introduced into the body and become activated only on contact with a given signal. For example, a drug delivery system consisting of dual hydrophilic and hydrophobic surrounding will lead to enhance the solubility of a poorly soluble drug. Additionally, a medicine may injure tissue, although this can be avoided with controlled drug release and drug delivery. A patient may be forced to take larger doses of a medication if it is eliminated from the body too quickly. However, by altering the medication's pharmacokinetics, drug delivery devices may decrease clearance.

Three facts serve as the foundation for the utilization of drug delivery:

(a) effective medication encapsulation,
(b) efficient drug delivery to the desired location of the body.
(c) drug's effective release there.

By tackling issues including poor solubility, instability, and restricted absorption, nanotechnology improves the bioavailability of medications. Nanoparticles greatly increase the solubility of poorly soluble medications, enabling them to dissolve more readily in biological fluids. For example, it has been demonstrated that lipid-based carriers and magnetic nanoparticles (MNPs) improve the solubility and therapeutic effectiveness of substances with low bioavailability, such as curcumin. Furthermore, a greater percentage of the medication reaches its target site undamaged thanks to nanocarriers, which shield medications from deterioration in the gastrointestinal tract or during systemic circulation (Gunasekaran et al., 2014).

Nanoparticles serve as potent agents in controlled drug delivery systems, allowing targeted administration and prolonged therapeutic activity. Such systems are based on mixture compounds to control drug release through diffusion, degradation, stimulation, complex designs, including multi-layered nanoparticles, offer additional control with passive or active sequential or sustained release profiles. Example of such materials are polymeric nanoparticles, whose composition can be designed so they undergo safe and controlled degradation in the body, releasing a drug at controlled rate. Chitosan-based nanoparticles (CSNPs) have received more attention due to the fact that it is a biocompatible form of biopolymer that ensures mummy in blood and enables bi-phasic drug delivery, which is characterized by burst drug release followed by a slow-release pattern (Băcăiță et al., 2023). The use of nanoparticles that can specifically target to the cells or tissues are gaining the attention and are resulting better medication delivery systems. This targeting can be accomplished in two ways: passive and active. Passive targeting exploits the natural relationship between the properties of nanoparticles and the physiology of human tissue for targeted delivery. This is exemplified by the passive targeting of tumor tissues via the tumor-specific enhanced permeation and retention (EPR) effect, wherein the abnormal structure of tumor blood vessels facilitates the accumulation of nanoparticles within the tumor core. This approach is particularly helpful for cancer therapy. This strategy yields high therapeutic concentrations of drugs which selectively accumulate in target tumor sites at low systemic exposure, minimizing toxicity to normal tissues. In contrast, active targeting requires modification of the nanoparticle surfaces and that they bear special ligands that can selectively bind to the target cell-specific receptors (peptides, antibodies, or small molecules). This increased specificity of drug delivery, as the nanoparticles directly interact with the relevant cells or tissues. For example, upon conjugation of antibodies toward tumor-specific antigens, nanoparticles deliver chemotherapeutic agents specifically to cancer cells and, thus, raise the concentration of drugs at the tumor site while saving the healthy tissues. Active targeting is particularly advantageous in overcoming obstacles such as the blood-brain barrier (BBB), where traditional delivery methods may fail. This focused strategy lessens adverse effects on healthy tissues and decreases drug loss (Yuhan et al., 2022).

Nanoparticles can be divided into carbon-based, inorganic and organic nanoparticles based on their chemical structure. Organic nanoparticles less than 100 nm are synthesized from proteins, carbohydrates, lipids, and other organic compounds. Inorganic nanoparticles are useful for applications like hydrophilicity, biocompatibility, non-toxicity and strong stability different from organic nanoparticles. Inorganic nanoparticles include metal salts, metal oxides and elemental metals. Graphene and its derivatives, fullerenes, and carbon nanotubes are examples of carbon-based nanomaterials (Joudeh et al., 2022).

Metal nanoparticle: These prepare metals, including palladium, platinum, silver and gold, from many chemical methods. However as they are nanoscale they have special optical and electrical properties. They can be used in chemical reactions as catalysts, imaging procedures and drug delivery systems. Examples of metal oxides nanoparticles are those composed of iron oxide (FeO_3), zinc oxide (ZnO), and titanium dioxide (TiO_2). Metal oxide nanoparticles are exploitable in many applications such as drug delivery, biological imaging, and environmental clean-up because of their photocatalytic and magnetic properties (Sánchez-López et al., 2020). Ceramic nanoparticles, generally produced using inorganic materials like phosphates or silicates, have high stability, chemical and heat resistance. In addition to coatings and catalysts, they can be used for drug delivery systems (Huang-Chiao et al., 2011).

Carbon-Based Nanoparticles: These comprise of graphene, fullerenes and carbon nanotubes. Carbon-based nanoparticles are robust, conductive and thermally stable, making them useful in materials science, electronics and biomedicine. Due to their exceptional power, electric conductivity, as well as thermal properties, carbon-based nanoparticles are preferred for working in materials, electronics, and biomedicine. Self-assembled amphiphilic compounds form Micelle nanoparticles. These compounds have both hydrophobic and hydrophilic parts. As a result, they can easily aggregate in aqueous systems. This mechanism forms a core–shell structure where the hydrophobic core encloses poorly soluble drugs while the hydrophilic shell stabilizes the micelle in water. In addition, micellar nanoparticles can be designed for stimuli-responsive properties, allowing for controlled release of the encapsulated nanomaterial at certain conditions such as increased temperature or altered pH levels. Both nanoparticulate liposomes and niosomes are lipid-based, but differ in stability and composition, and both are vesicular systems designed for drug delivery. Liposomes range includes their adaptability to host of different drugs, which is independent of solubility ion water, they can be either hydrophilic or hydrophobic, thanks to their phospholipid bilayer. They are useful for delivering therapeutic substances while reducing systemic side effects because they closely resemble biological membranes, which improves their capacity to interact with cells. Liposomes have a high achievability for solubility and are therefore beneficial for transdermal applications. Niosomes, on the other hand, are more chemically stable than liposomes because they are made using cholesterol and non-ionic surfactants. Polymeric Nanoparticles: These organic polymers can be constructed into nanospheres or nanocapsules. Nanocapsules have an oily core with the drug at the center, encased within a polymeric shell, whereas the nanospheres contain a polymeric matrix that retains drugs upon its surface. They are especially useful for controlled medicament release due to their biocompatibility and effective encapsulation. Their biocompatibility and efficacious encapsulation of medicinal substances make them especially valuable for applications involving controlled drug release. Dendrimers are distinguished by their distinct three-dimensional structure and a multitude of functional surface groups that can be customized for particular uses. Both hydrophilic and hydrophobic pharmaceuticals can be efficiently delivered thanks to their ability to encapsulate medications inside of them or affix therapeutic agents to their surface. Enhancing the solubility and stability of medications is one of dendrimers' main benefits in drug delivery; this is especially advantageous for poorly soluble substances. For example, it has been demonstrated that Poly(amidoamine) (PAMAM) dendrimers, via controlled release processes, greatly increase the bioavailability of a variety of medications (Liu et al., 2024).

In the Drug Delivery system, Particulates reduce the volume of dispersion and thus minimize the non-target tissues affected. Nevertheless, inadequate biodistribution persists as significant challenge because it can cause an unintended distribution and possible toxicity to normal tissues. One of the main implications nanotechnology and nanoscience are associated with, one of the most impactful being the potential for new therapeutic agents to be developed. These advancements are expected to allow the creation of drugs with improved effectiveness and less harmful side effects. Potential nanodrugs will function through highly specialized and well-understood processes (Vishwakarma et al., 2013).

5.2 *Nanotechnology in Diagnosis*

Nanotechnology is becoming an indispensable element in the treatment and diagnosis of many diseases as illustrated in Fig. 2. Its unique properties such as increased reactivity, better interaction with biological molecules, and large surface area make it ideal to diagnose a variety of diseases including genetic diseases, cancers and even infections. Nanotchnology has enabled those significant improvements in medical diagnostics, which in turn increased the accuracy, effectiveness, as well as case efficiency in diagnosis of various diseases (Jackson et al., 2017).

The great useful potential lies in the ability of nanometrology to analyze disease symptoms at the molecular scale, and thus opening great prospects for disease diagnosis. Nanometer-sized particles can easily break biological barriers, localize in diseased tissues and bind to specific targets such as viruses, proteins, or nucleic acids. This, attached with a molecular level understanding of the disease, makes diagnosis significantly more sensitive and specific. Nanotechnology has made detection easier through non-invasive or minimally invasive methods like molecular assays, biosensors, imaging, and smart pill cameras (Babu et al., 2014).

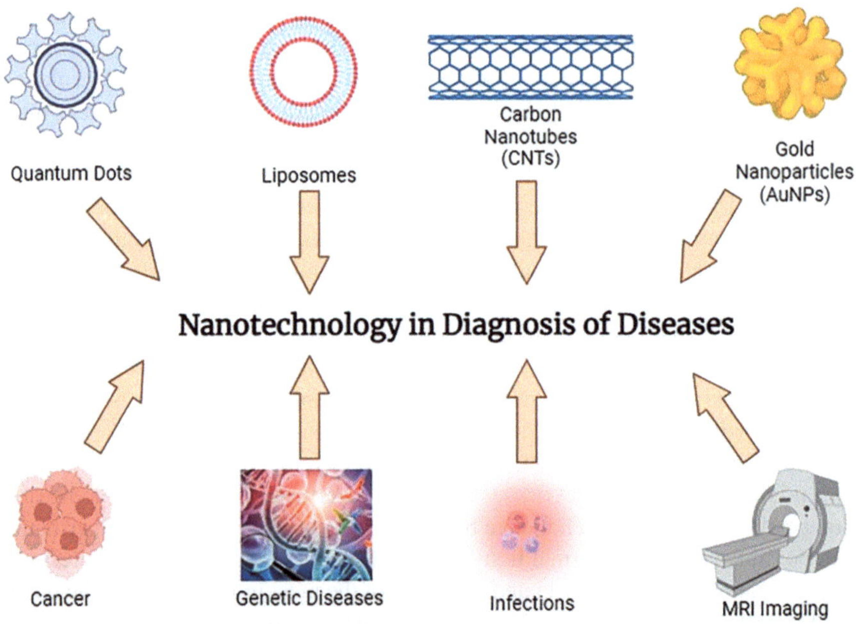

Fig. 2 Nanotechnology in diagnosis of diseases

With their differing properties and versatility, carbon nanotubes (CNTs), gold nanoparticles (AuNPs), quantum dots, and liposomes have become popular in therapeutic diagnostics. Gold nanoparticles offer the best attributes as compared to other types of nanoparticles, which makes them the popular choice despite their optical properties and biocompatibility not being the best. Their main use is in colorimetric assays for infection and biomarker identification. Magnetic nanoparticles improve imaging methods such as, which helps to detect cancer and infections (Liu et al., 2023).

Quantum Dots when exposed to light, these semiconductor nanoparticles glow. Quantum dots can be constructed to emit at different wavelengths, permitting the simultaneous identification of multiple biomarkers. Specifically, they are helpful for flow cytometry and fluorescence imaging. Due to the strong mechanical properties and excellent electrical conductivity that carbon nanotubes (CNTs) possess, they can be used as biosensors. They are useful in electrochemical sensors for the recognition of pathogens or disease-causing molecules due to their detection widespread molecular interactions. Drugs or genes can be encapsulated in liposomes or dendrimers, resulting in nanoparticles utilized in diagnostic procedures. Dendrimers are employed in gene delivery and biosensing applications, while liposomes are utilized in imaging applications (Allahou et al., 2021).

Nanotechnology changes the aspect of cancer diagnosis by allowing early detection of tumor biomarkers at very low levels during early stages of cancer. Magnetic nanoparticles may be utilized to label and eventually detect tumor cells as a result of their functionalization with antibodies directed against certain cancer markers. These also enable detection of circulating tumor DNA or exosomes from tumors in the blood. It helps in diagnosis of infectious diseases on the same principle, for direct and easy detection of specific pathogens like bacteria, virus, and fungus. Magnetic nanoparticles can thus enrich and isolate bacterial cells from body fluids for diagnosis. It transforms genetic screening through the most precise monitoring of mutations and disorders. In autoimmune diseases, magnetic nanoparticles can be used to identify the antibodies against the tissue targets; hence, they diagnose the disease at an earlier stage and provide real-time monitoring capability using biosensors. But all that will matter in the future is all about three things: safety concerns addressed, regulatory challenges taken care of, and accessibility made available (Eissa et al., 2024).

5.3 Nanotechnology in Tissue Restoration and Therapeutic Regeneration

Nanotechnology plays an important function in the development of regenerative therapy and tissue engineering, which are new approaches to repair and regenerate tissues. By producing materials at the nanoscale that mimic the biological environment, nanotechnology helps create scaffolds and biomaterials that promote cell growth, differentiation, and function in tissues (Christ et al., 2013).

The scaffolds created by nanotechnology have biochemical, mechanical, and electrical properties comparable to those of native tissues. Scaffolds created by nanopatterning, electrospinning, and self-assembly improve cell adhesion, proliferation, and differentiation. For example, gold nanoparticle-infused hydrogel scaffolds have been proved to enhance cardiomyocyte function; nanotopographical patterns will direct stem cell differentiation. At this scale, cellular behavior is determined by the interactions between cells and their microenvironment. Carbon nanotubes and nanoparticles are added to the scaffolds to optimize such interactions and thus enable high precision control over cellular responses. These advances contribute to regeneration of the tissue, for instance, myocardial infarction or neural injury with scaffolds assisting retention of cells and integration into host tissues. Novel materials have been developed for the creation of tissue-specific scaffolds aiding in bone, cartilage, nerve and skin tissue engineering. The introduction of nanostructured materials into scaffolds allows these structures to more closely approximate there in vivo counterparts for tissue regeneration. Extensive research is done on Stem cell-seeded scaffolds for bone and cartilage, but isolating stem cells can be an issue and there is also a need for further research on the efficiency of these scaffolds (Kim et al., 2014).

Iron Oxide Nanoparticles (IONPs) are emerging as an important tool for tissue engineering and cell therapy. Such nanoparticles allow for non-invasive therapy monitoring through the administration of bioactive chemicals or their magnetic properties that promote regeneration. Especially, IONPs are being customized for numerous uses, including cardiovascular, osteochondral, and neural defects. Incredible research is being done on the design of "smart" materials able to transmit signals to directly initiate regeneration. Nevertheless, developing ideal materials and assuring biocompatibility and safety remain challenging (Friedrich et al., 2021).

Although the huge potential of nanotechnology in human applications is well promising, safety of nanomaterials is the biggest concern. Before these systems translate to clinical usage, the toxicology of these materials must be rigorously characterized. Efforts of scientists and clinicians are crucial to facilitate the use of laboratory perceptions regarding regenerative therapy and tissue engineering in developing efficient treatments (Khang et al., 2008). Nanotech systems lead to the creation of advanced specialized scaffolds, improve cellular dynamics and develop cutting-edge biomaterials for regenerative medicines and also tissue engineering 20. However, in order to employ the remarkable potential of these technologies in the repair of injured tissue and organs, their safety, efficacy, and clinical translation must be thoroughly investigated (Engel et al., 2007).

5.4 Nanomedicine in Infectious Diseases

Nanomedicine is emerging as an effective weapon against infectious diseases, offering novel solutions for the long-standing problems of treatment and management. On a global level, morbidities and mortalities from infectious diseases, including tuberculosis, malaria and HIV are still considerable, especially in developing countries. These solutions are quite challenging due to solubility, bioavailability, drug resistance, and side effects in most traditional treatment regimes. Nanotechnology, with the help of nanoscale one possible strategy for overcoming these obstacles is the use of drug delivery systems (Unterweger et al., 2020).

Besides these, one major advantage of the nanocarriers is a great improvement in drug delivery to the infected site. For example, lipid-based nanocarriers can enhance the solubility and stability of certain drugs while delivering them directly to infected macrophages in TB. These systems provide double benefits-on-time, delivery-wise, and the drugs would surely reach their site of action or infected cell. Antibiofilm agents have now followed the recent emerging trend in treating pathogenic biofilm infections (Kirtane et al., 2021).

Additional problem arises from inflammation responding to some infections which at times are even more damaging compared to the infection itself. No other challenge seems harder than the increasing threat of multi-drug resistant microorganisms that underline further need for these innovative drug delivery technologies. Nanomedicine has ushered advanced technologies, such as targeted nanoparticles to improve the treatment of infections, while minimizing harm to healthy tissues and adverse effects that are distributed throughout the body (Marepally et al., 2022).

Additionally, nanotechnology plays a vital role in improving vaccines, which are a crucial defense against infectious diseases. Nanoparticles can assist as effective vaccine carriers, improving immunogenicity and bioavailability. In addition, nanomaterials are able to directly target pathogens and inflammation, which raises new therapeutic interception options (Rubey et al., 2021).

Many challenges remain in the clinical development of nanomedicine despite its promise. The ongoing challenges include those of ensuring biocompatibility and long-term safety of nanomaterials, enhancing the properties of nanoparticles (size, shape, surface charge, and so forth) and scaling of production for commercial purposes. In spite of these challenges, the prospects of nanomedicine applications in infectious disease therapy and prevention are considerable, with preclinical and clinical advances opening up pathways to next generation effector modalities (Peters et al., 2021).

5.5 Nanotechnology and Veterinary Medicine

Nanotechnology is revolutionizing the field of veterinary medicine, as it introduces novel approaches to treat and diagnose diseases in animals, facilitate drug delivery, augment vaccine effectiveness, and provide alternatives to conventional therapies.

Nanotechnology provides novel and innovative methods for the diagnosis and treatment of animal diseases; improved delivery of drugs; the enhancement of vaccine efficacy; and the opportunity for alternatives to traditional therapies. Due to their unique properties such as high surface area, small size, and customized surface chemistry, nanoparticles are emerging as popular candidates in numerous veterinary applications. Nonetheless, nanoparticles are still a major prospect in animal health, and any limitations on safety, toxicity and environmental effect, need to be addressed before their image in animal health would be achieved. Nanotechnology is one of veterinary medicine's most revolutionary aspects, and drug delivery is one of its most powerful roles. Nanoparticles can encapsulate drugs, making them more stable, bioavailable, and precise in their release mechanism. It allows targeted treatments that are more efficient and less damaging. Delivering the drug selectively to the tissue or organ makes treatment more potent (Ansari et al., 2024). This also helps in treating chronic diseases or infections for longer periods of time necessitating maintenance medication. Nanoparticle drug delivery systems ensure lesser pharmacophore doses to be administered in stage with less risk of pharmaceutical residues in animal products and also reduce the costs of overall therapy. Besides the advancing delivery of drugs, nanotechnology is now appearing to contribute significantly to the creation of vaccines. The use of nanoparticles as carriers makes Antigen delivery to the body's defenses enhanced, which translates to better immune responses to vaccines, hence protecting animals. Additionally, nanomaterials stabilize vaccines and allow for improved biological activity of vaccines and are also considered adjuvants which further magnify the immune response. This technology has great potential in improving livestock, pet, and other animals' health through the provision of better and more efficient vaccines (Ianiski et al., 2022).

Nanotechnology has advanced diagnostics. Nanoparticles are able to interact with specific biomarkers because of their small size, allowing for the early diagnosis of conditions such as infection or cancer. Such sensitive diagnostic agents would permit disease discovery at very early stages, allowing for very good treatment interventions directly related to the outcome of a patient. When it comes to taking care of health, the early diagnosis is extremely valuable; it can help prevent a serious disease from worsening. Diagnostic instruments employing nanoparticles may therefore be constructed for disease marker identification, affording quicker and more accurate results. Moreover, nanotechnology is contributing to overcome in creating advancements in healing wounds and regenerating tissue. Nanoparticles may repair damaged tissues and keep a well-oriented healing process, even enabling the regeneration of organs while replacing complex medical procedures such as organ transplantation. For example, carbon nanotubes could aid in recovering bone tissue, assisting animals in recovering from severe injuries. Because they are biocompatible and integrate easily into biological systems, nanoparticles might be an attractive way of managing injuries in animal species like horses and pets, where tissue damage can be quite severe, and recovery takes place over long periods (Prasad et al., 2021).

Another area where the role of nanotechnology requires immediate attention is the application of it as a potential solution to the increasingly apparent problem of antimicrobial resistance in veterinary medicine. In the world of animals, the overuse of antibiotics in agricultural settings still drives the development of resistant strains. This trend threatens both animals and humans with dire consequences. Traditional antibiotics may be replaced by antimicrobial therapies that include metallic nanoparticles and other types of nanoparticles. These particles can kill bacteria and are also less toxic and more effective. Antimicrobials have emerged as a promising source of growth for agriculture because they can effectively address pathogens such as Brucella, Mycobacterium bovis, and Streptococcus. This allows a veterinarian to successfully treat a chronic infection while contributing to the solution of the resistance problem (Danchuk et al., 2023).

Nanotechnology possesses vast potential in the area of veterinary medicine but faces abundant challenges. Complications related to safety and toxicity are most significant, because moreover they enter biological systems through unrecognized routes or may affect other unknown biological pathways. The accumulation of nanoparticles in tissues or across ecosystems brings a genuine concern from environmental perspectives. It has to be low-cost, large-scale production that runs on sustainability for access. With these challenges in mind, nanotechnology is being adopted in veterinary medicine by new substances in the form of quantum dots, magnetic nanoparticles, liposomes, and dendrimers. These nanoparticles have been more potent than conventional therapies because of having higher potency and lower toxicity and improved bioavailability. The continuous upgradation of nanotechnology is promising in gene therapy for veterinary applications for genetic diseases and augmenting immune response (Hill & Li, 2017).

As research in nanotechnology progresses, it is expected that its use in veterinary medicine will continue to expand. Countries like Brazil are starting to recognize the potential of nanotechnology in animal healthcare, although the field remains in its early stages. Ongoing research has already shown the potential for nanotechnology to significantly improve animal health, advance veterinary practices, and increase productivity in animal farming. With the evolution of nanotechnology, one can expect its application in veterinary medicine to broaden. Brazilian researchers are realizing that nanotechnology can have far reaching spheres of influence in the domain of animal health, although it is still an emerging field. So far, studied nanotechnology promised better health of animals, enhanced veterinary practices, and increased productivity in animal farming (Jafary et al., 2023). Nanotechnology is poised to revolutionize veterinary medicine by offering cutting-edge approaches to tissue repair, antimicrobial treatments, medication delivery, vaccine research, and diagnostics. The field grapples with safety, regulation, and production cost issues, as the potential initiatives are numerous. The issues such as antibiotic resistance and chronic diseases in animals may find solutions from nanotechnology. Continuing research, development, and the establishment of safety standards could see nanotechnology change the face of veterinary care, with improvement in animal health and eventual benefits to veterinarians and animal owners alike on a global scale (Youssef et al., 2019).

5.6 Nanobiotechnology and Gene Therapy

Nanobiotechnology is an multifaceted combination of nanotechnology and biotechnology originated from Richard Feynman's vision of building objects from bottom-up at CIT in 1959 whereas the actual recognition gained after 2 decades when Eric Drexler written about the concept of molecular nanotechnology in his book "Engines of Creation," where he speculated on nanobots capable of manipulating biological systems. Considering biological components function at the nanoscale, nanotechnology involves creating and assembling structures at the atomic, molecular, or macromolecular levels by successfully changing the basic structure of materials to improve, modulate, or change their properties.

Nanobiotechnology emerged as a specialized field by combining methods from nanotechnology, biology, and biochemistry. This will helpful in biological sciences by enhancing their physicochemical properties. The modified or functionalized nanoparticles can discharge the pharmaceutical or biomolecule at the targeted site whereby reducing the effect on other organs/tissue. The combination of nanotechnology and biotechnology tends to make various innovations with ease in the pharmaceutical field in respect of materials and methods. The tiny size of these makes it easy to penetrate the cells, interact cell organelles, and demonstrate unique effects (Karnwal et al., 2024).

An another budding approach in the biomedicine field is gene delivery which involves transferring genetic material into cells for gene therapy or scientific study. In recent decades, this science has made significant progress in treating genetic diseases that were previously incurable, including cancer and abnormalities of the central nervous system. The first successful gene therapy was invented in 1990 by Dr. W. French Anderson and his team, a young girl with adenosine deaminase (ADA) insufficiency, a rare immunological illness, was treated by introducing an appropriate replica of the ADA gene into her cells. Subsequently, in the 2010s, researchers like Jennifer Doudna and Emmanuelle Charpentier transformed the area of gene therapy with the creation of sophisticated gene editing tools, especially CRISPR-Cas9 (Mendes et al., 2022). Inactivating (or "knocking out") a defective gene that isn't working properly, substituting a healthy gene for a mutant one, and introducing foreign NA into the cells to shield them from illnesses are the three major methods of gene therapy. Inactive viruses are frequently used in gene therapy for transferring genetic material into cells. Despite its effectiveness, this strategy may result in mutations or immunological reactions. In addition, it necessitates a great deal of vector testing and the use of cells to evaluate vector properties, which lengthens manufacturing timelines and increases variability and may denaturize genetic material. Thus, the selection of appropriate gene delivery plays a crucial role for the treatment of cancer. With the help of unique characteristic to interact at the molecular level of nanotechnology the biotechnology and gene therapy becomes more promising (Hamimed et al., 2022; Zhang et al., 2024).

Both gene therapy and nanobiotechnology indicate outstanding potential in the treatment of a several kinds of illnesses, including cancer, hereditary, and chronic ailments. With their novel techniques to gene editing, drug delivery, and illness treatment, both disciplines increase accuracy, efficacy, and minimize adverse effect. Nanoparticles are used to deliver therapeutic genes to correct genetic defects such as lipid nanoparticles. Nano-Insulin—Insulin is encapsulated in nanoparticles to enable controlled, long-acting release, potentially reducing the need for frequent injections in diabetic patients. These are used to deliver the CFTR gene in cystic fibrosis patients to restore normal gene function. Nano-Insulin—Insulin is encapsulated in nanoparticles to enable controlled, long-acting release, potentially reducing the need for frequent injections in diabetic patients. Nanoparticles of methotrexate are commonly used in rheumatoid arthritis, and nanoparticles can enhance its delivery to inflamed joints, improving therapeutic outcomes (Dutt et al., 2023). Researchers are exploring nanoparticle-based vaccines that carry HIV proteins to stimulate a stronger immune response. A type of gene therapy that modifies a patient's T cells to diagnose and attack tumor cells. A CAR-T cell therapy approved for certain types of leukemia and lymphoma, wherein the T lymphocytes of the patient are genetically altered to specifically target cancer cells. Zolgensma (onasemnogene abeparvovec) therapy for spinal muscular atrophy (SMA) that stops further degradation by giving motor neurons a copy of the SMN1 gene. Patients with ischemic heart disease can benefit from angiogenesis, or the creation of new blood vessels, by receiving vascular endothelial growth factor (VEGF) gene therapy. Strimvelis (adenosine deaminase gene therapy)—Used for treating adenosine deaminase deficiency, a condition that leads to severe combined immunodeficiency (SCID), by delivering the ADA gene to the patient's cells. There are ongoing studies to use the CRISPR-Cas9 gene editing technology to repair genetic defects in the pancreas, which may ensure the regeneration of insulin-producing cells (Gentile et al., 2022). Nanobiotechnology and gene therapy are reinventing the treatment of various disease conditions. These fields now provide hope for more effective, personalized treatments because they aid in improving drug delivery, enable the therapeutic targeting of treatments, and also offer a potential cure through gene modification. Such is possibly achieved through nanoparticle-based technology for gene editing and drug delivery to patients in reduced side effects, improved efficacy, and even a cure for chronic and genetic diseases (Mendes et al., 2022).

5.7 *Green Nanotechnology-Driven Drug Delivery Assemblies*

Green nanotechnology is termed for the synthesis, design, and application of nanomaterials using eco-friendly processes. The aims include reducing the ecological footprint, lessening harmful chemicals, and encouraging the usage of renewable resources and health and environmentally safe methods. Drug delivery assemblies driven by green nanotechnology is an encouraging strategy within the pharmaceuticals and medication fraternity, especially in regulated and targeted systems of drug delivery. Eco-friendly, biodegradable, and biocompatible nanoparticles are used in these assemblies to enhance the effectiveness of therapy while minimizing side effects and environmental impact. This can include creating nanoparticles using bacteria, natural polymers, or plant-based extracts. Such techniques reduce the hazardous chemicals, which are mostly used to make nanoparticles, and are thus environmentally and human friendly (Kanwar et al., 2019). The use of green nanotechnology makes sure that nanoparticles used in drug delivery assemblies are biodegradable and biocompatible. This ensures that the body degrades the nanoparticles once the medicine has reached the target location without any likelihood of long-term toxicity or tissue accumulation. Drug delivery nanoparticles could be formed using biodegradable polymers like chitosan, a naturally occurring polysaccharide, or PLGA, which represents the abbreviation for poly(lactic-co-glycolic acid). The biodegradable polymers do not degrade into toxic metabolites thus they are beneficial for the controlled and extended drug release. This will allow for the specific attachment of gold nanoparticles functionalized with targeting molecules such as antibodies or peptides to tumor cells, thus improving the efficiency of drug delivery and reducing side effects from conventional chemotherapy. Green-synthesized chitosan nanoparticles can be used for the delivery of insulin in a sustained-release fashion to control blood sugar levels of diabetic individuals (Kumar et al., 2017).

A safer, environmental-friendly choice of viral delivery methods is by using green-synthesized nanoparticles, wherein nanoparticles are synthesized from eco-friendly materials, such as chitosan and alginate. They can take the form of transferring RNA molecules like siRNA into target cells for gene editing or silencing. Green-synthesized silver nanoparticles have intense antimicrobial property and can also be used as targeting agents of bacterial infections. The nanospheres have applications as topical remedies or within drug delivery carriers in the management of infections, for example, wound infections and urinary tract infections. Nanoparticle-based vaccines acting as antigen delivery vehicles of influenza virus and have been able to enhance immune responses and efficiency of the vaccines. Drug delivery assemblies powered by green nanotechnology: This novel method of increasing medication and biologic delivery uses an environmental-friendly way. The applications of these systems have some benefits, like having less toxicity and higher accuracy in the drug therapy, resulting in better clinical outcomes, utilizing renewable production processes, and utilizing environmental-friendly material components. Medicine is likely to undergo a green revolution with the integration of nanotechnology and gene therapy and drug delivery (Khoobchandani et al., 2020; Pearce et al., 2023).

5.8 Surgical Nanorobotics and Nano-Bioelectric Medicine

Recent advancements in surgical nanorobotics and nanobioelectric medicine have significantly accelerated the progress of medical science, enabling unprecedented precision in diagnosis, targeted therapy, and minimally invasive surgical interventions. Nanorobots are nano-electromechanical devices made to precisely complete a certain task at nanoscale dimensions. Its size is what gives it an advantage over traditional medicine. The behavior of bacteria serves as a biological model for the construction of nanorobots. Some of the various components that constitute a nanorobot's design include onboard sensors, motors, manipulators, power supply, and molecular computers. Professor Toshio Fukuda, the father of micro/nanorobots, developed the first nanorobot computing system in history for single-cell research and manipulation in the early 2000s. Since carbon is strong and inert in the form of fullerene and diamond, it is the primary ingredient used in nanorobots. All of the remaining elements, which are silicon, fluorine, sulfur, nitrogen, oxygen, and hydrogen, can be utilized at the nanoscale. This nanorobot can perform a smart chemotherapy where it could concentrate on neoplastic-specific cells and tissues, shielding healthy cells nearby from the toxicity of chemotherapy drugs and could help facilitate early dissolution of the cancer efficiently through medication administration. When used as drug transporters for timely dose administration, nanorobots enable chemical compounds to remain in the bloodstream for as long as necessary, providing the anticipated pharmacokinetic properties for chemotherapy in anticancer treatments (Aggarwal et al., 2022). The intravenous injection of nanorobots opens the door to their clinical application in diagnosis, treatment, and surgery. It is likely that the patient will receive an intravenous injection of the nanorobots. In addition to absorption, metabolism, and excretion, pharmacokinetics of chemotherapy also involves a rest interval that allows the body to re-establish itself before the next chemotherapy session. Most of the patients with small tumors are treated in 2-week cycles. Nanomedicine and in vivo diagnostics can also establish technology that will be able to function inside the human body to detect and quantify harmful substances and tumor cells, diagnose illnesses sooner, and possibly other applications. In terms of surgery, a surgical nanorobot guided or controlled by a human surgeon may function as a semi-autonomous on-site surgeon when introduced into the body by an intravenous channel or cavities. Surgical nanorobotics is the practice of performing medical procedures, especially surgeries, using nanorobots, which are microscopic robots with a scale of nanometers (1–100 nm). These nanorobots are designed to work inside the human body, performing precise operations such as tissue healing, targeted medicine distribution, and even intricate surgery (Patil et al., 2016).

Surgical nanorobots are revolutionizing the fields of precision and minimally invasive surgery and significantly reducing the need for traditional invasive treatments by performing activities with remarkable accuracy at the cellular or molecular level. These nanorobots can selectively target specific organs or tissues without causing much harm to the adjacent healthy tissue. Therefore, in the case of cancer surgery, it is achievable with nanorobots to administer chemotherapy drugs, which reduces the side effects and ensures that the treatment process has a maximum effect. Among the major applications of surgical nanorobots is targeted drug delivery. Nanorobots reduce toxic effects and minimize waste since drugs will reach affected tissues or organs, meaning that drugs will work precisely where they are applied. For example, nanorobots can be designed to detect cancer cells based on biomarkers found on their surface and to deliver drugs or other therapeutic agents directly to the tumor, thereby improving the treatment outcomes (Yang et al., 2023).

Moreover, nanorobots may be required for healing tissue and cells. They can assist in healing wounds, regenerating damaged tissues within organs such as the liver, heart, or nerves, and repair tissues at the molecular level. For example, nanorobots might be able to repair damaged nerve fibers and promote regeneration for patients with diseases that were thought to be impossible to cure through spinal cord injuries. In addition, nanorobots for surgery are capable of being used for diagnostic and biopsy purposes. Such devices can be introduced into tissue or blood circulation that carries diagnostic substances, which are capable of detecting cancer at very early stages or allows retrieval of samples for biopsy. Compared to conventional techniques, nanorobots would enable far earlier and more precise diagnoses by detecting molecular markers in the blood or tissues.

Although this technique has a number of advantages, there are a number of complications that come in the way of the development of surgical nanorobots. It is essential that the nanorobots not cause unintentional damage to humans or elicit undesirable reactions within the body. Hence, safety and control become essential issues. The key technological challenge that arises here would be the fact that nanorobots must be produced in volumes large enough for therapeutic application. It will raise some issues that are also a bit of regulation and ethics concerns: the treatment of permission problem from the patient, how privacy issues regarding patient data may occur, and so on. In spite of the above-mentioned difficulties, surgical nanorobots have great potential for advancing surgery and its results, and further research will continue to enhance these technologies into broader clinical applications. So nano-bioelectric medicine can be defined as a type of medicine which modifies the electrical characteristics of biological tissues in order to achieve some therapeutic effect with the aid of nanotechnology. This includes the use of nanoscale materials and devices in order to change the electrical signals in the cells, tissues, or organs for the management of many medical conditions like chronic illnesses, injuries to the nervous system, and even psychological problems (Mahima et al., 2024).

Nano-bioelectric medicine represents a promising new field in medicine that utilizes electrical signals to control or stimulate the nervous system in the treatment of various conditions including Parkinson's disease, epilepsy, and chronic pain. The most advanced technique that has been developed is the one using nanowires, the most innovative up to date for stimulation at particular areas within the central nervous system. This device controls the neural activity far better than even deep brain stimulators, meaning its interventions will be more specifically designed for certain disorders in neurology. Another very important area of nano-bioelectric medicine is tissue regeneration through electrical stimulation. Electrical signals are very important in promoting healing and regeneration in damaged tissues. Nano-bioelectric devices may be used for the application of electrical stimulation, which can aid in the encouragement of new cells to grow or reduce inflammation. For example, the regeneration of bone cells might be stimulated through the use of nanostructured scaffolds coupled with electrical signals that help achieve faster recovery when there is fracture or bone defect. The drugs are released in a controlled manner in a targeted way, so electroporation is under research for the treatment of cancers and gene therapy. Nano-bioelectric medicine can be used in the management of chronic pain since electrical signals may block or modulate pain signals within the nervous system, an alternative to painkillers with no invasion into the body. For example, nanoparticle-based pain relief modulates the electrical activity of the pain receptors through nanoscale materials, making it a targeted approach to the management of pain with fewer side effects than traditional methods. Nano-bioelectric devices also have the possibility of curing mental health disorders such as depression and anxiety by very accurately tuning brain activity (Yoon et al., 2020).

One is transcranial magnetic stimulation (TMS), which can be advanced by nanotechnology to better target and more effectively stimulate the involved regions of the brain. Although highly promising, nano-bioelectric medicine presents a number of challenges of important interest. Among them, there is an essential one about biocompatibility; the devices, that are meant to provide biomedical treatments or functionalities, shall not cause undesirable biological effects. This would necessitate careful extended investigations into long-term safety and effectiveness before possible broad clinical usage. Second, a fully acceptable nano-bioelectric device will need to integrate compatibly with already available medical equipment and other healthcare infrastructures. This brings with it the ability to obtain precise control of brain and nervous system activity, yet again with ethical issues on privacy and consent, combined with the danger of misapplication. All this will be met stride by stride as this field continues to grow to actualize the potential of nano-bioelectric medicine (Jiang et al., 2020).

5.9 Implantable Medical Nanogenerators

These devices, exploiting body movements in the form of breathing, heartbeats, or muscle contraction, can actually produce electricity in order to feed the sensors, pacemakers, drug delivery system, and other types of implantable devices, thereby ushering in medical treatments that can be more effective, sustainable, and long-lived. The nanogenerators for implantations work on either triboelectric nanogenerator or piezoelectric nanogenerator principle another as shown in Fig. 3. These mechanisms convert the mechanical energy to electrical energy via deformation of zinc oxide, graphene, or piezoelectric polymers once they come in contact with one (Baburaj et al., 2024; Wang et al., 2022).

Implantable medical nanogenerators represent one of the revolutionary technologies within medicine, and such devices will provide the capability of powering implantable devices and sensors without requiring external power sources. These devices could reduce the necessity for many invasive procedures while helping to advance patient outcomes through providing continuous self-sustaining power for critical medical applications. While challenges like biocompatibility, energy efficiency, and durability are still to be overcome, implantable nanogenerators have a

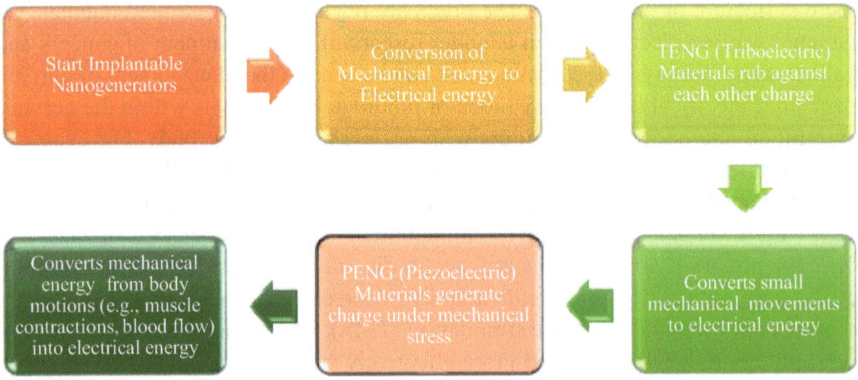

Fig. 3 The process from the basic concept of implantable nanogenerators converting mechanical energy into electrical energy, through the two mechanisms of TENGs and PENGs

great future in the near future of personalized and sustainable healthcare (Jian et al., 2024; Liu et al., 2019).

5.10 *Nanotechnology in Dentistry*

Nanotechnology involves the application of materials and devices on a nanoscale of 1–100 nm. Nanotechnology has changed the world for many areas; it also applied in the field of dentistry. Integration of nanotechnology in dental practices opened new innovative ways for restorations, preventive care, diagnostics, and treatment modalities which have greatly enhanced patient outcomes (Sachin et al., 2024). One of the principal applications is found in dental restorations, where nanocomposites have incorporated nanoparticles into the restorative material, thus further enhancing mechanical strength, durability, and wear. These materials, in addition, provide superior aesthetic qualities and superior polish retention to traditional composites. Similarly, Nano-adhesives containing the nanoparticles in its formulation offer additional bonding strength both to enamel as well as dentin, ensuring that restorations are that much more predictable and longer in service. In preventive dentistry, nanotechnology has played an important role, especially with the introduction of nano-hydroxyapatite. It mimics the tooth enamel in natural structure and has been extensively used for the remineralisation of enamel, in reducing sensitivity, and for filling up small cracks in teeth. Antimicrobial nanoparticles like silver, zinc oxide, and titanium dioxide are now being incorporated into toothpaste and mouthwash products. They prevent the growth of bacteria and also the formation of biofilm; thus, they play a critical role in preventing dental caries and gum diseases (Kumbhar et al., 2022).

Nanorobotics is one of the most promising areas in dental applications. The futurism of nanobots in removing plaque, microscopic robotic devices that will navigate through the oral cavity to remove plaque and tartar from the teeth and gums with unmatched accuracy. In addition, nanorobots hold much promise for targeted drug delivery: exact administration of therapeutic agents to the site of disease for minimum systemic side effects and increased effectiveness of treatment. Dentistry diagnostics have gained much with the nanotechnology discovery. Nanosensors are highly sensitive devices that identify specific biomarkers in saliva or tissues, identifying caries, periodontal disease, or oral cancer, hence detecting those diseases early and enhancing diagnostic accuracy with advanced imaging techniques such as X-rays, MRI, or optical coherence tomography, further improved with the help of nanoparticles (Arvind et al., 2025).

In orthodontics, nanotechnology contributes in terms of applying nanocoatings and nanoparticle-based adhesives. The orthodontic wires and brackets applied with nanocoatings diminish friction and increase durability, making the treatments more efficient and comfortable for patients. The bonding strength of brackets to tooth surfaces is improved with nanoparticle-based adhesives and this reduces the possibility of deboning in the process of orthodontic procedures. Nanostructured coatings and nanoceramics are also seen as advancements in the field of dental implants and prosthetics. Coatings containing nanoparticles of titanium or zirconia improve Osseo integration and tissue regeneration and reduce the rate of implant failure. Nanoceramics crowns, bridges, and veneers offer increased strength, abrasion resistance, and aesthetic value, establishing a new benchmark in dental prosthetics Nanoparticles and nanoscaffolds have had a great impact on periodontal therapy. Silver and zinc oxide nanoparticles in periodontal gels and mouthwashes play their role in targeting periodontal pathogens, helping improve gum health. Nanoscaffolds are used for the reconstruction of periodontal tissues like bone and ligament, which makes them functional for the treatment of advanced gum diseases. The agents based on nanotechnology have also improved teeth whitening, allowing for more uniform and effective whitening results with less tooth sensitivity, a common side effect of conventional bleaching treatments. The advantages of nanotechnology in dentistry are very numerous. It allows for greater precision and efficiency in dental treatments, enhances the mechanical and aesthetic properties of restorative materials, allows for targeted therapies with minimal side effects, promotes faster healing and better tissue regeneration, and facilitates early disease detection and preventive care. Despite its high potential, however, a number of challenges and limitations should be addressed: it is pricey for nanotechnology-based materials and treatments, only limited clinical trials with long-term safety data exist, nanoparticles might have toxicity problems, regulatory matters may arise, and advanced training would be necessary for dental practitioners. Still, integrating nanotechnology into dentistry offers great potential to transform this field and help advance oral health care worldwide (Manjeet et al., 2025).

6 Toxicology and Safety Analysis of Nanotechnologies

The application of nanotechnology in dental clinical practice has become the apprehensive point concerning their safety and potential toxicity, since the physical and chemical materials properties are quite unique at nanoscale. Such materials can have adverse interactions with biological systems and can also pose toxicological risks. Some nanoparticles such as silver and zinc oxide were shown to present cytotoxic activity at high concentration while others may be genotoxic since they involve damage to the DNA and a step toward the occurrence of mutations. Secondly, nanoparticles stimulate reactive oxygen species which can produce oxidative stress along with cellular damage. Extended exposure to these substances can lead to bioaccumulation in the tissues, making their long-term impact a cause for concern (Zhang et al., 2024). Exposure in dental settings has varied routes, including the inhalation pathway when they are formed or applied, ingestion due to occasional exposure to dental components, and dermal contact leading to skin irritation or penetration into deeper tissues. In this context, it is essential to minimize the toxic effects to human health and ensure biocompatibility of nanomaterials by carrying out extensive preclinical studies that reduce the concentration of nanomaterials for safe therapeutic applications. It should include safety procedures for the use and applications of nanomaterials and must become the standard to minimize occupational exposure; regulation is required to avoid public exposure where it cannot be ensured that the public which would benefit from using nanotechnology in dentistry are not exposed to its application or its practice. Long-term studies are necessary to track the long-term effects of these materials, but evaluation of the environmental impact of nanoparticle disposal from dental practices is equally crucial (Domb et al., 2021).

7 Challenges and Ethical Considerations in the Use of Nanotechnology in Medicines and Healthcare

While nanotechnology provides exciting opportunities in medicine and healthcare as specified in Fig. 4, its application necessitates certain considerations, challenges, and ethical issues, which must be addressed to guarantee its safe and equitable adoption. NPs could unproductively interact with the biological system to cause toxicity, inflammation, and oxidative stress. Another mechanism of toxicity would be that nanoparticles reside for long periods in tissues that would otherwise be biocompatible, which would lead to bioaccumulation and chronic adverse effects. This situation is further complicated by the lack of evidence on safety regarding long-term exposure both to humans and the environment. Regulatory and standardization problems come into play, as well, because regulatory frameworks in use are partially unsuited for nanotechnology products, and global standards for evaluation and validation of these applications in healthcare do not exist to date. The manufacture

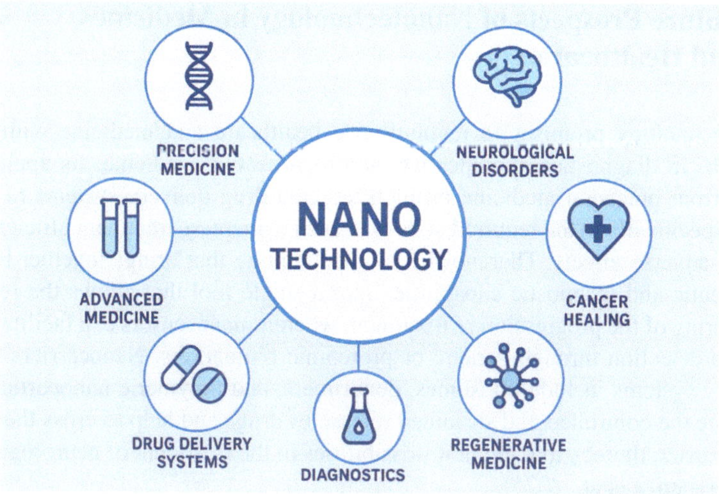

Fig. 4 Applications of nanotechnology in healthcare

of nanomaterials and their reduced scale become a major impediment because of high production costs, inherent problems in quality control, and difficulties in scaling laboratory methods to industrial levels. Other complications arise due to worries over the environment, such as how these nanomaterials might be disposed of and their possible ecotoxicological effect. Even more technically, stability of the nanoparticles and the ability to target them precisely to diseased tissues without interfering with the functioning of healthy cells remain two huge roadblocks. More significant ethical dilemmas come into play in terms of patient safety and informed consent, along with issues regarding how fairly and in what fashion such new nanotechnology-based medical procedures will be made available to everyone thus creating further healthcare inequities. Wearable nanotechnology would also bring issues of privacy and data security (Sachin et al., 2024).

Another that can't be avoided is potential dual-use issues like bioterrorism, creating a requirement for more stringent controls. Of at least equal concern is the social and ethical dilemma of the ethics of nanotechnological applications regarding human enhancement when it comes to equity and community acceptance. Then there's also the question of environmental justice: this needs great care because affected communities are very vulnerable. To answer the challenge, it should be required to have appropriate regulatory frameworks in place, routes for public access to information and interdisciplinary collaboration. Ethical supervision should be ensured so that developments and applications of nanotechnology will be responsibly, equitably undertaken. Aside from that, safeguards to the environment such as nanoparticles that are biodegradable can be prioritized to minimize damage to the ecology. If these problems are identified now and addressed in time, nanotechnology can find an implementation in health care that is responsible, sustainable, and beneficial to all (Ebbesen et al., 2006).

8 Future Prospects of Nanotechnology in Medicines and Healthcare

Nanotechnology promises to revolutionize healthcare and medicine with breakthroughs in diagnostics, therapeutics, and regenerative medicine. Its applications range from precision medicine through targeted drug delivery systems to deliver drugs specifically to the required cells or tissues to increase treatment efficiency and reduce adverse effects. Theranostics is a technology that brings together both the therapeutic and diagnostic capabilities into a single tool that allows the real-time monitoring of the progression of treatment, whereas nanosensors can facilitate early disease detection through genetic or proteomic biomarkers. Nanocarriers in drug delivery systems include liposomes, dendrimers, and polymeric nanoparticles that facilitate the controlled and sustained release of drugs and help to cross the blood–brain barrier, thereby making new possibilities in the treatment of neurological disorders (Hamza et al., 2025).

Multifunctional delivery systems also provide combination therapies. Nanotechnology is a cornerstone of regenerative medicine, where nanomaterials such as nanofibers, hydrogels, and scaffolds stimulate tissue regeneration, and nanoparticles facilitate tracking, improving, or delivering stem cells to the body site. In cancer therapy, nanoparticles functionalized with ligands allow targeting of cancer cells with minimum damage to the healthy cells, whereas gold nanoparticles and other nanostructures allow for localized destruction of cancer cells by hyperthermia and photothermal therapy (Sachin et al., 2024).

Nanotechnology also finds applications in immune modulation for cancer therapy. For infectious diseases, diagnostics and antimicrobial nanoparticles are advocated to present solutions in broad-spectrum medicine; nano-based vaccine delivery systems enhance the immune response. Cardiac-related applications involve nanoparticles for targeted removal of plaque formation, regeneration of cardiac tissue, and improved imaging. Finally, nanotechnology advances diagnostics through nanosensors that monitor biomarkers in real time, while enhanced imaging agents improve the resolution on MRI, PET, and CT scans. In neurology, nanotechnology enables drugs to cross the blood-brain barrier and support neuroregeneration (Elyor et al., 2024).

Nanorobotics, such as nanobots in precision surgeries and smart wearables for continuous monitoring of health parameters, are emerging technologies that have the potential to improve patient care. Nanotechnology also provides the solutions for sustainability and accessibility as it can enable low-cost manufacture, thereby bringing healthcare within an affordable reach for people, while point-of-care devices will bring healthcare to resource-poor areas or remote regions. Challenges related to safety, regulation, and equity must still be addressed to achieve full realization of nanotechnology and ensure its global accessibility (Sivasubramanian et al., 2014).

9 Conclusion

In a nutshell, nanotechnology is an extraordinary field that can, in the future, revolutionize industries like medicine, electronics, and environmental science. Unique features like an enhanced surface area and quantum phenomena are what enable breakthroughs in areas like drug delivery, diagnostics.

In addition, nanomaterials can also trigger tissue regeneration and improve the efficiency of stem cell therapies within the field of regenerative medicine, and nanostructures in cancer therapy will help to deliver treatment in a targeted and localized fashion with minimal destruction of healthy cells. There is also promising application in veterinary medicine in animal disease treatment and improvement of drug delivery.

Although such revolutionary leaps exist, numerous problems have to be addressed and solved for safe and effective utilization in the field of medicine and health care. The toxicology, biocompatibility, and environmental effects of nanoparticles must be understood through broad preclinical and long-term studies. Patient safety, informed consent, and fair access concerns related to ethics must also remain on the frontline as nanotechnology is implemented in clinical settings. There will be frameworks for regulations and standardized practice in the harnessing of nanotechnology benefits and ensuring its risks are reduced to a bare minimum. Precision medicine and targeted drug delivery in healthcare to improve treatment outcome with minimal side effects are now possible.

Vast potential still exists, but safety, biocompatibility, frameworks of regulation, and ethical concerns persist. Cross-disciplinary approaches towards further research and development will ensure that nanotechnology is properly introduced into clinical environments in an efficient and effective way having encountered these challenges, nanotechnology shall provide individually customized, environmentally sound, and available health care systems that improve care for patients to satisfy world medical needs. Impacts on the human and animals may grow from time to time to more potent treatments, perfect diagnostic methods, and enhanced regeneration treatments.

References

Abid, H., Mohd, J., Ravipratap, S., Shanay, R., & Rajiv, S. (2023). Applications of nanotechnology in medical field: A brief review. *Global Health Journal, 7*(2), 70–77.

Aggarwal, M., & Kumar, S. (2022a). The use of nanorobotics in the treatment therapy of cancer and its future aspects: A review. *Cureus, 14*(9), 29366.

Aggarwal, M., & Kumar, S. (2022b). The use of nanorobotics in the treatment therapy of cancer and its future aspects: A review. *Cureus, 14*(9), e29366.

Allahou, L. W., Madani, S. Y., & Seifalian, A. (2021). Investigating the application of liposomes as drug delivery systems for the diagnosis and treatment of cancer. *International Journal of Biomaterials, 4*, 1–16.

Ansari, S., Rajpurohit, V., & Dabas, R. (2024). Application of nanotechnology in veterinary medicine-an overview. *International Journal of Livestock Research, 14*(12), 11–19.
Arvind, K. S. C., Parihar, A. (2025). Smart Ways of Biomaterial Designing, Synthesis and Characterization – Prospects of Enhanced Application from Labs to Clinics. CRC Press.
Babu, A., et al. (2014). Nanodrug delivery systems: A promising technology for detection, diagnosis, and treatment of cancer. *AAPS PharmSciTech, 15*(3), 709–721.
Baburaj, A., et al. (2024). Biodegradable based TENGs for self-sustaining implantable medical devices. *Nano Energy, 127*, 109785.
Băcăiţă, E. S., Raţă, D. M., Cadinoiu, A. N., Ghizdovăţ, V., Agop, M., & Luca, A.-C. (2023). Drug release from nanoparticles (polymeric nanocapsules and liposomes) mimed through a multifractal tunnelling-type effect. *Polymers, 15*(4), 1018.
Bhatia, S. (2016). Nanoparticles types, classification, characterization, fabrication methods and drug delivery applications. In *Natural polymer drug delivery systems*. Springer.
Bhushan, B. (2015). Governance, policy, and legislation of nanotechnology: A perspective. *Microsystem Technologies, 21*, 1137–1155.
Bhushan, B. (2016). *Encyclopedia of nanotechnology* (2nd ed.). Springer.
Bhushan, B. (2017). Introduction to nanotechnology. In B. Bhushan (Ed.), *Springer handbook of nanotechnology. Springer handbooks*. Springer.
Cavalcanti, A., Shirinzadeh, B., Freitas, R. A., Jr., & Hogg, T. (2008). Nanorobot architecture for medical target identification. *Nanotechnology, 19*(1), 015103.
Christ, G. J., et al. (2013). The pharmacology of regenerative medicine. *Pharmacological Reviews, 65*(3), 1091–1133.
Danchuk, O., et al. (2023). Meeting contemporary challenges: Development of nanomaterials for veterinary medicine. *Pharmaceutics, 15*, 2326.
Domb, A. J., Sharifzadeh, G., Nahum, V., & Hosseinkhani, H. (2021). Safety evaluation of nanotechnology products. *Pharmaceutics, 13*(10), 1615.
Dutt, Y., Pandey, R., Dutt, M., et al. (2023). Therapeutic applications of nanobiotechnology. *Journal of Nanbiotechnology, 21*, 148.
Ebbesen, M., & Jensen, T. (2006). Nanomedicine: Techniques, potentials, and ethical implications. *Journal of Biomedicine & Biotechnology, 5*, 51516.
Eissa, M., et al. (2024). Nanosensors for early detection and diagnosis of cancer: A review of recent advances. *Journal of Cancer Research and Reviews, 1*(1), 1–13.
Elyor, B., Hussain, C. M., & Tüzün, B. (2024). *Multifunctional. magnetic nanoparticles in therapy, biology, and pharmacy - Synthesis, fundamentals andapplications*. CRC Press.
Engel, E., et al. (2007). Nanotechnology in regenerative medicine: The materials side. *Trends in Biotechnology, 26*(1), 39–47.
Friedrich, R. P., Cicha, I., & Alexiou, C. (2021). Iron oxide nanoparticles in regenerative medicine and tissue engineering. *Nanomaterials, 11*, 1–72.
Gentile, P., & Tonda-Turo, C. (2022). Editorial: Methods in nanobiotechnology. *Frontiers in Bioengineering and Biotechnology, 10*, 1067952. https://doi.org/10.3389/fbioe.2022.1067952
Gunasekaran, T., Haile, T., Nigusse, T., & Dhanaraju, M. D. (2014). Nanotechnology: An effective tool for enhancing bioavailability and bioactivity of phytomedicine. *Asian Pacific Journal of Tropical Biomedicine, 4*(Suppl 1), S1–S7.
Hamimed, S., Jabberi, M., & Chatti, A. (2022). Nanotechnology in drug and gene delivery. *Naunyn-Schmiedeberg's Archives of Pharmacology, 395*, 769–787.
Hamza HM, Malik MM, Asad M, Ali S, Awan AA. (2025). Advances in orthopedic implants: the role of nanotechnology in enhancing performance and longevity. *Regen Med Rep, 2*(1):15–21.
Hill, E. K., & Li, J. (2017). Current and future prospects for nanotechnology in animal production. *Journal of Animal Science and Biotechnology, 8*, 1–13.
Huang-Chiao, H., Barua, S., Sharma, G., Dey, S. K., & Rege, K. (2011). Inorganic nanoparticles for cancer imaging and therapy. *Journal of Controlled Release, 155*(3), 344–357.
Ianiski, L. B., et al. (2022). Nanotechnology in veterinary medicine: A review. *Ciência Rural, 52*(6), 1–12.

Jackson, T. C., Patani, B. O., & Ekpa, D. E. (2017). Nanotechnology in diagnosis: A review. *Advances in Nanoparticles, 6*(3), 93–102.

Jafary, F., Motamedi, S., & Karimi, I. (2023). Veterinary nanomedicine: Pros and cons. *Veterinary Medicine and Science, 9*, 494–506.

Jian, G., et al. (2024). Biodegradable triboelectric nanogenerator as a implantable power source for embedded medicine devices. *NPG Asia Materials, 16*(1), 12.

Jiang, D., et al. (2020). Emerging implantable energy harvesters and self-powered implantable medical electronics. *ACS Nano, 14*(6), 6436–6448.

Joudeh, N., & Linke, D. (2022). Nanoparticle classification, physicochemical properties, characterization, and applications: A comprehensive review for biologists. *Journal of Nanbiotechnology, 20*, 262.

Kanwar, R., Rathee, J., Salunke, D., et al. (2019). Green nanotechnology-driven drug delivery assemblies. *ACS Omega, 4*(5), 8804–8815.

Karahmet Sher, E., Alebić, M., Marković Boras, M., Boškailo, E., Karahmet Farhat, E., Karahmet, A., Pavlović, B., Sher, F., & Lekić, L. (2024). Nanotechnology in medicine revolutionizing drug delivery for cancer and viral infection treatments. *International Journal of Pharmaceutics, 660*, 124345.

Karnwal, A., Sharma, V., Kumar, G., Jassim, A. Y., Dohroo, A., & Sivanesan, I. (2024). Transforming medicine with Nanobiotechnology: Nanocarriers and their biomedical applications. *Pharmaceutics, 16*(9), 1114.

Karnwal, A., Sharma, V., Kumar, G., et al. (2024). Transforming medicine with nanobiotechnology: Nanocarriers and their biomedical applications. *Pharmaceutics, 16*(9), 1114.

Khang, D., et al. (2008). Nanotechnology for regenerative medicine. *Biomedical Microdevices, 12*, 575–587.

Khoobchandani, M., Katti, K., Karikachery, A., et al. (2020). New approaches in breast cancer therapy through green nanotechnology and nano-ayurvedic medicine—Pre-clinical and pilot human clinical investigations. *International Journal of Nanomedicine, 15*, 181–197.

Kim, E. S., et al. (2014). Emerging nanotechnology approaches in tissue engineering and regenerative medicine. *International Journal of Nanomedicine, 9*(1), 1–5.

Kirtane, A. R., et al. (2021). Nanotechnology approaches for global infectious diseases. *Nature Nanotechnology, 16*, 369–384.

Kumar, B., Jalodia, K., Kumar, P., et al. (2017). Recent advances in nanoparticle-mediated drug delivery. *Journal of Drug Delivery Science and Technology, 41*, 260–268.

Kumbhar, A., Chaudhari, P., & Karimunnisa, S. (2022). Nanocomposite based drug delivery for periodontal disease. *International Journal of Health Sciences, 6*(S9), 2982–2995.

Liu, R., et al. (2023). Advances of nanoparticles as drug delivery systems for disease diagnosis and treatment. *Chinese Chemical Letters, 34*(2), 107518.

Liu, Y., Liang, Y., Yuhong, J., Xin, P., Han, J. L., Du, Y., Yu, X., Zhu, R., Zhang, M., Chen, W., & Ma, Y. (2024). Advances in nanotechnology for enhancing the solubility and bioavailability of poorly soluble drugs. *Drug Design, Development and Therapy, 18*, 1469–1495.

Liu, Z., et al. (2019). Wearable and implantable triboelectric Nanogenerators. *Advanced Functional Materials, 29*(20), 1808820.

Mahima, A., & Vaibhav, G. (2024). Nanorobots in medicine: Advancing healthcare through molecular engineering: A comprehensive review. *IgMin Research, 2*(11), 938–949.

Manjeet Singh, Ashish Kumar Singh, Balaram Pani (2025). Layered Nanomaterials for Solution-Processed Optoelectronics, CRC Press.

Marepally, S., Sahu, T., & Tyagi, R. K. (2022). Nanomedicine in infectious diseases: Drug delivery and vaccines. *Frontiers in Pharmacology., 13*, 1–3.

Mendes, B., Conniot, J., & Avital, A. (2022). Nanodelivery of nucleic acids. *Nature Reviews Methods Primers, 2*, 24.

Nicholas, W. (2020). The Applications and Implications of Nanotechnology. American University, School of International Service, Center for Security and Technology. Available at: https://www.american.edu/sis/centers/security-technology/the-applications-and-implications-of-nanotechnology.cfm

Patil, L. B., et al. (2016). Review on-novel approaches in nanorobotics. *Asian Journal of Pharmaceutical Research, 6*(4), 217.

Pearce, K., Thipe, V., Henkel, R., et al. (2023). Green nanotechnology as an innovative drug delivery approach for Typha capensis and Naringenin—New class of phytochemical embedded biocompatible gold nanoparticles in prostate cancer therapy. *Journal of Drug Delivery Science and Technology, 80*, 104100.

Peters, E. B., & Banerjee, R. (2021). Special issue: Nanomedicine advances in infectious disease. *ACS Biomaterials Science & Engineering, 7*, 1722–1724.

Prasad, R. D., et al. (2021). A review on concept of nanotechnology in veterinary medicine. *ES Food and Agroforestry, 4*, 28–60.

Rubey, K. M., & Brenner, J. S. (2021). Nanomedicine to fight infectious disease. *Advanced Drug Delivery Reviews, 179*, 113996.

Sachin Kumar, Ajit Kumar Verma, Amna Mirza. (2024). Chapter 10 *Artificial Intelligence and Climate Change Mitigation*, Springer Science and Business Media LLC.

Sánchez-López, E., Gomes, D., Esteruelas, G., Bonilla, L., Lopez-Machado, A. L., Galindo, R., Cano, A., Espina, M., Ettcheto, M., Camins, A., et al. (2020). Metal-based nanoparticles as antimicrobial agents: An overview. *Nanomaterials, 10*, 292.

Shyam, L. S., Hawaldar, R., Titus, E., Gracio, J., Singh, M. K. (2012). Biomedical Engineering – Technical Applications in Medicine. In: Hudak R, Penhaker M, Majernik J (eds) InTechOpen, 6 September 2012.

Singh, A. K., & Chandel, A. P. (2025). *Smart ways of biomaterial designing synthesis and characterization - Prospects of enhanced application from labs to clinics*. CRC Press.

Sivasubramanian, M., Hsia, Y., & Lo, L. W. (2014). Nanoparticle-facilitated functional and molecular imaging for the early detection of cancer. *Frontiers in Molecular Biosciences, 1*, 15.

Suhag, D., Thakur, P., & Thakur, A. (2023). Introduction to nanotechnology. In D. Suhag, A. Thakur, & P. Thakur (Eds.), *Integrated nanomaterials and their applications*. Springer.

Unterweger, H., et al. (2020). Nanomedicine for infectious diseases. *Nanomedicine, 15*(13), 1263–1267.

Vishwakarma, K., Vishwakarma, O. P., & Bhatele, M. (2013). A brief review on role of nanotechnology in medical sciences. In V. Kumar & M. Bhatele (Eds.), *Proceedings of All India Seminar on biomedical engineering 2012 (AISOBE 2012). Lecture notes in bioengineering*. Springer.

Wang, C., Shi, Q., & Lee, C. (2022). Advanced implantable biomedical devices enabled by triboelectric Nanogenerators. *Nanomaterials, 12*(8), 1366.

Yang, Y., & Jiao, P. (2023). Nanomaterials and nanotechnology for biomedical soft robots. *Materials, 17*, 100338.

Yoon, H. J., & Kim, S. W. (2020). Nanogenerators to power implantable medical systems. *Joule, 4*(7), 1398–1407.

Youssef, F. S., et al. (2019). Application of some nanoparticles in the field of veterinary medicine. *International Journal of Veterinary Science and Medicine, 7*(1), 78–93.

Yuan, T., Yao, Z., Yang, W., Tao, S., Li, B., Wang, Y., Su, Z., Li, S. (2022). Application of Nanoparticles in Tumour Targeted Drug Delivery and Vaccine. Front. *Nanotechnol.* 4:948705. https://doi.org/10.3389/fnano.2022.948705

Zhang, W., Hou, Y., Yin, S., et al. (2024). Advanced gene nanocarriers/scaffolds in nonviral-mediated delivery system for tissue regeneration and repair. *Journal of Nanbiotechnology, 22*, 376.

Nanotechnology in Energy and Environment

Jibrin Muhammad Yelwa, Joseph Usman, Ibrahim Umar, Alhagie Dremmah, Jibrin Yusuf Yahaya, Tari Joel Honda, and Haruna Musa

J. M. Yelwa (✉)
Scientific and Industrial Research Department, National Research Institute for Chemical Technology, Zaria, Kaduna State, Nigeria

Department of Pure and Industrial Chemistry, Bayero University Kano, Kano, Kano State, Nigeria
e-mail: jibrin.yelwa@narict.gov.ng

J. Usman
Department of Electrical and Electronics Engineering, Abubakar Tafawa Balewa University, Bauchi, Nigeria

I. Umar
Department of Chemistry, Aliko Dangote University of Science and Technology, Wudil, Kano State, Nigeria
e-mail: umaribrahim@kustwudil.edu.ng

A. Dremmah
Chemistry Unit, Division of Physical and Natural Sciences, School of Arts and Sciences, University of The Gambia, Serekunda, The Gambia

J. Y. Yahaya
Scientific and Industrial Research Department, National Research Institute for Chemical Technology, Zaria, Kaduna State, Nigeria

T. J. Honda
Department of Chemistry, Modibbo Adama University, Yola, Nigeria

H. Musa
Department of Pure and Industrial Chemistry, Bayero University Kano, Kano, Kano State, Nigeria
e-mail: hmusa.chm@buk.edu.ng

S. Kulkarni (ed.), *A Journey from Nanotechnology to Picotechnology*, Lecture Notes in Nanoscale Science and Technology 47,
https://doi.org/10.1007/978-3-032-07228-3_6

1 Introduction

Nanotechnology has its scientific bases in handling matter at atomic, molecular, and nanometric levels, with conventional size from 1 to 100 nm. During the last few decades, nanotechnology applications have largely increased, especially in the fields of energy and environment. With improved material characteristics, offering novel ways to tackle long-standing problems, nanotechnology holds transformative potential in many of the key issues related to energy production, energy storage, energy efficiency, and environmental sustainability (Gao et al., 2013).

These developments focused on nanomaterials, including carbon nanotubes, graphene, and nanoparticles. They are characterized by properties different from those of bulk materials, including high surface area, enhanced reactivity, extraordinary mechanical strength, etc. These make them potential candidates for energy and environmental applications (Aricò et al., 2005). Some of these properties have been harnessed in various applications, improving energy storage systems and water purification technologies.

Nanotechnology promises a lot for the energy sector, such as better production, storage, or efficacy of renewable energy. In the future, presumably, nanomaterials will majorly improve the performance of solar cells by improving the absorption of light and electron mobility. In one of the fastest-growing technologies, for instance, perovskite solar cells, the introduction of nanoparticles has already been able to show improved efficiencies (Ramsurn & Gupta, 2013). Nanotechnology also allows for the development of more efficient batteries and supercapacitors by providing increased storage and durability for energy that will further enable both renewable energy systems and electric vehicles (Liang et al., 2009).

Similarly, nanotechnology plays an important role in energy efficiency. Nanomaterials have been introduced to thermoelectric materials, which transform heat into electricity and probably also provide the basis for waste heat recovery in the future (Zhao et al., 2019). Additionally, in hydrogen production, nanomaterials are used to improve catalysts for water splitting with a goal of producing hydrogen as a clean energy carrier (Khin et al., 2012).

Its applications to the environment range from pollution control to water treatment and waste management. Nanoparticles can be employed for effective adsorption of pollutants in water, air, and soil instead of conventional methods. For example, nanoparticles of iron and other materials were found to degrade organic pollutants in water as an effective solution for water purification at industrial levels (Wang et al., 2015). To their part, nanomaterials have variably been used to improve pollutant disintegration and the remediation of contaminated soils (Khin et al., 2012).

Nanotechnology now enables the creation of a new generation of filtration material that can capture even fine particulate matter and toxic gases, thus creating very much improved air quality within the air pollution zone (Khin et al., 2012). In addition, nanotechnology is being explored for capture and storage technology, CCS, where nanoparticles may be used in the absorption of CO_2 from industrial emissions, to mitigate climate change (Wang et al., 2014).

But at the same time, it is true that despite all the pros, nanotechnology has significant challenges in fully scale applications related to energy and the environment. In terms of large-scale industrial production, environmental hazards, toxicological issues, economic viability, are all significant points related to the nanomaterials (Simpa et al., 2024). Besides, long-term stability and their reliability of nanomaterials in practical application need to be fully understood (Banin et al., 2021).

This chapter seeks to discuss the immense potential of nanotechnology by contributing to clean and sustainable energy solutions and environmental protection. With ongoing research and overcoming such challenges, nanotechnology is expected to help shape the future in both energy and environmental fields.

2 Nanotechnology in Energy

Nanotechnology is a relatively an emerging field that has renewed many industries and in energy production sector. Because of the increased global demand for sustainable and clean energy, nanotechnology offers exciting opportunity for increasing energy production, storage, and conversion with minimum environmental pollution.

The following section explains various types of energy applications where nanotechnology is playing a major role, including renewable energy, energy storage, and energy conversion, while explaining a wide variety of nanomaterials shaping the future of the energy sector.

2.1 Energy Applications of Nanotechnology

2.1.1 Renewable Energy

Nanotechnology has been able to advance the growth of renewable energy resources like solar, wind, and hydrogen fuels by increasing the efficacy of energy systems and cutting down their cost.

Solar Energy

Nanomaterials including nanoparticles, quantum dots, and thin films are finding their applications in increasing the efficiency of photovoltaic cells. Quantum dots refer to semiconductor nanocrystals that have the ability to absorb and emit light. Quantum dots have the potential for highly efficient solar cells by broadening the spectrum of sunlight that can be absorbed, thereby increasing the conversion of sunlight into electricity. An investigation on quantum dot-based solar cells shows that it can achieve higher efficiency than conventional silicon-based solar cells

(Rezakazemi et al., 2021). Besides, other nanomaterials, such as highly efficient light absorbers, including perovskites, have increasingly been applied in place of silicon within the solar cells and have reduced the cost of production without altering the performance (Pandey et al., 2019).

Wind Energy

Nanotechnology is also being applied to improve the materials used in wind turbines. Carbon nanotubes and graphene allow greater mechanical strength and great flexibility that so far increases the reliability of the blades of turbines. For example, CNTs may be used in preparing some of the composites found in building wind turbine blades owing to increasing their structural strength but not affecting their overall weight substantially. These result in more robust turbines that can produce higher amounts of energy and for long periods. Graphene-based coatings applied on them minimize wear and tear, thereby prolonging the lifespan of the turbines (Eldada, 2011).

Hydrogen Energy

Hydrogen, a clean fuel, is considered as an alternative to fossil fuels. Nanotechnology can significantly improve hydrogen production, storage, and conversion. Nanocatalysts, such as Pt-nanoparticles, are used in fuel cells and act as catalysts for the hydrogen oxidation reaction. These catalysts highly enhance the efficiency of hydrogen fuel cells, which have an important role in the advancement of zero-emission power systems. In addition, nanomaterials are being researched for hydrogen storage; porous materials including metal-organic frameworks have shown high surface areas, which are very useful for gas adsorption, playing a central role in the development of more compact and efficient hydrogen storage systems (Sahaym & Norton, 2008).

Figure 1 depicts the role of nanotechnology toward better efficiency and cost-reduction in various renewable energy forms, including solar, hydrogen, and wind. It shows how nano-enabled materials, which include new solar cells, nanocoatings for wind turbines, and even nanocatalysts for hydrogen processing, all come together to amplify energy conversion and storage.

2.1.2 Energy Storage

Most important, nanotechnology can make vital contributions to the performance of energy storage systems, be it in batteries or supercapacitors which are crucial not only just for renewable energy applications but also for electric vehicle technology.

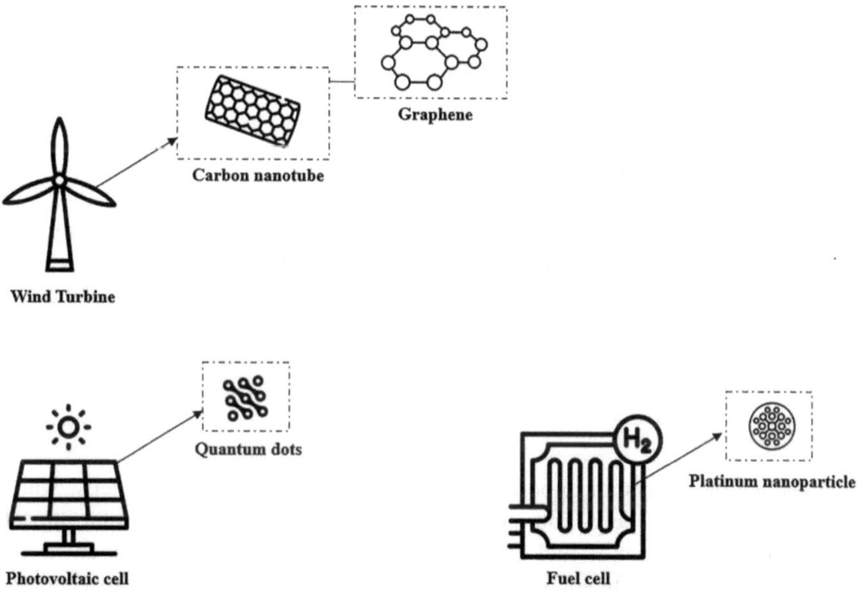

Fig. 1 Nanotechnology in renewable energy applications

Batteries

In developing advanced lithium-ion and solid-state batteries, nanomaterials play a vital role in enhancing efficiency, charge capacity, and durability. The incorporation of silicon nanoparticles in anode material is underway to replace graphite in lithium-ion batteries. Silicon-based anodes provide far better energy density than the currently used graphite-based ones, translating in to long battery life combined with quite fast charging times.

This research also covers nanostructured material, adding lithium iron phosphate ($LiFePO_4$), which will make contributions to reaching overall performance and safety for batteries by reducing risks of overheating and explosions. Researchers have been pointing out that nanostructuring would offer the ability for batteries to sustain capacity over a big number of charge cycles (Hudak, 2014).

Supercapacitors

Supercapacitors can benefit so much from nano-materials including graphene or carbon nanotubes, which have exceptional properties like high surface area, conductivity, and good mechanical strength—all three reinforcing energy density, charge/discharge rates, and lifecycle of supercapacitors. Hybrid energy storage systems that combine batteries and supercapacitors are also under research. While supercapacitors can provide very fast bursts of energy during peak demand, the batteries can supply sustained energy over a period of time. Integrating nanomaterials into these devices will add to the improvement in the energy storage capacity and efficiency of supercapacitors (Maksoud et al., 2020).

2.1.3 Energy Conversion

While energy conversion systems are one of the core parts for efficient energy conversion of stored energy into usable power, nanotechnology is definitely causing an evolution. The areas that have tremendous development are in fuel cells and thermoelectric materials.

Fuel Cells

Nanotechnology plays an important role in the advancement of fuel cells, which transform chemical energy into electrical energy with less environmental impact. Among the challenges of fuel cell technology, one of the most important is the improvement of the efficiency of catalysts used in the electrochemical reactions, and nanocatalysts hold the key to solving this issue. Nanoparticles, mainly platinum and platinum alloys, demonstrate higher catalytic activity by using less material and reducing costs. Nanostructured electrodes will be contributing to better efficiency because it provides a greater surface area over which the reaction can take place in fuel cells. On the other hand, nanomaterials were found capable of offering a decrease in the working temperature of fuel cells, which may enable wider use of these devices in more versatile applications (Pandiyan & Prabaharan, 2020).

Thermoelectric Materials

These materials are the materials that transform heat into electricity. Nanotechnology enhances their efficiency with the improvement of their thermoelectric properties. Various research studies involve nanostructured material nanowires, quantum dots, and nanocomposites to increase the Seebeck coefficient, which is a measure for thermoelectric efficiency, and reduce thermal conductivity in such a way that much of heat conversion into electricity would be realized with high efficiency. Most recent research studies show that an introduction of nanomaterials into thermoelectric device configurations really enhances their performance and enables more opportunities for waste heat recovery and energy harvesting, from industrial processes (Zhang et al., 2020).

2.1.4 Nano-Materials in Energy Sector

Several nanomaterials are involved in the advancement of efficient energy systems, including nanoparticles, nanotubes, quantum dots, graphene, fullerenes, perovskites, and nanocatalysts.

Nanoparticles and Nanotubes

In energy systems, nanoparticles such as gold, silver, and platinum are used as catalysts while carbon nanotubes are employed in enhancing mechanical strength, energy storage devices, and fuel cells. With their small size, large surface area, and electrical conductivity, these materials have a lot of unique properties making them highly effective in energy applications.

Graphene and Fullerenes

Graphene is one of the first-generation nanomaterials, well studied, and presents exceptional conductivity, both electrical and thermal, with considerable mechanical strength due to its crystalline hexagonal lattice monoatomic carbon thin film. Also, graphene in solar cell applications, hydrogen storage, and as supercapacitors have undergone extended use lately. Fullerenes, or buckyball, are spherical forms taken by carbon nanoparticles with exciting properties that take place in solar applications, mainly being organic photovoltaics.

Perovskites

Perovskite solar cells represent a new generation in renewable energy systems, featuring high efficiency while keeping relatively low production costs. This material has shown record efficiencies in converting sunlight into electricity, and their production is more cost-effective compared to traditional silicon-based solar cells.

Catalysis and Membranes

Nanocatalysts find their applications in various energy systems, including fuel cells and hydrogen production. These catalysts increase the rates of reaction and improve efficiency in energy conversion processes. Membranes based on nanomaterials also contribute to increasing the effectiveness of energy storage and conversion by allowing the passing of ions or molecules while inhibiting the passage of others.

It is important to highlight the huge potential that nanotechnology has demonstrated in the energy sector, starting with renewable sources, including energy storage, right to the technology of energy conversion. The new nanomaterials like nanoparticles, nanotubes, quantum dots, graphene, and perovskites would possibly change how we produce, store, and transform energy toward much more efficient and sustainable systems. With each step forward in research, the integration of nanotechnology into energy systems promises to be instrumental in overcoming the world's growing needs for energy while addressing environmental concerns.

3 Renewable Energy Nanotechnology

Nanotechnology has emerged as one of the most innovative fields in renewable energy development because of its vast potential to improve efficiency, cost-effectiveness, and environmental sustainability in energy systems. This is because nanotechnology has the distinctive capability to manipulate matter at the atomic or molecular level, thereby providing specific benefits in energy generation, storage, and conversion. Application of nanotechnology in various renewable energy resources such as solar, wind, and hydrogen energy has been reviewed in the following section along with discussions on important nanomaterials involved, advantages thereof, and challenges offered.

3.1 Solar Energy

3.1.1 Photovoltaics

Solar energy has become one of the most abundant and renewable source transformations into very powerful forms with the help of nanotechnology. Among various innovations in solar cells, nanostructured solar cells, especially perovskite solar cells and quantum dot solar cells, draw much attention. The efficacy of perovskite-based solar cells has garnered greater attention lately because of their high efficiency at low production cost and easiness of fabrication (Banin et al., 2021; Mao et al., 2012). Researchers managed to boost the efficacy of perovskite solar cells by integrating nanomaterials, such as nanostructured electron transport layers, which significantly enhance the performance of the cells as represented in Fig. 2.

Figure 2 depicts the inner structure and working mechanism of a perovskite solar cell. It shows the key layers of the solar system, such as the transparent electrode, perovskite layer, and electron transport layer, and how these layers operate together to produce electricity from sunlight. Nanomaterials, in particular, perovskites with a tunable bandgap, are emphasized for their high absorption efficiency and relatively low cost of production. The figure also explains how nanostructured interfaces assist in charge separation while minimizing recombination losses. This

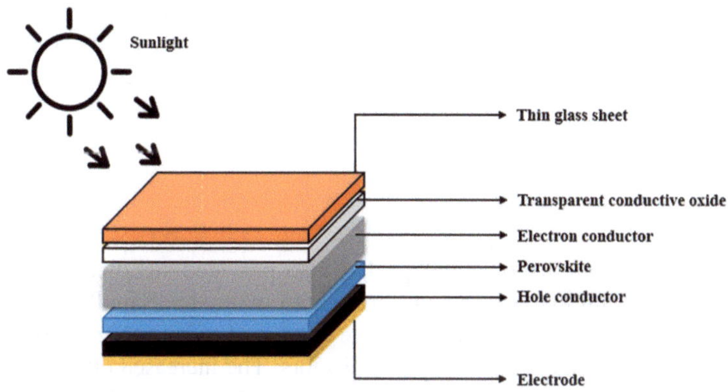

Fig. 2 A perovskite solar cell schematic diagram

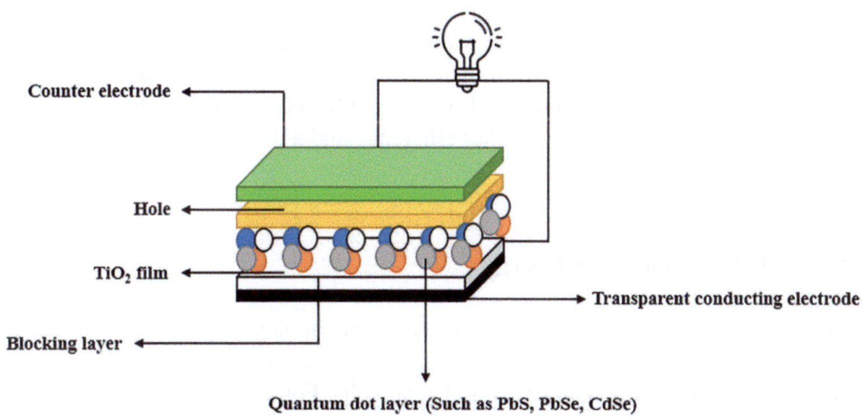

Fig. 3 A quantum dot solar cell schematic diagram

schematic is instrumental in showing how several aspects of nanotechnology, especially material design, can offer extreme leverage toward the prompt evolution and commercialization prospects of next-generation solar photovoltaic systems.

The main use of these nanomaterials recently is in quantum dot solar cells. The cells employ semiconductor material in nanoparticle form. Therefore, because of this aspect, they possess unique optical properties that allow better light absorption. Quantum dots have also made it possible for the better optimization of the absorption of light at higher wavelengths to enhance overall efficiencies (Abed-Mottaleb et al., 2011).

Figure 3 illustrates a quantum dot solar cell, emphasizing the ability of semiconductor nanocrystals to capture and convert solar energy. The layer structure is pictured, indicating that quantum dots act as a light absorber, followed by the absorption of photons, creation of electron-hole pairs, and eventual separation of charge. Quantum dots widen the absorption spectrum and allow for bandgap tuning, which greatly contributes to efficiency enhancement over existing silicon cells. This figure explain how nanotechnology aids in utilizing new materials with size-dependent optical properties to achieve more efficient and possibly less expensive, solar energy conversion options.

3.1.2 Solar Thermal Applications

Besides photovoltaics, nanotechnology has also provided significant benefits for solar thermal systems that collect and store heat from the sun. Nanofluids, which are suspensions of nanoparticles in a fluid, have been used to improve the thermal conductivity of heat transfer fluids in solar collectors. The increased efficiency in heat transfer allows these systems to result in greater capture and better transfer of energy, thus improving the overall performance of solar thermal energy systems (Bhong et al., 2024).

Another very important field of application where nanocoatings are finding their place is in the enhancement of heat capture and storage in solar thermal applications. These coatings are made from nanomaterials, solar radiation absorption and resistance of collectors are improved, which could reduce thermal losses and, in turn, give solar thermal systems higher efficiency and a better economic prospect in the long run.

3.1.3 Challenges in Solar Energy

Notwithstanding the promising advances in nanotechnology, a number of obstacles are yet to be conquered as far as solar energy is concerned. Stability is still the number one issue concerning nanostructured solar cells. For example, perovskite-based solar cells degrade with humidity and temperature. Presently, research is very active for the enhancement of their stability in such a way that the performance can last for a longer time, particularly in commercial aspects. Thirdly, nanostructured solar cells are hardly scalable: although the lab-scale fabrication is already successful, it is very challenging to mass-produce high-quality nanomaterials at low costs (Abed-Mottaleb et al., 2011).

3.2 *Wind Energy*

3.2.1 Nano-Coatings

Wind energy is another renewable energy source where nanotechnology is making a great impact. The wind turbine blades are exposed to harsh environmental conditions that might cause wear and tear, which reduces their efficiency and operational lifespan. Nanocoatings were fabricated to enhance durability and efficiency of turbine blades. These coatings, derived from nanomaterials, ensure an improved resistance to corrosion, wear, and tear, and even UV degradation, thus giving turbine blades a much improved performance and longer service life (Abed-Mottaleb et al., 2011).

3.2.2 Nano-Sensors for Predictive Maintenance

Apart from nanocoatings, nanosensors are also being integrated into wind turbine systems for predictive maintenance and optimization. Nano-sensors can monitor the shape/structures of the turbine blades or any stress, fatigue wear, and other real situations, transmitting real-time records to the operators for facilitating proactive maintenance before failure happened. Nanosensors were utilized to ensure maximum operational and economic efficiency by reducing expenditure made on maintenance and preventing inactivity (Jordi et al., 2024).

3.3 Hydrogen Energy

3.3.1 Nanomaterials for Hydrogen Production

Hydrogen energy is clean and sustainable source of power in nature. Nanotechnology plays an important role in the development of systems in hydrogen production, mainly the use of photocatalysts in the splitting of water. Water splitting is a process where water molecules are split into hydrogen and oxygen using sunlight. Nanostructured materials, such as TiO_2, while modified at the nanolevel, showed enhanced photocatalytic activity toward effective water splitting. More recently, newer nanostructures in the form of nanowires and nanotubes are also being targeted by researchers with regard to their application in hydrogen production mechanisms (Mao et al., 2012).

3.3.2 Storage and Transportation

Another critical bottleneck in the hydrogen energy sector is associated with hydrogen storage. Thus, nanotechnology is used to develop light-weight and high-density hydrogen storing systems that are not only essential for transportation but also in practical applications. For this, nanostructures such as MOFs, and carbon nanotubes are being highly explored due to their capacity in the absorption and storage of a large amount of hydrogen at relatively low pressures, making the storage and transport much safer or efficient. These nanostructures developed are promising to overcome the intrinsic limitations of hydrogen storage currently and allow the practical use of hydrogen as a clean energy source (Pomerantseva et al., 2019).

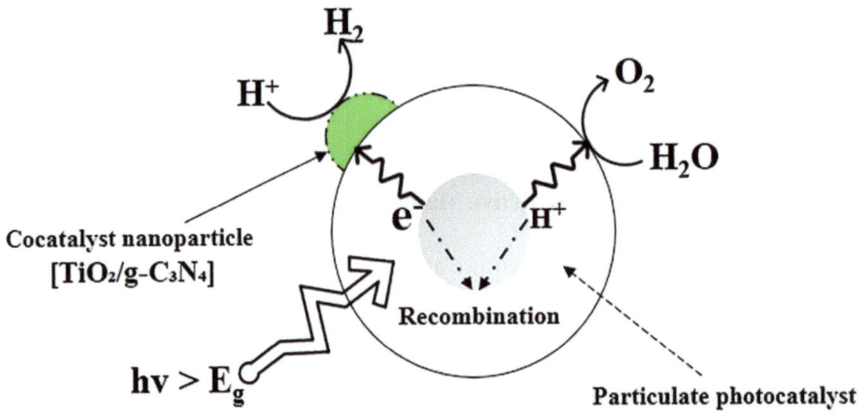

Fig. 4 Mechanism of photocatalytic water splitting for hydrogen production

Figure 4 shows the mechanisms of photocatalytic water splitting for hydrogen production in nanostructured materials, such as titania (TiO_2) and graphitic carbon nitride (g-C_3N_4). The photocatalysts presented herein are able to split water into H_2 and O_2 under sunlight-a source considered clean and renewable for the generation of hydrogen energy. One may recognize therein an important contribution of nanotechnology in enhancing such activity.

3.3.3 Fuel Cells

Nanocatalysts also play a critical role in fuel cells in terms of the conversion of hydrogen into electric energy. At the present time, platinum is the most common catalyst in fuel cells. However, its high cost prevents the wide diffusion of hydrogen fuel cells. Therefore, researchers are exploring at nanocatalysts made from metals like nickel, cobalt, and iron, which are widely available and cheaper than platinum. These nanocatalysts will reduce the costs of the fuel cells without losing their efficiencies (Banin et al., 2021).

4 Nanotechnology for Energy Storage

Nanotechnology has so far transformed many industries, especially in energy storage technologies. This revolution is notable in areas such as batteries, supercapacitors, and grid storage, where nanotechnology could improve energy efficiency, prolong device life, and reduce environmental impact. The following section discusses the role of nanotechnology in energy storage, especially in terms of its impact on lithium-ion batteries, emerging battery technologies, supercapacitors, and large-scale energy storage systems.

4.1 *Nanotechnology in Batteries*

4.1.1 Lithium-Ion Batteries (Li-Ion)

Lithium-ion batteries are the dominant energy storage in user electronics, electric vehicles (EVs), and renewable energy storage applications. Despite their widespread use, traditional lithium-ion batteries are confronted with several challenges regarding energy density as well as charging speed during which critical lifestyle. Nanotechnology has enhanced current reconfigurable analog lithium-ion batteries through advanced cathode, anode current collector, and electrolytic separator sheet materials.

Nanostructured electrodes, such as graphene-based materials, carbon nanotubes, and silicon nanoparticles, were found to improve significantly the performance of lithium-ion batteries. These materials possess larger surface areas, enhanced conductivity, and better mechanical properties, highly useful in improving energy storage capacity, reducing charging times, and enhancing the overall longevity of batteries. For instance, researchers have incorporated nanostructured silicon into the anode of lithium-ion batteries, improving their energy capacity while reducing the expansion and contraction of the silicon that often leads to performance degradation during cycling (Wu et al., 2010).

4.1.2 New Battery Technologies

In addition to Li-ion batteries, a set of next-generation energy storage systems is under development. These include solid-state batteries, lithium-sulfur (Li-S), and lithium-air ($Li\text{-}O_2$) batteries. Solid-state batteries replace the liquid electrolyte with a solid one and are considered much safer and more energy-dense than traditional Li-ion batteries. Nanotechnology can help enhance the ionic conductivity of solid-state electrolytes, which is one critical element in making large-scale use of solid-state batteries practical.

In principle, lithium-sulfur batteries have a much higher theoretical energy density than lithium-ion batteries but they suffer from very poor conductivity and rapid degradation of the sulfur cathode. For the enhancement of conductivity and ensuring cycle stability, conductive carbon nanostructures have been used within sulfur cathodes. Similarly, lithium-air batteries possess a far higher energy density compared to lithium-ion batteries, but their widespread adoption has been limited by issues related to air cathode stability and efficiency. Nanocatalysts are being developed with the intention of increasing the efficiency of such systems and prolonging their lifetime (Nazar, 2022).

4.2 Nanotechnology in Supercapacitors

Supercapacitor is another very promising and complementary device to the battery for energy storage. This maintains very high-power density, faster charge/discharge cycles, and very long-life cycles. However, in respect of energy density, energy stored in supercapacitors is much lower when compared with that in batteries, which limits their application in some areas. Nanotechnology has been instrumental in enhancing the energy density and overall performance of supercapacitors.

4.2.1 Nanostructured Carbon Materials

Graphene and carbon nanotubes (CNTs) are the latest advancement and development in supercapacitors. Some of the attributes that make graphene a uniquely adapted material for supercapacitor applications include its excellent conductivity, two-dimensional structure, and high surface area. These materials added within the electrodes of these supercapacitors improve their energy storage density, charge/discharge efficiency, and cycling stability. Carbon nanotubes, with high electrical conductivity and mechanical strength, have also been integrated with the aim of improving supercapacitor performance by enhancing conductivity in the electrodes and increasing the capacitance of charge storing (Li & Wei, 2013).

4.2.2 Hybrid Supercapacitors

Hybrid supercapacitors are a new, promising class of energy storage devices that merge properties from both batteries and supercapacitors. These devices offer both the high energy density typical of batteries and the high-power density typical of supercapacitors. As an example, hybrid devices may be based on the combination of battery-type electrode-like lithium-based anodes, which are excellent examples-with a capacitor-type electrode, such as graphene or carbon nanotubes. It offers the combination of advantages that each brings such as faster times of charging, longer life, and higher energy storage capacity. Nanotechnology contributes to electrode material optimization in hybrid supercapacitors for better general performance (Noor et al., 2022).

4.3 Nanotechnology in Grid Storage and Distribution

Currently, demand is increasingly high with regard to the advancement of efficient and large-scale energy storage solutions in integrating most renewable energy sources, like solar and wind, into the electrical grid. Nanotechnology is the best solution so as to come up with efficiency and reliability improvements in the grid storage system. This is important, particularly in balancing the intermittent nature of renewable energy generation.

4.3.1 Nanotechnology for Smart Grid Application

The idea of smart grid systems is basically the optimization of efficient electricity distribution and energy with betterment in the power system. Nanotechnology therefore takes a prime role in smart grid systems through the development of advanced sensors, energy-efficient transmission lines, and energy storage solutions. Nanosensors have the ability to monitor energy usage, fault detection, and optimization of energy flow in the grid such that electricity is effectively utilized. It has also used nanomaterials including graphene and carbon nanotubes in enhancing the transmission lines' conductivity to their efficiency by reducing energy losses in the grid (Eldada, 2011).

4.3.2 Large-Scale Storage Solutions

Large-scale energy storage seems to be a very promising solution in technologies such as flow batteries. Systems like these can be further improved using nanotechnology to enhance performance by developing efficient and stable electrolyte material. Nanostructured material approaches currently being applied to carbon-based electrodes have demonstrated great potential for increasing conductivity and energy storage capacity in flow batteries, making them suitable for large-scale and long-duration storage applications (Xu et al., 2020). Nanotechnology is being applied to develop next-generation energy storage systems, including batteries, supercapacitors, and large-scale storage systems.

Nanotechnology has the capability to enable the constituent material performance of these devices toward higher energy densities, faster charging times, longer cycle life, and overall higher efficiency; therefore, nanotechnology can be considered very promising in energy storage. Due to the fact that research is continuous, which also entails new nanomaterials and technologies, the potential of nanotechnology on energy storage is immense. It will most probably have a critical role in transitioning into sustainable and renewable energy systems.

A summary of some of the nanomaterials used, including quantum dots, graphene, and carbon nanotubes, together with their specific applications in energy, is given in Table 1, detailing how different nanomaterials envisage a wide range of energy technologies owing to their unique properties.

Table 1 Nanomaterials and their applications in energy

Nanomaterial	Applications in energy	Properties	References
Quantum dots	Photovoltaic cells, LEDs, display technologies	Size-dependent optical properties, tunable bandgaps, high quantum yield	Bressi et al. (2021), Wang et al. (2015)
Graphene	Supercapacitors, batteries, fuel cells, photovoltaics, sensors	High electrical conductivity, high surface area, mechanical strength, thermal stability	Bouclé and Herlin-Boime (2016), Bressi et al. (2021)
Carbon nanotubes (CNTs)	Batteries, supercapacitors, fuel cells, structural materials	High electrical conductivity, high surface area, mechanical strength, chemical stability	Liu, Chang, et al. (2012)
Perovskites	Solar cells (high-efficiency photovoltaics), LEDs, photodetectors	High-power conversion efficiency, cost-effective, tunable electronic properties	Zhou et al. (2022)
Silicon nanowires	Lithium-ion batteries (anode materials), high-capacity storage devices	High capacity for lithium storage, structural stability, fast lithium diffusion	Bressi et al. (2021), Gao et al. (2013)
Metal nanoparticles (e.g., platinum)	Fuel cells (catalysts for electrochemical reactions), hydrogen production	High surface area, enhanced catalytic activity, stability in harsh environments	Kim et al. (2021), Xiao et al. (2024)
Metal-organic Frameworks (MOFs)	CO_2 capture and storage, hydrogen storage, catalysis	High surface area, tunable pore size, selective adsorption	Xiao et al. (2024), Yao et al. (2023)
Nanostructured oxides (e.g., TiO_2)	Photocatalysts for hydrogen production, environmental remediation	High photocatalytic activity, low cost, stability under UV light	Chen et al. (2023)
Nanoscale metal hydrides	Hydrogen storage systems, fuel cells, portable power devices	High hydrogen storage capacity, reversible absorption/desorption properties	Tejashwini et al. (2024)

5 Nanotechnology in Protection of the Environment

Most of the developments in nanotechnology have made significant strides toward application in many industries; perhaps one of the most promising uses, however, might be environmental protection. So far, it has been possible to elaborate on some very efficient approaches to control pollution, manage waste, and even capture carbon by using nanomaterials. The section that follows describes how nanotechnology contributes to such important areas of concern.

5.1 Pollution Control and Remediation

5.1.1 Nanomaterials for Purification of Air and Water

Nanotechnology, therefore, have immense potential to contribute to cleaning air and water pollution. Nanocatalysts, nanosorbents, and nanomembranes are some of the materials whose application has revolutionized processes in this area of purification. Nanocatalysts are materials of at least one dimension with size in nanometer scale used in acceleration of chemical reactions which will break down pollutants in both water and air. For instance, nanocatalysts based on metal oxide forms have shown promising degradation of VOCs and other noxious gases (Jadhav et al., 2011). These nanocatalysts, when compared to traditional catalysts, are advantageous because high surface area and reactivity are certain features that make them highly effective in the degradation of pollutants at very low concentrations too.

Another breakthrough in environmental cleanup involves nanosorbents or nanomaterials designed to adsorb contaminants. For instance, nanoparticles such as activated carbon and metal oxide nanoparticles have excellent adsorption capabilities and are very effective in the removal of heavy metals, dyes, and other toxicants in various types of wastewater treatments (El-sayed, 2020). These can be engineered toward specific pollutants and thus make better water treatment possible.

Nanomembranes are thin films or coatings through which impurities get filtered out of the air or water. A typical example of these membranes includes nanomaterials such as carbon nanotubes and graphene oxide, which filter some pollutants out while allowing some to pass through. These have high surface areas that have ensured contaminant trapping higher than by traditional materials, hence ideal for use in a large-scale water purification system (Bouclé & Herlin-Boime, 2016).

Figure 5 shows the application of different nanomaterials that can help reduce air pollution. These depict nanocatalysts, nanomembranes, and adsorbents used for the capture or destruction of harmful gaseous substances and particulate matter. Due to their high surface area and reactivity, these nanoparticles are ideal for filtration and catalytic degradation processes. The depiction further highlights the role nanotechnology plays in clean air and ties to broader environmental sustenance concepts by enumerating the practical applications employed in urban and industrial pollution control systems.

5.1.2 Nanoparticles in Oil Spill Cleanup

Oil spills are one of the worst environmental hazards possible, but nanotechnology has been able to provide ways for quicker cleaning. Among the highly promising nanomaterials for the fast removal of oil from the water surface, magnetic nanoparticles significantly rank top. Magnetic nanoparticles based on iron oxide have the tendency to adsorb large amounts of oil and, at the same time, can easily be separated from water with the help of an external magnetic field (Stolarczyk et al., 2018).

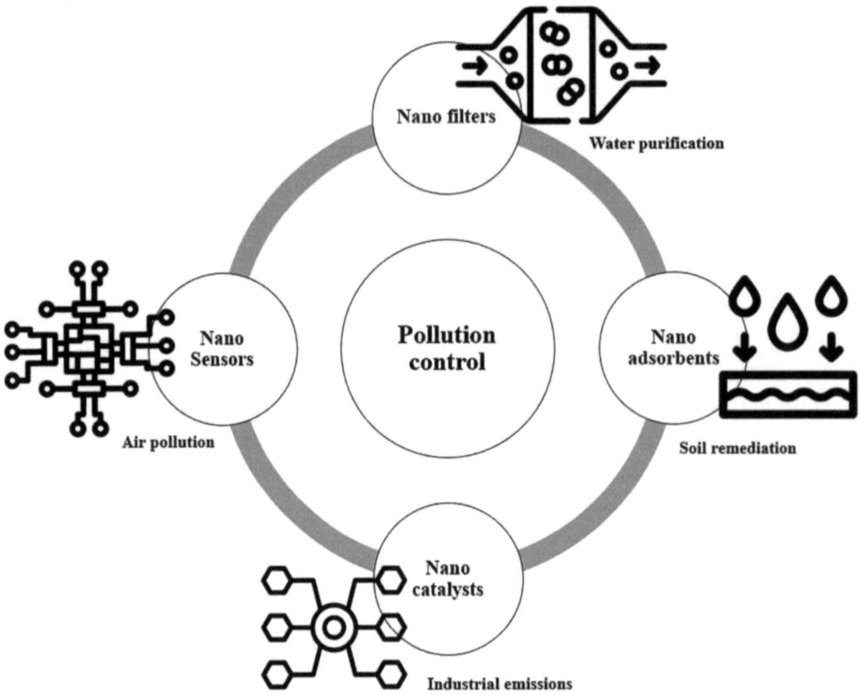

Fig. 5 Nanomaterials for air pollution control

The primary idea of this technique is not only the increase in the rate of recovery but also reducing the risk of pollution during cleaning. Another thing is that the magnetic nanomaterials are reusable, which makes them economically efficient in case of considerable oil spills.

5.2 Waste Management and Recycling

5.2.1 Nanotechnology in E-Waste Recycling

E-waste has been a growing concern due to an increase in volumes coming from discarded electronic equipment, which is made up of many valuable materials that can also be hazardous. Nanotechnology now offers some new approaches to effective rare earth metal recovery from e-waste. These nanomaterials involve gold, silver, and copper nanoparticles, which could be used in the extraction processes that would recover such valuable materials unavailable or hardly accessible by conventional methods of recycling. Nanomaterials can also be utilized in the extraction of precious metals from low-grade ores and, hence, render mining greener (Liu, Xue, et al., 2012).

5.2.2 Nanomaterials in Biodegradation

The second important approach in waste management involves biodegradation, mainly when the organic pollutant is not readily degradable. It is achieved by enhancing the degradation rate of organic pollutants through the application of nanomaterials. For example, nanozymes are nanomaterials engineered in such a way that they express properties like enzymes, and may, therefore, catalyze harmful chemicals into harmless forms in the environment at rates higher than natural enzymes do (Simpa et al., 2024). These nanozymes can be engineered for the purpose of targeting specific pollutants, including pesticides and herbicides, which in turn promote their rapid degradation, reducing their toxic effects. Nanomaterials can also be coupled with bacteria or fungi in bioremediation processes, increasing the effectiveness of pollutant degradation.

5.3 Carbon Capture and Storage (CCS)

Nanomaterials applied in DAC systems for removing carbon dioxide from the atmosphere are depicted in Fig. 6. It shows the use of advanced nanostructured sorbents like MOFs and amine-functionalized materials with strong CO_2 selectivity and

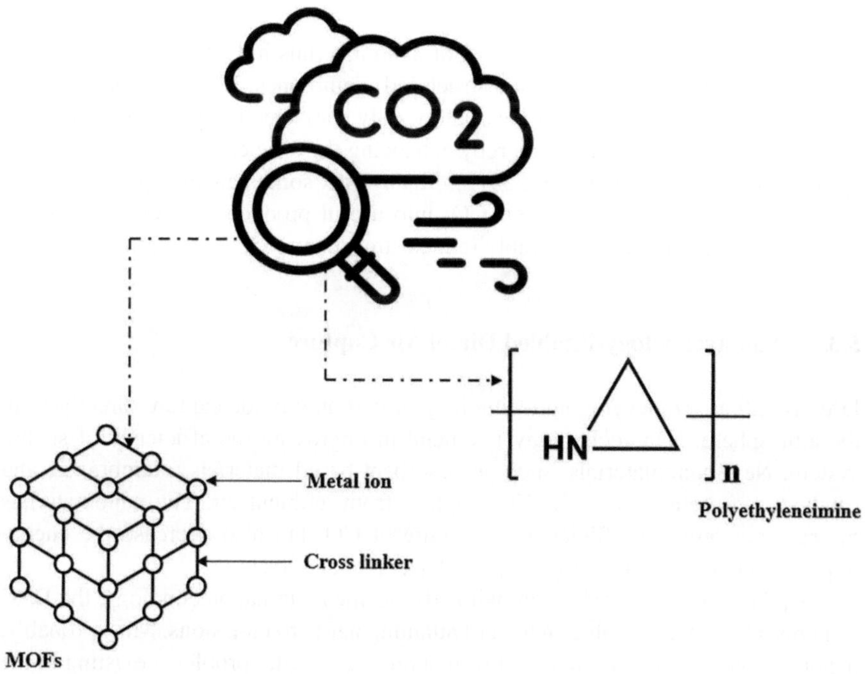

Fig. 6 Nanomaterials in direct air capture systems

adsorption capacity. Also, each step is outlined: air intake, binding of CO_2, and either storage or use. This illustration is critical to show how nanotechnology is applied to combat climate change, with nanoscale materials making scalable, energy-efficient solutions possible to decrease concentrations of greenhouse gases along with carbon-neutral or negative emission technologies.

5.3.1 Nanomaterials for Carbon Dioxide Capture

It has become a problem that in the last few years has taken development attention to the Carbon Capture and Storage technologies. Under development, nanomaterials capture CO_2 from industrial emissions, reducing greenhouse gas concentration in the atmosphere. Like all the other forms of adsorbents nanoporous material-adsorption of CO_2 is greatly enhanced by their having very highly extensive surface area on which a large amount of gas molecules of CO_2 would adsorb, among reasons (Chen et al., 2023). Examples of these are metal-organic frameworks, like others, in their use at large while being considered and targeted for technological innovations because their selective capability of separating carbon dioxide into industrial flue gases (Tong & Sekizkardes, 2021).

5.3.2 Membrane and Catalyst Nanomaterials

Besides adsorbents, other applications of nanomaterials in CO_2 separation include membrane-based separations. Nanostructured membranes, like graphene oxide or carbon nanotube-based membranes, allow only the selective permeation of CO_2 while blocking all other gases, thereby enhancing the efficiency of the CO_2 capture systems significantly (Afif et al., 2019). Meanwhile, some nanocatalysts have been synthesized in attempts to convert CO_2 into useful products, namely methane or methanol, which could considerably reduce atmospheric CO_2 concentration levels.

5.3.3 Nanotechnology-Enabled Direct Air Capture

DAC is still a new developing technology of carbon dioxide capture directly from the atmosphere. Nanotechnology is crucial in improving the efficiency of such a system. New nanomaterials, such as adsorbent-based materials, membranes, and catalysts, are being tested for CO_2 capture from ambient air. The nanomaterials increase not only the efficiency of capture of CO_2 but also decrease the energy requirements for systems of DAC (Rezakazemi et al., 2021).

It is, therefore, reasonable that with advancements in nanotechnology, the DAC can present a practical solution toward attaining net-zero emissions. Most probably, nanotechnology will save mankind from a number of vital problems existing in the contemporary world facing ecological catastrophes. Beginning with air and water cleaning, cleaning up oil spilled in the sea, ending with electronic wastes recycling—there are a number of fields that nanomaterials act better and more effectively than the so-calledtraditional ways by all means. Moreover, in terms of CC, advanced development of nanomaterials for carbon capture also shows great perspectives. Continuing to innovate with nanotechnology holds great promise for significant positive impacts on environmental protection.

6 Nanotechnology for Sustainable Development

Nanotechnology is a relatively emerging sector that holds great promise toward the solution of many global challenges included within the United Nations Sustainable Development Goals. Considering unique properties of nanomaterials and nanoscale processes, nanotechnology can contribute to sustainable solutions in diverse sectors, starting with clean energy production and ranging up to environmental protection. This section discusses nanotechnology with respect to economic and social consequences in the context of achieving SDG 7 (Affordable and Clean Energy), SDG 13 (Climate Action), SDG 14 and 15 (Life Below Water and Life on Land). Furthermore, it discusses the challenges and opportunities of cost-benefit analysis as well as the necessary policy and regulation frameworks for an ethical and safe deployment of nanotechnology.

6.1 Nanotechnology and the UN SDGs

6.1.1 SDG 7: Affordable and Clean Energy

The most promising applications of nanotechnology include energy production and storage, especially from renewable energy sources. Nanomaterials enhance the efficiency of solar cells, batteries, and fuel cells contributing to the development of cleaner and cheaper energy solutions.

Solar Energy

Nanotechnology has so far enabled immense improvements in photovoltaic (PV) technology. Nanostructured materials, such as quantum dots and perovskite solar cells, exhibit higher light absorption and energy conversion efficiency compared to conventional silicon-based solar cells. Evidence from research has shown that nanomaterials will reduce the cost of production while improving the performance of solar panels and, as such, make solar energy more available and accessible in both developed and developing regions (Deshmukh et al., 2021).

Energy Storage

The advancement of renewable energy resources is causing an increased demand for effective energy storage systems. Nanotechnology significantly aids in the advancement of high-capacity batteries and supercapacitors. Graphene and Carbon nanotubes were found to improve energy density and lifetime of batteries for efficient energy storage in solar and wind energy conversion (Gao et al., 2013).

Energy Efficiency

The second major contribution of nanotechnology has involved making energy utilization more efficient for buildings and at industrial levels. Their applications have emerged in buildings in developing nano-coatings, which minimize the heat loss and the development of excessive amounts of heat inside them. In addition to this aspect, another potential application field that opens for nanosensors relates to real-time energy monitoring and optimization (He et al., 2020).

6.1.2 SDG 13: Climate Action

Nanotechnology can affect two major aspects of the mitigation of climate change, such as reduction in greenhouse gas emissions and improvement of carbon capture and energy efficiency.

Carbon Capture and Storage

It enhances the effectiveness of the carbon capture technologies by increasing the surface area that is able to absorb CO_2. Examples include carbon-based nanomaterials, which have been identified to possess excellent CO_2 absorption characteristics. These could be integrated into existing systems for carbon capture and, in that way, enhance the efficiency of these systems and make them less energy-intensive (Saeed et al., 2024).

Climate Modeling and Monitoring

The application of nanosensors is immense in environmental monitoring, such as air and water quality, and the level of atmospheric carbon. These sensors are able to provide current data on the dynamics of climate change and help make informed policy decisions (Liu et al., 2010).

6.1.3 SDG 14 and 15: Life Below Water and Life on Land

Nanotechnology is also applied to conserve the environment, especially aquatic and terrestrial ecosystems.

Water Purification

Availability of clean water is one of the important challenges concerning SDG 14-LBW and SDG 15-LOL. Different nanomaterials have been noted to show a great potential and excellent performance within water remediation through removal of heavy metals, pesticides, and pathogens by carbon nanotubes and nanocomposites (Jadhav et al., 2011). These nanotechnology-based filtration systems are highly efficient and economically viable; hence, they are attractive in water purification for both developed and developing regions (Jadhav et al., 2011).

Remediation of Pollution

It is clear that nanotechnology is effective in environmental remediation. Speccifically, nanoparticles have been utilized for the degradation of oil spills, agricultural runoff, and other such pollutants at the molecular level. Thereafter, this has posed a very important influence in preserving aquatic ecosystems and raising better soil health (Zhao et al., 2012).

Biodiversity Conservation

Nanotechnology can further help in the conservation of biodiversity through advanced methods of observation. For example, in an advanced way, it manages chemical pollutants that may bring along contamination of ecosystems. The early detection and intervention at the site of harmful contamination are, consequently enhanced. Beyond that, there is also a promising use in nanotechnology for the preparation of green pesticides and fertilizers to reduce any impacts related to their application in agriculture (Bouclé & Herlin-Boime, 2016).

6.2 *Economic and Social Implications*

While nanotechnology has a huge potential to achieve SDGs, its economic and social considerations need attention to ensure that these technologies are deployed in an equitable and responsible manner.

6.2.1 Cost-Benefit Analysis

The biggest challenges to implementing nanotechnology for sustainable development include making its benefits accessible to all, particularly in countries of poor development.

The Price of Nanomaterials

Although the manufacturing costs of some nanomaterials have significantly gone down over the last years, high initial costs of advanced systems based on nanotechnology are still a serious obstacle for their wide-scale implementation. For example, nanostructured solar cells, while much more efficient, are still more expensive in production and hence less accessible in low-income regions (Liu, Chang, et al., 2012).

Long-term Savings

Despite the high initial costs, the nanotechnology-enabled solutions may bring some long-term economic benefits. For example, the application of nanomaterials in energy generation and storage can help in reducing energy costs in the future, while nanotechnology-based water purification systems make clean water cheaper. The cost analysis of such technologies indicates that in the long term, these could be far more economically viable than traditional solutions, provided scaling up is done according to merit (Shapira & Youtie, 2015).

6.2.2 Politics and Regulation

Due to rapid development in nanotechnology requires effective policy and corresponding regulatory frameworks to ensure not only safety but also moral deployments, particularly in energy utilization and environmental perspectives.

Safety Issues

One of the major issues in relation to nanotechnology includes health and environmental impacts. Because of their minute sizes, nanoparticles easily pass through biological membranes and can lead to the increase in toxicity and long-term exposure risks that may pose a challenge to human health in the future. It is based on these premises that regulatory agencies must set safety standards in relation to the manufacture and applications of nanomaterials toward the energy and water-purifying industries (Zhang et al., 2020).

Ethical Considerations

Nanotechnology could potentially increase the current inequalities, especially if most of the benefits were to be concentrated in industrialized nations or a few giant corporations. Policymakers should ensure that the deployment of nanotechnology is in line with ethical considerations, including fairness, equity, and environmental justice. International cooperation will be necessary for the development of policies that foster responsible use of nanotechnology to benefit all sectors of society.

Regulatory Frameworks

There is a need for collaboration among governments in laying down standards and regulations related to nanotechnology at the global level. Such a framework would detail the safe development, testing, and use of nanomaterials, together with mechanisms for monitoring and enforcement. This would ensure that any potential risks due to nanotechnology are reduced to a minimum while ensuring maximum benefits toward sustainable development (Liu et al., 2010).

7 Challenges and Future Prospects

Besides the advancement of energy technologies, especially nanotechnology, renewable energy systems, and nanomaterials, different critical issues and interesting perspectives appear. Those challenges span from technical and economic ones to environmental aspects of ongoing and future trends in materials and technologies that may bring quite new perspectives in the working of energy systems during the coming decades.

Some of the major issues associated with nanomaterials in energy applications are instability, scalability of production, health and environmental risks, high cost, and regulatory challenges. Specific research and current solutions which try to solve each issue are also summarized in Table 2.

7.1 Technical Challenges

7.1.1 Stability, Scalability, and Integration into the Existing Systems

Stability and scalability of the advanced energy technologies using nanomaterials are considered to be a major technical challenge. Thus, carbon nanotubes and graphene display exceptional properties at a small scale, and it is very difficult to keep these properties on a much larger scale applicable in realistic applications. There are plenty of great ideas for new technologies, such as nanophotovoltaics that can

Table 2 Challenges and solutions in nanotechnology for energy applications

Challenge	Description	Solutions or current research efforts	References
Stability of nanomaterials	Nanomaterials often degrade or lose efficiency under real-world conditions (e.g., moisture, UV light)	Development of advanced encapsulation techniques and material engineering to enhance stability	Song and Zhu (2024)
Scalability of production	Difficulty in producing nanomaterials consistently at a commercial scale without compromising properties	Research into cost-effective and scalable synthesis methods, including green chemistry approaches	Zhang et al. (2021)
Health and environmental risks	Concerns over toxicity, bioaccumulation, and long-term environmental impact of nanomaterials	Establishing standardized protocols for handling, use, and disposal of nanomaterials	Chen et al. (2023)
High cost of nanomaterial synthesis	High cost of synthesizing and processing nanomaterials, such as graphene and perovskites	Focus on scaling up production and reducing costs through mass production and improved fabrication techniques	Pomerantseva et al. (2019)
Market penetration and adoption	Slow market adoption due to lack of established supply chains, manufacturing capabilities, and awareness	Increasing market awareness and building infrastructure to support supply chains	Zhou et al. (2022)
Regulatory and policy challenges	Need for guidelines to ensure the safe and ethical use of nanomaterials considering potential risks	International cooperation to harmonize standards and facilitate responsible development	Gupta et al. (2025)

convert sunlight into energy with nanoscale materials, but scaling technologies remains an issue to this date (Banin et al., 2021). If these systems are ever to find viable applications commercially, they have to be cost-scalable, production-scalable, and deployable across industries. Other issues are integration with the existing infrastructure: most of the existent energy systems were not originally designed for advanced technologies represented by nanofluidic energy systems (Liu, Xue, et al., 2012).

7.1.2 Health, Safety, and Environmental Risks

While nanomaterials indisputably confer many benefits, one cannot disregard the associated health and safety risks or even possible environmental risks quite so easily. This is because of the fact that certain nanomaterials could possess hazardous as they could interact with biological systems in unexpected ways (Liu et al., 2010). Besides, there is another special risk regarding their possible environmental degradation or the disposal of nanomaterials.

It is now important and urgent to carry out due diligent testing and regulation of these nanomaterials to control their introduction into the production chain in a manner aimed at ensuring they do no long-term ecological harm. Certainly, a lot more needs to be learnt about the potential safety of nanomaterials and most importantly about each stage in the nanoproduct lifecycle: production, maintenance, and disposal.

7.1.3 Economic and Market Barriers

Cost of Production and Market Penetration

Any new technology will, in general, have specific economic factors that will decide its future. Advanced energy technologies generally involve high production costs, particularly for nanomaterials. For instance, nanostructured material manufacturing in solar cells or batteries is still expensive due to the complex fabrication process involved (Tejashwini et al., 2024). Market penetration is another critical issue. Although technologies such as perovskite solar cells hold tremendous promise in the lab, translating these advancements into mass-market products that can be cost-competitive with already established energy solutions such as silicon solar cells continues to be a major barrier (Roy et al., 2022). Overcoming these economic barriers requires innovation in manufacturing processes and a strong push toward reducing costs.

Intellectual Property and Innovation Challenges

Other critical barriers to the fast development and deployment of nanotechnology in energy systems include various intellectual property issues. It is believed that a mere potential patent dispute over a new material or technology may cause significant preventing of the velocity of innovation, with access to various critical advancements becoming restricted (Bastani & Fernandez, 2002). Furthermore, the rate at which this field is evolving is such that tracking of patent rights, especially in areas involving nanotechnology, becomes cumbersome and costly both for researchers and companies. The only way such IP issues could be sorted out is through cooperation internationally and through smooth patenting mechanisms; this would help accelerate the pace of innovation.

7.2 *Future Trends*

7.2.1 Nano-Energy Materials Development: Graphene and Perovskites

The most exciting energy technology prospect enables power through advanced nanoenergy material development. Graphene, a material known for its strength, conductivity, and flexibility, has shown promise in different applications, from energy storage systems to solar cells. Researchers are focusing on overcoming the challenges of mass production of high-quality graphene and integrating it into functional energy devices (Liu, Chang, et al., 2012). Similarly, perovskite solar cells also emerged as a frontrunner for next-generation photovoltaics because of their high efficiencies at relatively low manufacturing costs (Liang et al., 2009). These rapid advances in properties and production methods for these materials suggest that they can play an important role in renewable energy systems of the future.

7.2.2 Emerging Technologies

Nanofluidics

Nanofluidics, dealing with the manipulation of fluids at the nanoscale, extends applications in energy harvesting, water purification, and energy storage. These nanoscale channels greatly enhance both passive and active heat transfer/energy conversion processes, thus giving rise to improved energy system efficiency in devices (Pomerantseva et al., 2019).

Nanophotovoltaics

Another area of emerging technology involves nanophotovoltaics, studying different properties of nanomaterials with the purpose of using them to enhance the efficiencies of photovoltaic cells. Research on nanostructured materials, including quantum dots, is being explored for their superior capability in sunlight absorption and energy conversion compared to conventional materials (Semonin et al., 2012). These emerging technologies are envisioned to enhance efficacy and cut the cost of many renewable energy sources, among them solar power.

These emerging technologies are envisioned to improve the efficiency and reduce the cost for many renewable energy sources, including solar power.

7.2.3 The Next Frontier in Energy and Environmental Applications

Going forward, picotechnology, or the study and application of material and systems at the picometer scale, holds immense promise to disrupt energy systems. It is envisioned that by manipulating materials at even smaller scales, researchers will achieve unprecedented levels of precision and efficiency in energy conversion, storage, and transmission. It is at this scale that quantum effects start to become much more pronounced, thus enabling innovations like quantum computing, energy storage, and energy transmission. These will make systems orders of magnitude more efficient compared to the current technologies and will open completely new perspectives for energy and environmental applications in general (Tan et al., 2012).

8 Conclusion

Nanotechnology has been a revolutionary force in the energy sector, which has brought about newer solutions to the problems of energy generation, storage, and environmental sustainability. From the improvement of renewable energy systems, including solar and wind, to enhancing energy storage with nanostructured batteries and supercapacitors, nanotechnology offers key advances in different applications. Furthermore, recently, the emergence of new nanomaterials, such as graphene, quantum dots, and perovskite, paved the way not only toward much better energy systems but also allowed for groundbreaking advancements in energy conversion and storage technologies.

All things considered, notwithstanding this great potential of nanotechnology, a great multitude of challenges does exist in the path to leveraging Nanotechnology. Technical barriers regarding scalability, stability, and integrative capacity of nanotechnology through and across existing energy infrastructure inhibit—what many consider—the progress of rapid commercialization of most nanotechnology-based solutions. Health, safety, environmental concerns regarding the use of these nanomaterials offer much danger that needs forethought regulation, plus testing before deployment. Additionally, there are a myriad array of economic and market issues on high production cost among another challenge of intellectual property.

Nanotechnology has the potential to support sustainable development in accordance with global efforts toward the attainment of the United Nations' Sustainable Development Goals, particularly in the aspect of clean energy, climate action, and preservation of life on land and in water. Nanomaterials, if integrated through energy systems, pollution, and waste, are able to reduce carbon emissions, help in dwindling the use of fuels, and resource utilization at an efficient level.

References

Abed-Mottaleb, M. S., Byrne, J. A., & Chakarov, D. (2011). Nanotechnology for solar energy. *International Journal of Photoenergy, 2011*, 1–2. https://doi.org/10.1155/2011/194146

Afif, A., Rahman, S. M., Azad, A. T., Zaini, J. H., Islam, A., & Azad, A. (2019). Advanced materials and technologies for hybrid supercapacitors for energy storage: A review. *Journal of Energy Storage, 25*, 100852. https://doi.org/10.1016/j.est.2019.100852

Aricò, A., Bruce, P., Scrosati, B., Tarascon, J., & van Schalkwijk, W. (2005). Nanostructured materials for advanced energy conversion and storage devices. *Nature Materials, 4*(5), 366–377. https://doi.org/10.1038/nmat1368

Banin, U., Waiskopf, N., Hammarström, L., Boschloo, G., Freitag, M., Johansson, E. M. J., Sá, J., Tian, H., Johnston, M. B., Herz, L. M., Milot, R. L., Kanatzidis, M. G., Ke, W., Spanopoulos, I., Kohlstedt, K. L., Schatz, G. C., Lewis, N., Meyer, T., Nozik, A. J., Beard, M. C., & Brudvig, G. W. (2021). Nanotechnology for catalysis and solar energy conversion. *Nanotechnology, 32*(4), 042003. https://doi.org/10.1088/1361-6528/abbce8

Bastani, B., & Fernandez, D. (2002). Intellectual property rights in nanotechnology. *Thin Solid Films, 420*, 472–477. https://doi.org/10.1016/S0040-6090(02)00843-X

Bhong, M., Singh, R., Kumar, P., & Saraswat, M. (2024). Renewable energy integration for urban sustainability: A nanomaterial perspective. *E3S Web of Conferences, 511*(4), 01034. https://doi.org/10.1051/e3sconf/202451101034

Bouclé, J., & Herlin-Boime, N. (2016). The benefits of graphene for hybrid perovskite solar cells. *Synthetic Metals, 222*, 3–16. https://doi.org/10.1016/J.SYNTHMET.2016.03.030

Bressi, V., Ferlazzo, A., Iannazzo, D., & Espro, C. (2021). Graphene quantum dots by eco-friendly green synthesis for electrochemical sensing. *Nanomaterials, 11*, 1120. https://doi.org/10.3390/nano11051120

Chen, H., Zheng, Y., Li, J., Li, L., & Wang, X. (2023). AI for nanomaterials development in clean energy and carbon capture, utilization and storage (CCUS). *ACS Nano, 17*, 9763–9792. https://doi.org/10.1021/acsnano.3c01062

Deshmukh, M. A., Park, S. J., Hedau, B. S., & Ha, T. (2021). Recent progress in solar cells based on carbon nanomaterials. *Solar Energy, 220*, 953–990. https://doi.org/10.1016/J.SOLENER.2021.04.001

Eldada, L. (2011). Nanotechnologies for efficient solar and wind energy harvesting and storage in smart-grid and transportation applications. *Journal of Nanophotonics, 5*, 051704. https://doi.org/10.1117/1.3574149

El-sayed, M. E. A. (2020). Nanoadsorbents for water and wastewater remediation. *Science of the Total Environment, 739*, 139903. https://doi.org/10.1016/j.scitotenv.2020.139903

Gao, M., Xu, Y., Jiang, J., & Yu, S. (2013). Nanostructured metal chalcogenides: Synthesis, modification, and applications in energy conversion and storage devices. *Chemical Society Reviews, 42*(7), 2986–3017. https://doi.org/10.1039/c2cs35310e

Gupta, G., Bisht, Y. S., Saluja, A., Lakshmi, A. R., & Vanam, M. (2025, April). Nanotechnology in renewable energy: A comprehensive review with experimental analysis. In *AIP Conference Proceedings* (Vol. 3157, No. 1, p. 120048). AIP Publishing LLC. https://doi.org/10.1063/5.0271287

He, Y., Zhang, Y., Zhang, C., & Zhou, H. (2020). Energy-saving potential of 3D printed concrete building with integrated living wall. *Energy and Buildings, 222*, 110110. https://doi.org/10.1016/j.enbuild.2020.110110

Hudak, N. S. (2014). Nanostructured electrode materials for lithium-ion batteries. In G. Pistoia (Ed.), *Lithium-ion batteries* (pp. 57–82). Elsevier. https://doi.org/10.1016/B978-0-444-59513-3.00004-2

Jadhav, M., Todkar, A., Gambhire, V. R., & Sawant, S. (2011). Nanotechnology for powerful solar energy. *International Journal of Advanced Biotechnology and Research, 2*(1), 208–212.

Jordi, Z. J., Erasmo, C., & Rafael, C. (2024). Multi-criteria optimal design of small wind turbine blades based on deep learning methods. *Energy, 293*, 130625. https://doi.org/10.1016/j.energy.2024.130625

Khin, M. M., Nair, A. S., Babu, V. J., Murugan, R., & Ramakrishna, S. (2012). A review on nanomaterials for environmental remediation. *Energy and Environmental Science, 5*(8), 8075–8109. https://doi.org/10.1039/c2ee21818f

Kim, I. Y., Ko, J., Ahn, T. Y., Cheong, H. W., & Yoon, Y. S. (2021). Energy materials for energy conversion and storage: Focus on research conducted in Korea. *Journal of the Korean Ceramic Society, 58*(6), 645–661. https://doi.org/10.1007/s43207-021-00152-2

Li, X., & Wei, B. (2013). Supercapacitors based on nanostructured carbon. *Nano Energy, 2*(2), 159–173. https://doi.org/10.1016/j.nanoen.2012.09.008

Liang, M., Luo, B., & Zhi, L. (2009). Application of graphene and graphene-based materials in clean energy-related devices. *International Journal of Energy Research, 33*, 1161–1170. https://doi.org/10.1002/er.1598

Liu, C. J., Burghaus, U., Besenbacher, F., & Wang, Z. L. (2010). Preparation and characterization of nanomaterials for sustainable energy production. *ACS Nano, 4*(10), 5517–5526. https://doi.org/10.1021/nn102420c

Liu, D., Chang, D., Baek, J. B., & Lu, W. (2012). Carbon nanomaterials for advanced energy conversion and storage. *Small, 8*(8), 1130–1166. https://doi.org/10.1002/smll.201101594

Liu, J., Xue, Y., Zhang, M., & Dai, L. (2012). Graphene-based materials for energy applications. *MRS Bulletin, 37*, 1265–1272. https://doi.org/10.1557/MRS.2012.179

Maksoud, M. A. A., Fahim, R. A., Shalan, A., et al. (2020). Advanced materials and technologies for supercapacitors used in energy conversion and storage: A review. *Environmental Chemistry Letters, 19*, 375–439. https://doi.org/10.1007/s10311-020-01075-w

Mao, S., Shen, S., & Guo, L. (2012). Nanomaterials for renewable hydrogen production, storage, and utilization. *Progress in Natural Science: Materials International, 22*(6), 522–534. https://doi.org/10.1016/j.pnsc.2012.12.003

Nazar, L. (2022). Nanotechnology for industrial electrochemical energy storage. ECS Meeting Abstract MA2022-01 623. https://doi.org/10.1149/ma2022-017623mtgabs

Noor, A., Fatima, N., Tahir, M. B., Pervaiz, M., & Tanveer, M. (2022). Neoteric advances in affordable photocatalytic energy conversion and storage with perovskites nanocrystals. *International Journal of Energy Research, 46*, 24516–24527. https://doi.org/10.1002/er.8772

Pandey, R., Vats, G., Yun, J., Bowen, C., Ho-Baillie, A., Seidel, J., & Seok, S. (2019). Mutual insight on ferroelectrics and hybrid halide perovskites: A platform for future multifunctional energy conversion. *Advanced Materials, 31*, 7376. https://doi.org/10.1002/adma.201807376

Pandiyan, G. K., & Prabaharan, T. (2020). Implementation of nanotechnology in fuel cells. *Materials Today: Proceedings, 33*, 2681–2685. https://doi.org/10.1016/j.matpr.2020.01.368

Pomerantseva, E., Bonaccorso, F., Feng, X., Cui, Y., & Gogotsi, Y. (2019). Energy storage: The future enabled by nanomaterials. *Science, 366*, 8285. https://doi.org/10.1126/science.aan8285

Ramsurn, H., & Gupta, R. B. (2013). Nanotechnology in solar and biofuels. *ACS Sustainable Chemistry & Engineering, 1*(7), 779–797. https://doi.org/10.1021/SC400046Y

Rezakazemi, M., Kumar, B., & Sadasivuni, K. K. (2021). Preface on "nanomaterials for energy conversion and storage systems". *Emergent Materials, 4*, 387–388. https://doi.org/10.1007/s42247-021-00227-0

Roy, P., Ghosh, A., Barclay, F., Khare, A., & Cuce, E. (2022). Perovskite solar cells: A review of the recent advances. *Coatings, 12*(8), 1089. https://doi.org/10.3390/coatings12081089

Saeed, A., Munir, S., Gull, N., & Khan, S. M. (2024). Nanomaterials for carbon capture and their conversion to useful products for sustainable energy production. In K. Rizwan & M. Bilal (Eds.), *Woodhead series in bioenergy. Nanomaterials in biomass conversion* (pp. 369–395). Woodhead Publishing. https://doi.org/10.1016/B978-0-443-13500-2.00015-8

Sahaym, U., & Norton, M. G. (2008). Advances in the application of nanotechnology in enabling a 'hydrogen economy'. *Journal of Materials Science, 43*(17), 5395–5429. https://doi.org/10.1007/s10853-008-2749-0

Semonin, O. E., Luther, J. M., & Beard, M. C. (2012). Quantum dots for next-generation photovoltaics. *Materials Today, 15*, 508–515. https://doi.org/10.1016/S1369-7021(12)70220-1

Shapira, P., & Youtie, J. (2015). The economic contributions of nanotechnology to green and sustainable growth. In V. Basiuk & E. Basiuk (Eds.), *Green processes for nanotechnology*. Springer. https://doi.org/10.1007/978-3-319-15461-9_15

Simpa, P., Solomon, N. O., Adenekan, O. A., & Obasi, S. C. (2024). Nanotechnology's potential in advancing renewable energy solutions. *Engineering Science & Technology Journal, 5*, 1695–1710. https://doi.org/10.51594/estj.v5i5.1137

Song, Y., & Zhu, W. (2024). Application of nanomaterials in solar cells. *MATEC Web of Conferences, 404*, 03007. https://doi.org/10.1051/matecconf/202440403007

Stolarczyk, J. K., Bhattacharyya, S., Polavarapu, L., & Feldmann, J. (2018). Challenges and prospects in solar water splitting and CO2 reduction with inorganic and hybrid nanostructures. *ACS Catalysis, 8*(4), 3602–3635. https://doi.org/10.1021/acscatal.8b00791

Tan, C. W., Tan, K. H., Ong, Y., Mohamed, A., Zein, S., & Tan, S. H. (2012). Energy and environmental applications of carbon nanotubes. *Environmental Chemistry Letters, 10*(4), 265–273. https://doi.org/10.1007/s10311-012-0356-4

Tejashwini, D. M., Deepa, H. M., Deshmukh, V. V., & Naik, Y. V. (2024). Challenges and opportunities in nanomaterials for energy storage. In R. Naik, H. P. Nagaswarupa, H. C. Ananda Murthy, & M. E. T. Sillanpää (Eds.), *Advanced nanomaterials for energy storage devices. Nanostructure science and technology*. Springer. https://doi.org/10.1007/978-3-031-74730-4_10

Tong, Z., & Sekizkardes, A. K. (2021). Recent developments in high-performance membranes for CO_2 separation. *Membranes, 11*(2), 156. https://doi.org/10.3390/membranes11020156

Wang, H., Xu, Z., Yi, H., Wei, H., Guo, Z., & Wang, X. (2014). One-step preparation of single-crystalline Fe_2O_3 particles/graphene composite hydrogels as high performance anode materials for supercapacitors. *Nano Energy, 7*, 86–96. https://doi.org/10.1016/j.nanoen.2014.04.009

Wang, W., Tadé, M., & Shao, Z. (2015). Research progress of perovskite materials in photocatalysis- and photovoltaics-related energy conversion and environmental treatment. *Chemical Society Reviews, 44*(15), 5371–5408. https://doi.org/10.1039/c5cs00113g

Wu, Z., Ren, W., Wen, L., Gao, L., Zhao, J., Chen, Z., Zhou, G., Li, F., & Cheng, H. M. (2010). Graphene-anchored Co3O4 nanoparticles as anode of lithium ion batteries with enhanced reversible capacity and cyclic performance. *ACS Nano, 4*(6), 3187–3194. https://doi.org/10.1021/nn100740x

Xiao, C., Tian, J., Chen, Q., & Hong, M. (2024). Water-stable metal–organic frameworks (MOFs): Rational construction and carbon dioxide capture. *Chemical Science, 15*(5), 1570–1610. https://doi.org/10.1039/D3SC06076D

Xu, T., Shen, W., Huang, W., & Lu, X. J. M. T. N. (2020). Fullerene micro/nanostructures: Controlled synthesis and energy applications. *Materials Today Nano, 11*, 100081. https://doi.org/10.1016/j.mtnano.2020.100081

Yao, Y., Zhao, X., Chang, G., Yang, X., & Chen, B. (2023). Hierarchically porous metal–organic frameworks: Synthetic strategies and applications. *Small Structures, 4*(1), 2200187. https://doi.org/10.1002/sstr.202200187

Zhang, Y., Zhang, Q., & Chen, G. (2020). Carbon and carbon composites for thermoelectric applications. *Carbon Energy, 2*(3), 408–436. https://doi.org/10.1002/cey2.68

Zhang, M., Miao, Z., & Fan, L. (2021). Reduced-order analytical models of grid-connected solar photovoltaic systems for low-frequency oscillation analysis. *IEEE Transactions on Sustainable Energy, 12*(3), 1662–1671. https://doi.org/10.1109/TSTE.2021.3061296

Zhao, G., Wen, T., Chen, C., & Wang, X. (2012). Synthesis of graphene-based nanomaterials and their application in energy-related and environmental-related areas. *RSC Advances, 2*(24), 9286–9303. https://doi.org/10.1039/c2ra20990j

Zhao, W., Chen, I., & Huang, F. (2019). Toward large-scale water treatment using nanomaterials. *Nano Today*. https://doi.org/10.1016/j.nantod.2019.05.003

Zhou, B., Gao, R., Zou, J. J., & Yang, H. (2022). Surface design strategy of catalysts for water electrolysis. *Small, 18*(27), 2202336. https://doi.org/10.1002/smll.202202336

Nanotechnology in Computing and Communication

Nitiraj V. Kulkarni, Sampathi Sunitha, Jagadish V. Tawade, and Ashish Katke

1 Introduction

The study and use of materials and structures at an extraordinarily small scale more precisely, at the nanometre level is known as nanotechnology. A nanometre is one billionth of a metre, and atoms and molecules can be manipulated with direct manipulation. This level of precision will empower scientists and engineers to develop materials and devices with new functionalities. The roots of nanotechnology trace back to the celebrated lecture by Richard Feynman in 1959, where he proposed that it was theoretically practicable to arrange atoms into precise configurations to form compact, powerful tools (Baum, 2003). Even though the ideas are theoretical, as technology developed over time, they progressively became useful applications. Nanotechnology is frequently used to create quicker, more effective, and smaller technologies in various industries, including computing and communication (Allhoff et al., 2010).

Nanotechnology is one such development that has changed the ways scientists has been able to manipulate materials at atomic levels. The milestone was reached in 1981 when Gerd Binnig and Heinrich Rohrer invented the Scanning Tunnelling Microscope or STM. This invention allowed researchers to view and manipulate individual atoms for the first time (Binnig & Rohrer, 1986). Their success was recognized by the Nobel Prize and laid down the foundation for the nanotechnology of

N. V. Kulkarni · J. V. Tawade
Faculty of Science and Technology, Vishwakarma University, Pune, Maharashtra, India
e-mail: research@nitirajkulkarni.in; jagadish.tawade@vupune.ac.in

S. Sunitha (✉) · A. Katke
School of Pharmacy, Vishwakarma University, Pune, Maharashtra, India
e-mail: s.sunitha@vupune.ac.in

S. Kulkarni (ed.), *A Journey from Nanotechnology to Picotechnology*, Lecture Notes in Nanoscale Science and Technology 47,
https://doi.org/10.1007/978-3-032-07228-3_7

today. Soon afterward, fullerenes where discovered, characterized as molecular structures resembling howl spheres and carbon nanotubes, particularly noted for their remarkable strength and conductivity. These developments greatly implied that possibility of improving nanoscale materials (Kroto et al., 1985; Drexler, 1986). This progress prompted further exploration and innovation. Over the years, improvements in tools and techniques for nanoscale manipulation have led to substantial applications in medicine, energy, and computing (Buzea et al., 2007).

Nanotechnology adds a key component in communication and computing. It makes it easier to create high-performance small computers that use less energy in computation. It is for this reason that the invention is still valid even when Moore's Law states that a chip's transistor count doubles approximately every 2 years (Buzea et al., 2007). Transistors need to be miniaturized if the growing demands of new technologies like machine learning and artificial intelligence are to be met better. Nanotechnology also paved the way for faster and more efficient communication networks. Advanced materials such as carbon nanotubes and graphene improve data transmission efficiency by using minimal energy. (Shinn 2012, Goodman et al., 2005). Nanotechnology is also a requirement for making secure systems for data and making safer and faster communication.

This chapter will examine the revolutionary impact nanotechnology is going to have on computing and communication. The book opens with an introductory outline of the basics of nanotechnology as well as important historical landmarks in the timeline. Subsequent chapters delineate the tools and techniques applied in the fabrication of nanoscale devices, with great interest in how such devices may be designed to speed up processing and communication networks. It discusses scalability and reliability issues within an environment of ethical and environmental implications of these new technologies.

2 Fundamentals of Nanotechnology

2.1 *Basic Principles of Nanoscience*

Nanoscience is the study of materials and phenomena at the nanoscale, between 1 and 100 nm. Structures in this range clearly show some behaviour and characteristics quite different from larger structures. The main reasons for these differences are: (a) increased surface-to-volume and (b) quantum effects.

At the nanoscale, surface area-to-volume ratio increases dramatically. For instance, atoms in a nanoparticle are much bigger compared to those found in bulk material. There is increased reactivity as well as interaction between different species. This is why catalytic nanosized versions, such as platinum nanoparticles, work much better in chemical reactions than their bulk counterpart counterparts made of the same material (Adams & Baughman., 2005, Buzea et al., 2007).

Quantum effects are very crucial in nanoscience. Excitations at the nanoscale can no longer be described using classical physics but by quantum mechanics when electron behaviour in materials is of concern. This has consequential effects on electronic, magnetic, and optical properties. (Elishakoff et al., 2013), For example, semiconductor particles at the nanoscale emit different colors due to quantum confinement (Elishakoff et al., 2013; Goodman et al., 2005).

More so, nanoscience focuses on the exact control of material structures and has the capability to change its properties for a particular need. These principles lay grounds for developing revolutionary nanomaterials and devices used in electronics, medicine, and in energy storage industries among others.

2.2 *Nanomaterials and their Unique Properties*

Features of nanomaterials include small size and large surface areas, which are synthesized at the nanoscale. Examples of nanomaterials include nanoparticles, nanotubes, nanowires, and thin films that offer several advantages in their applications.

Graphene is one of the most used nanomaterials with a single sheet of carbon atoms forming a hexagonal pattern. It is highly electrically conductive, mechanically strong, and thermally stable, hence applicable in flexible electronics and energy storage (Kroto et al., 1985). Analogous carbon nanotubes that specifically suggest a tubular form from a sheet of graphene have various applications because of their excellent tensile strength and electrical property.

Nanomaterials also have unique optical properties. For example, gold nanoparticles exhibit phenomenon of colour change with the changing size of nanoparticles. Such phenomenon is termed as surface plasmon resonance. Already, this property has been effectively used in biomedical imaging and drug delivery systems (Cho et al., 2015).

Another property of nanomaterials is superparamagnetism, a phenomenon that is observed in magnetic nanoparticles when the size is below a certain level. This property has application in magnetic resonance imaging and targeted drug delivery (Belkin et al., 2015, Leary et al., 2006).

With nanomaterials, accompanied by small dimensions, they can even penetrate the biological systems. It will bring unparalleled medical applications: drug carriers at the nanoscale to improve efficiency and allow for more precise treatments. However, nanomaterials raise concerns about the possible toxicity and environmental effects since it underlines the need for proper research and regulation.

2.3 *Nanoscale Dimensions and Impacts on Physics and Chemistry*

At the nanoscale, a change in the physical and chemical properties is seen as they significantly depend on the size. The reduced dimensions of nanomaterials enlarge the impact of surface effects, quantum confinement, and altered intermolecular interactions.

From a physics point of view, nanoscale materials often exhibit superior mechanical properties. For example, nanostructured materials usually possess higher strength and hardness values than their large-scale analogs, primarily because they are less prone to structural defects. An excellent example is carbon nanotubes, which are around 100 times stronger than steel but much lighter (Monthioux & Kuznetsov, 2006).

In chemistry, substances at the nanoscale often manifest increased chemical reactivity said to result from their high surface energy. This is especially visible in nanoscale catalysts, which work more efficiently than similar conventional catalysts because they have a larger surface area, providing more active sites for chemical reactions (Buzea et al., 2007). Such confinement also changes the energy band structure of the semiconductor nanoparticles, making their electronic and optical properties tuneable, and a principle that underlies the operation of quantum dots in cutting-edge display technologies.

Thermal properties are also nanoscale dependent. Nanofluids are the liquid suspensions of nanosized particles and usually exhibit higher thermal conductivity values that are utilized for enhanced heat transfer purposes. The melting points of nanomaterials may vary enormously compared to their respective bulk counterparts; for instance, gold nanoparticles exhibit a melting temperature that is lower than that of the bulk gold (Allhoff et al., 2010).

These nanoscale properties enable the development of a family of materials with significantly distinct properties and have vast applications in energy storage, sensors, and advanced computing.

2.4 Fabrication Techniques at the Nanoscale

The production of nanoscale materials and devices involves controlling their structure and composition, as well as the assembly. The two primary strategies in nanoscale fabrication include top-down techniques along with bottom-up approaches.

These top-down approaches begin with materials having large sizes, which are successively decreased in scale to nanoscales. Some such approaches are photolithography and electron beam lithography. Though photolithography is important for the mass production of microprocessors and integrated circuits, as envisioned by Binnig & Rohrer, (1986), the main disadvantages associated with the top-down approaches are problems related to scaling up the process and also material loss.

In contrast, most bottom-up approaches usually start by fabricating materials from atom or molecule to atom or molecule. Thus, in this regard, examples of this approach include CVD and self-assembly. CVD is usually used for thin films and nanostructure synthesis like graphene and carbon nanotubes where, in this case, the material is built from the vapour phase onto a substrate. Conversely, self-assembly makes use of natural forces like van der Waals interactions to assemble molecules into ordered structures (Goodman et al., 2005).

Other advanced instruments such as STM and AFM are required in nanoscale fabrication. In the case of STM, atomic manipulation is possible at the Nano level, whereas in the case of AFM, topographical imaging of surface surfaces can be obtained. All these technologies make possible devices with controlled conditions at the nanoscale (Binnig & Rohrer, 1986).

Hybrid approaches combining top-down and bottom-up are emerging as a solution to the limitations of both methods. For example, the hybrid approach between lithography and self-assembly further improves nanoscale fabrication's accuracy and scalability.

Advances in fabrication technology allow for developments in complex and functional nanostructures and enhance applications in electronics, energy storage devices, and medical devices.

3 Atomic-Scale Engineering Techniques

3.1 Scanning Tunnelling Microscopy (STM)

STM was the most recent technological breakthrough for nanotechnology in 1981 by Gerd Binnig and Heinrich Rohrer. This advancement of a new instrument dramatically changed how we visualize at the nanoscale, allowing the imaging of surfaces with atomic-level resolution (Binnig & Rohrer, 1986). The basic principle behind STM is quantum tunnelling, which enables electrons to move from the sharp tip of the STM to the conductive and semiconductive surface by passing through a potential energy barrier. The technique produces an extremely accurate atomic map by exploiting this link between the distance and tunnelling current.

An STM tip, typically made up of tungsten or platinum-iridium, is placed at angstrom-scale distances from the sample surface. A bias voltage applied creates a tunnelling current that varies with the alteration of surface characteristics. It maintains real-time control over the tunnelling current through height variation that varies its tip position and thereby maintains control over the height of the scanning tunnelling microscope, keeping the tunnelling current constant while allowing three-dimensional mapping of atomic details across the surface. This excellent resolution allows one to examine and even manipulate individual atoms, thereby providing crucial insight into material properties that drive innovation in nanotechnology.

3.1.1 Principles of STM

Surface mapping, feedback-based control, and quantum tunnelling are the fundamental ideas of scanning tunnelling microscopy. Across the use of quantum tunnelling, electrons can pass across a vacuum gap that is regulated by the laws of quantum theory. The exponential dependence of the Tunnelling current on the separation distance between the tip and the sample provides very high sensitivity to changes at the nanometre scale. The control system continuously adjusts the tip position and achieves a constant Tunnelling current independent of the surface irregularities, thus offering a basis for achieving atomic-scale maps of the surface with high resolution. Although it has limitations; STM can only be applied to conductive surfaces, and it is typically used ultra-high vacuum or inert conditions in order to minimize interference and, therefore, typically is constrained by the demands of biological or non-conductive materials, etc. (Binnig & Rohrer, 1986).

3.1.2 Applications of STM in Nanotechnology

The basic principles of STM lie in quantum tunnelling, control through feedback, and surface mapping. Quantum tunnelling enables electrons to cross a vacuum gap controlled by principles of quantum theory. The Tunnelling current's exponential dependency on the distance between the tip and the sample provides extremely high sensitivity to nanometre scale alterations. In spite of surface abnormalities, the feedback control system maintains the Tunnelling current constant and continuously modifies the tip's position, allowing for the creation of intricate atomic-scale surface maps. However, it has limitations; STM can only be applied to conductive surfaces, and it is usually used in ultra-high vacuum or inert conditions in order to suppress interference and, therefore, usually is bound by the requirements of biological or non-conductive materials, etc. (Binnig & Rohrer, 1986).

3.2 Atomic Force Microscopy (AFM)

Atomic Force Microscopy (AFM) was first reported in 1986 and offers improved imaging resolution at the atomic scale for non-conductive samples. AFM differs from STM in that it is based on the mechanical forces acting between the tip and the sample's surface rather than relying on quantum Tunnelling. The cantilever bends due to these mechanical forces, which include electrostatic and van der Waals forces. These deflections are monitored by a laser beam reflecting off the cantilever and onto a photodetector (Boukallel et al., 2007).

AFM is carried out in various modes: contact, non-contact, and tapping modes, of which contact mode is best suited for hard surfaces, whereas the two others are better suited for soft or delicate samples to avoid scratching. The flexibility of AFM enables it to create a three-dimensional image of the surface topography and assess mechanical properties like elasticity and adhesion (Gaharwar et al., 2014).

3.2.1 Principles of AFM

AFM is founded on the extremely sensitive measurement of tip-sample interactions so that the main body might be for example a cantilever attached to a fine tip that scans the surface and detects the subsequent deflection triggered by forces from the surface, which are further processed to yield a very detailed topographical representation. In addition to imaging, AFM has been applied for the determination of nanoscale mechanical properties, like stiffness and adhesion, which proved indispensable for designing materials with potential for use on the nanoscale level (Goodman et al., 2005). An important strength of AFM lies in the ability of the technique to probe non-conducting material; therefore, the scope of applications is many times broader than that of STM. The second aspect is that AFM can be used in vacuum conditions different from air or liquid and, therefore, increases general flexibility of performance.

3.2.2 Role of AFM in the Design of Nanoscale Devices

AFM plays an important role in the design and characterization of nanoscale devices. Information about its surface roughness, thickness, and mechanical properties are highly critical to the development of the material. It finds extensive usage in nanolithography in patterning nanoscale features on surfaces. This mechanical handling of surfaces in building circuits and electronic components like transistors and memory devices basically forms the basis of this process (Buzea et al., 2007).

In addition, AFM is a very important application in nanomedicine where the mechanics of biological samples like cells and tissues are analysed. This aspect is vital for nanoscale drug delivery systems and diagnostic devices. The capability of assessing properties such as mechanical and topographical characteristics makes it a very valuable tool in research and innovation at the nanoscale (Cavalcanti et al., 2007).

3.3 Comparison of STM and AFM for Atomic-Scale Fabrication

Scanning Tunnelling Microscopy (STM) and Atomic Force Microscopy (AFM) are two such techniques that have contributed greatly to the advancement of nanoscale fabrication. Even though both techniques do offer atomic resolution and have great applications in nanotechnology, they function on different principles and have their applicability and limitations.

Scanning Tunnelling Microscopy (STM) is based on the principle of quantum tunnelling and is best suited for imaging and manipulating conductive or semiconducting surfaces. It gets a phenomenal resolution to observe even single atoms by measuring the tunnelling current between the STM tip and the sample surface. Its usage is limited to conductive materials, and ambient conditions must be very stable, thus frequently forcing ultra-high vacuum or inert gas environment (Binnig & Rohrer, 1986).

Atomic Force Microscopy (AFM) works based on measuring the mechanical forces between the tip and the sample surface. It is, therefore, capable of working with conductive and non-conductive materials. AFM can be used in various

Table 1 Comparison summary between STM and AFM for Atomic-Scale Fabrication

Feature	Scanning tunnelling microscopy (STM)	Atomic force microscopy (AFM)
Principle	Quantum tunnelling current	Mechanical forces between tip and sample
Sample type	Conductive or semiconductive materials	Conductive and non-conductive materials
Resolution	Atomic resolution	High resolution, but typically lower than STM
Operating environment	Requires vacuum or inert conditions	Operates in air, vacuum, or liquids
Applications	Surface imaging, atomic manipulation, spectroscopy	Topography, mechanical property measurements
Strengths	Atomic-level precision, electronic property analysis	Versatility in material type and property measurement
Limitations	Limited to conductive surfaces, sensitive to noise	Slower scan speeds, potential sample damage

environments: in air, liquid, or a vacuum, so it is very versatile for analysing very broad samples, including biological specimens. While AFM provides high resolution, it generally obtains slightly lower resolution than STM when observing atomic-scale features (Boukallel et al., 2007).

STM is suited for atomic manipulation applications, including positioning single atoms or forming quantum dots, because of its great application in nanoscale fabrication. Quantum dots are crucial for the development of single-electron transistors. On the other hand, AFM is applied in nanolithography; the tip is used to pattern materials or modifies surfaces at the nanoscale. Both techniques play different but complementary roles in nanoscale fabrication, as outlined in Table 1.

3.4 *Applications in Constructing Nanoscale Transistors and Circuits*

Major growth in nanoscale transistors and circuits relies on the functionalities of scanning tunnelling microscopy (STM) and Atomic Force Microscopy. Such tools are advanced enough to provide the precision and control needed for manipulating atoms and molecules, allowing greater advancement in miniaturized electronic technologies.

STM plays a crucial role in realizing the principles of quantum dots and single-electron transistors. It allows for manipulating individual atoms at surfaces, thus enabling the creation of structures that exhibit quantum phenomena, including tunnelling and energy quantization. Such phenomena are fundamental to progressing quantum computing as well as energy-efficient technologies. For example, STM has made it possible to build atomic-scale logic gates and memory devices with the help of careful atom-by-atom assembly of atoms for optimal information storage and processing. Goodman et al. (2005) cite STM as the tool essential for conductive material through operations with non-conductive materials and is the key instrument in the fabrication of flexible and hybrid electronic devices. It is heavily applied in nanolithography to transfer circuit designs onto substrates such as polymers and other non-conductive materials. The capacity of AFM to measure and manipulate surface properties offers a significant key to attaining mechanical and electrical stability at the nanoscale. For instance, AFM significantly contributed to developing carbon nanotube-based transistors that are lightweight, energy-efficient, and can operate at high speeds (Buzea et al., 2007).

Both STM and AFM must include the nanoscale components within the circuitry, ensuring that Moore's Law remains valid for the future. The atomic-level accuracy that STM provides is crucial to manipulating materials, while the flexibility of AFM ensures that strong and diverse electronic systems are produced. Together, these techniques have facilitated the pace at which transistors and circuits became smaller, faster, and more efficient, all of which continue to evolve in computers, communication devices, and energy storage.

4 Nanotechnology's Impact on Moore's Law

4.1 What Is Moore's Law and its Relevance

Moore's law formalized by Gordon Moore, co-inventor of Intel, in 1965, is an empirical observation that implies the number of transistors on a microchip roughly doubles every couple of years. This phenomenon effectively leads to an immense increase in computational power and simultaneously reduces the cost per transistor. Although it is not a physical law, this pattern has been a guiding principle in the semiconductor industry for many years. The phenomenon known as the scaling effect, explained by Moore's Law, played the driving force for technological advancement in computing, allowing the creation of smaller faster and less powerful machines. For ease of textual representation Moore's Law can be written mathematically as:

$$N(t) = N_0 \cdot 2^{\frac{t}{T}}$$

where $N(t)$ the number of transistors at time t, N_0 The initial number of transistors, T the doubling time (approximately 2 years).

This exponential growth model underscores even the rapid development of technology. For example, the Intel 4004 processor 1971 contained only 2300 transistors, whereas some modern processors contained billions. Moore's Law created hardware design and facilitated new advances in software, storage, and networking to keep in step with a surge in computational capabilities.

The importance of Moore's Law is its ability to predict new technological advances and set industry standards. They have fuelled investments in research and development, compelling companies to innovate. However, as transistors get close to atomic scales, the continued validity of Moore's Law is confronted by major physical and economic barriers in this exponential growth curve. Issues with heat dissipation, quantum effects, and manufacturing costs have led the industry to look into nanotechnology to sustain this logarithmic increase path (Lyon & Hubler 2013).

4.2 Enhancing Processing Speeds through Nanotechnology

Nanotechnology has revolutionized processing speeds by effectively overcoming the constraints of conventional semiconductor technology. Using materials and designs at the nanoscale, it is now possible to fabricate transistors and circuits that perform at unprecedented speed and efficiency.

A critical development in nanotechnology is the achievement of materials with extraordinary electrical properties. Graphene is just one layer of carbon atoms, structured in a hexagonal configuration, possessing extraordinary electron mobility and extremely low resistance, where transistors are comfortably operated at terahertz frequencies higher than any silicon-based transistor. Carbon nanotubes-tube-shaped forms of graphene demonstrate superb electrical conductivity and are being integrated into high-speed processors.

Another recent advancement is the introduction of Fin Field-Effect Transistors, where nanoscale features are added to enhance their performance. Unlike conventional planar transistors, FinFETs have a three-dimensional structure, increasing the effective electron movement surface. This structure reduces power and enables faster switching and, thus, is suited for high-performance computing applications.

In addition, it has opened the door to quantum dot technology advancement. Quantum dots are very small (nanoscale) semiconductor particles with unique optical and electronic characteristics resulting from quantum confinement effects. Quantum dots are used in next-generation transistors to accelerate and reduce energy consumption. Consequently, nanowires and incorporating two-dimensional materials like MoS_2 have contributed to faster processing due to reduced resistance to optimize charge transport.

Advanced fabrication techniques such as extreme ultraviolet (EUV) lithography and atomic layer deposition (ALD) have been enabled by nanotechnology to develop nanoscale transistors. These steps allow precise patterning and material application, ensuring consistent manufacturing of high-speed devices.

4.3 Power Consumption Reduction in Nanoscale Devices

Modern electronics' power consumption was a critical issue concerning nanoscale dimensions. Nanotechnology provided innovative solutions to this problem by enabling the development of energy-efficient devices and systems.

Table 2 Innovations in reducing power consumption using nanotechnology

Innovation	Description	Impact on power consumption
Carbon nanotubes and graphene	Materials with low resistivity and high electron mobility	Reduced energy loss during electron transport
Tunnel field-effect transistors	Transistors leveraging quantum tunnelling	Ultra-low voltage operation
FinFETs	Three-dimensional transistor structure	Minimized leakage currents
Quantum dots	Nanoscale particles with discrete energy levels	Efficient data storage and retrieval
Nanowire interconnects	Low-resistance signal transmission paths	Reduced power loss in circuits

One of the avenues that would help curb power consumption is through outstanding electrical properties of materials. For example, carbon nanotubes (CNTs) and graphene have very low resistivity and high mobility, thus considerably reducing energy loss in electron transport. This is finding more applications at the nanoscale in the transistor, leading to a marked decrease in the power needed by these transistors (Buzea et al., 2007).

Another prominent innovation is the development of tunnel field-effect transistors, or TFETs. TFETs do not rely on the traditional thermionic emission mechanism used by transistors for switching them off and on; instead, they work on quantum Tunnelling, permitting operation at much lower voltages. This allows them to provide high performance while consuming little power, making them well-suited for portable and IoT devices.

Nanotechnology has significantly contributed to advancing device architectures together with new materials and transistor designs. The Fin Field-Effect Transistors use a three-dimensional structure to reduce leakage currents, improving gate control and off-power consumption. Moreover, using quantum dots in memory devices enables low-power data storage and retrieval through their discrete energy levels.

Moreover, nanotechnology has streamlined advancements in power management systems. Energy-efficient interconnects made of nanowires or two-dimensional materials reduce power loss during signal transmission in processors. Furthermore, nanoscale capacitors are improving power delivery efficiency to transistors due to increased energy densities. Table 2 outlines these developments.

4.4 Prolonging Moore's Law Through Advanced Nanotechnology Techniques

Nanotechnology brings new paths to extend Moore's Law because traditional scaling methods are reaching the physical and economic limits. These involve advanced fabrication techniques, innovative materials, and transistor design alternatives.

An important tool is extreme ultraviolet (EUV) lithography, where nanoscale features with less than 10 nm can be formed. This is achieved by using a shorter wavelength of light to enable higher resolution of the created nanoscale transistors. Atomic layer deposition also allows for atomic-level control of material growth, leading to consistency and accuracy in the device's manufacturing (Goodman et al., 2005).

Nanotechnology has generated several new architectures for transistors, namely gate-all-around (GAA) transistors, using nanowires or nanosheets as the channel. The nanowire or nanosheet channels offer far better electrostatic control than planar or FinFETs. GAA supports continuous scaling because of reduced leakage currents and superior performance in general.

Hybrid materials have emerged as one of the most important innovations that extend Moore's Law. These combine silicon with nanomaterials like graphene or molybdenum disulfide (MoS_2), enhancing transistors that now benefit from each component's advantages. Scalability, speed, and power consumption are all enhanced by hybrid systems. Moreover, nanotechnology allows the incorporation of several functions into a single device. For instance, nanoscale 3D circuit stacking boosts density and performance without expanding chip space. By doing this, the difficulties associated with traditional two-dimensional scaling can be overcome, and layers of logic, memory, and interconnects can be stacked compactly.

4.5 Implications for Future Processor Technologies

Nanotechnology is a critical necessity in the ongoing miniaturization of devices; this remains one of the most significant factors impacting the consumer electronics sector. Nanoscale CPUs, mobile devices, gadgets, and Internet of Things devices are allowed to be more powerful and compact than earlier generations with continuously improved energy efficiency. This extends battery life, especially is crucial for portable gadgets.

These nanoscale processors offer enhanced data processing speed at lower energy consumption, especially for high-performance computing applications. In particular, artificial intelligence and machine learning fields must have very rapid speed and efficiency in effective data analysis and pattern recognition. Some notable advancements in this line of research include the development of quantum dot-based processors and the integration of nanoscale components into neuromorphic computing systems.

Nanotechnology is essential to the continued miniaturization that has been a hallmark of the consumer electronics industry. The creation of nanoscale processors that are energy efficient will make smartphones, wearables, and IoT devices both more powerful and greatly smaller than their current forms and, hence, prolong the lifetimes of portable electronics operations.

It also helps in having more sustainable approaches by decreasing power consumption. Nanoscale devices produce less waste and have lower energy consumption, which can help minimize the carbon footprint associated with computing.

Nanoscale production process advancements prioritize resource optimization and promote sustainability.

5 Nanotechnology in Communication Systems

Nanotechnology's revolutionary advancements in speed, energy efficiency, and security are transforming communication networks. The nanotechnology advancements through new materials and elements at the nanoscale enhance traditional communication networks' functionality and foster innovations, including quantum communication and the Internet of Things (IoT).

5.1 Overview of Communication Systems and Nanotechnology

Communication systems play a critical role within the modern infrastructure of society, providing a means of transferring information by many different means, from copper and optical fibres to wireless networks and satellite communications. These systems require effective communication, processing, and data storage, thereby requiring sophisticated materials to meet the demand for rapidity, bandwidth, and high efficiency in energy use.

Nanotechnology allows the advancement of communication systems using materials and structures on the atomic and molecular scale. For example, graphene and carbon nanotubes incorporated into devices have affected signals' reception and processing capabilities. They stand to be superior alternatives to traditional materials such as copper and silicon because they have outstanding electrical conductivity, excellent thermal characteristics, and impressive mechanical strength.

Nanotechnology not only enhances existing technologies but also develops new modes of communication. Quantum dots and nanophotonic devices enable data transmission at extremely high speeds while maintaining a low-energy efficiency. Additionally, nanoscale antennas and sensors boost the performance of wireless communication through elevated frequencies and higher energy efficiency. Nanotechnology also enables miniaturization, creating lighter and multifunctional communications devices.

5.2 Development of High-Speed, Low-Energy Networks

A very important application of nanotechnology in the communication industry is the development of very fast and energy-efficient networks. Although traditional communication systems have achieved impressive capabilities, they are limited to these specifications: having a bandwidth limit and energy consumption, most especially in how rapid data traffic grows. Nanotechnology offers solutions to issues related to the constraints above using new kinds of materials and architectures for devices.

5.2.1 Graphene-Based Communication Devices

The atomic architecture of graphene, an allotrope of carbon, is a monolayer hexagonal lattice configuration considered a milestone development in today's communication technology.

Its two-dimensional structure displays impressive electronic transport properties, especially its unmatched electron mobility and negligible electrical resistance. These intrinsic properties allow faster signal transmission, so graphene is one of the best candidates for designing high-performance electronic components, especially in advanced transistors and interconnect technologies. These intrinsic properties allow faster signal transmission, so graphene is one of the best candidates for designing high-performance electronic components, especially in advanced transistors and interconnect technologies. For example, graphene-based modulators and photodetectors have been incorporated into optical communications systems, allowing data transfer at rates higher than terabits per second (Goodman et al., 2005).

5.2.2 Nanophotonic Systems

Nanophotonic interaction between light waves and nanoscale materials is exploited to provide extremely high data transmission speeds. Innovations in components like photonic crystals and plasmonic waveguides reduce energy losses while guaranteeing exceptional speed, which makes them particularly helpful for applications in data centres and long-distance optical networks where effective light-based communication is essential (Hubler, 2010).

5.2.3 Carbon Nanotube Interconnects

Carbon nanotubes are a type of cutting-edge alternative to traditional copper interconnects. Since CNTs have superior electrical conductivity and thermal stability, signal delay times and power losses are much reduced, further improving network performance in high-frequency applications.

5.2.4 Low-Energy Nanoscale Transceivers

Nanomaterial-based transceivers, such as indium arsenide nanowire-based ones, provide efficient signal transmission and reception with a power consumption of at least one order of magnitude lower than usual transceivers. It is very attractive for

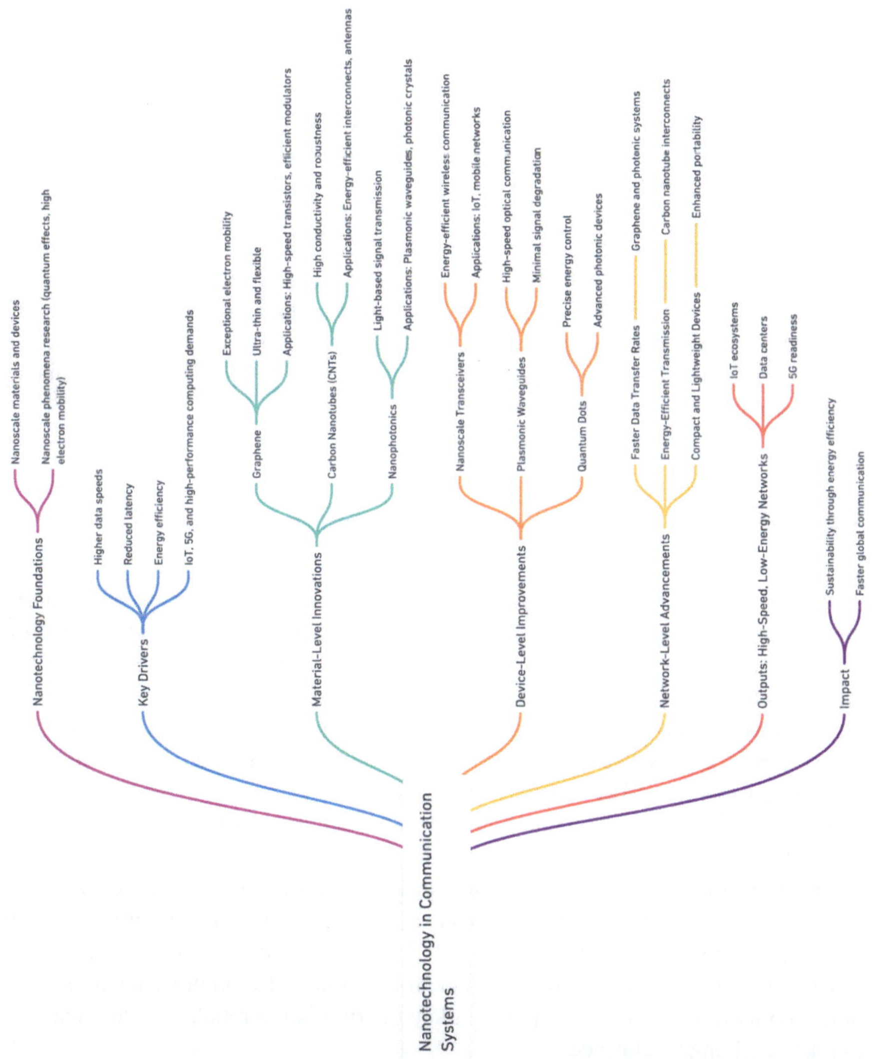

Fig. 1 Overview of nanotechnology in communication systems

low-power wireless networks and IoT systems where energy consumption should be minimal. The impact of these advances is reflected in Table 1, and Fig. 1.

5.3 Applications in Quantum Communication

Quantum communication is a new technological framework based on complex quantum mechanical principles. It brings forth previously unattainable methods for information transfer that conventional communication systems could not achieve.

For instance, quantum communication relies on quantum bits or qubits instead of binary digital representation. Qubits are far ahead of classical computing in that they represent probabilistic superpositions.

This new communication approach exploits the two fundamental quantum mechanical phenomena: quantum superposition, entanglement, and the no-cloning theorem. These quantum mechanical properties provide a far more secure and computationally efficient communications system than classical systems.

Being an important technological interface, nanotechnology considerably accelerates the progress of quantum communication. It gives the high-performance nanoscale materials needed for generating, manipulating, and measuring quantum states; thus, as one would expect, research in modern communication has reached a very vibrant junction between quantum mechanics and nanotechnology, promising some revolutionary technological change.

5.4 Principles of Quantum Communication

Quantum communication is based on three fundamental principles of quantum mechanics: quantum superposition, entanglement, and the no-cloning theorem. Quantum superposition allows qubits to exist in many states simultaneously, meaning that one can encode more information than is possible classically with bits. In the latter instance, a single qubit may simultaneously be a zero, a one, or both. This characteristic allows faster and more dependable data transport by significantly increasing capacity and efficiency.

Entanglement is another important principle, according to which two particles are so correlated that the state of one particle immediately changes the state of the other, regardless of the distance between them. The basic attribute of the entangled particles determines the features of the encrypted communication algorithms. Such particles' dynamic correlation is impacted by measuring or interference, which causes the communicating parties to get alarmed. Entangled photons are among the most common applications in quantum key distribution, a reliable, secure encryption key exchange technique.

Another important theory in quantum communication is the "No-Cloning theorem." The theory states that creating or designing a duplicate of an unknown quantum state is impossible.

The concept is necessary to prevent data modification while it is being transmitted. An observable error will be made in attempts to intercept and replicate quantum information, making security breaches immediately traceable. Quantum Key Distribution (QKD) is a good example in this regard. It enables secure and encrypted key exchange based on the principle of measurement sensitivity. The most well-known QKD protocol is BB84. It works on the basic principle that any attempt to measure a quantum state changes its characteristics, thereby exposing eavesdropping attempts.

5.5 The Role of Nanotechnology in Quantum Systems

Nanotechnology is significant as it fundamentally depends on the advancement and application of quantum communication. Acting at the nanoscale generally enables remarkable precision in quantum manipulation, transmission, and detection while facilitating the creation of nanoscale devices and materials.

For example, quantum dots are single-photon emitters that play a vital role in many encoding processes involved in quantum information. For quantum communication protocols involving the detection of photons, these semiconductor nanoparticles give high-fidelity on-demand emission with high efficiency. Furthermore, quantum dots make it easier to manipulate photons because of their optical characteristics. As a result, they perform exceptionally well in applications such as Quantum Key Distribution.

A very impressive advance made possible by nanotechnology is the appearance of nanophotonic waveguides. The effective propagation of quantum states over long distances with little loss is made possible by these novel structures, which efficiently restrict and guide light at the nanoscale. Nanophotonic waveguides, made via nanoscale lithography, are essential for creating the quantum networks needed to swiftly and safely transfer entangled photons. Nanotechnology plays a crucial role in developing and implementing quantum communication systems while catalysing the emergence of nanoscale devices and materials that allow for the precise manipulation, transmission, and detection of quantum states with exceptional efficiency.

Plasmonic nanostructures utilize light coupling with free electrons in metals to ensure the high efficiency of quantum communication systems. They confine and manipulate light at scales of or below its wavelength, enhancing the generation and detection of quantum states. Such plasmonic nanostructures thus offer outstanding advantages in advanced quantum devices by compact dimensions and superior performance.

Decoding quantum information is impossible without nanowire-based single-photon detectors. The efficiency with which the above detectors can identify single photons with a certain noise level makes them highly valuable in quantum key distribution systems. Compact and sensitive detection enables the incorporation of nanowires in QKD systems.

Nanoscale quantum memories are also intended for the temporary storage of quantum information with the purpose of synchronizing the quantum network. These devices usually apply nanostructured materials doped with rare-earth elements to reach high efficiency and stability. Nanotechnology encourages seamless and scalable operations once these advanced components are integrated into the quantum communication system.

Incorporating nanotechnology in quantum systems has pushed the field considerably ahead as concepts once purely theoretical are brought into tangible applications. Materials and device fabrication advances have efficiently overcome many challenges in quantum communication, including signal attenuation, scalability,

Table 3 Nanotechnology's role in quantum communication

Quantum communication component	Nanotechnology contribution	Impact
Quantum dots	Single-photon emitters with tuneable properties	High fidelity and efficiency in encoding information
Nanophotonic waveguides	Nanoscale light confinement and guidance	Minimal loss in transmitting quantum states
Plasmonic nanostructures	Sub-wavelength manipulation of light	Enhanced generation and detection of quantum states
Nanowire single-photon detectors	Highly sensitive photon detection	Reliable decoding of quantum information
Nanoscale quantum memories	Temporary storage of quantum information	Synchronization of distributed quantum networks

and dependability. Enhanced developments at the nanoscale devices are expected further to enhance the effectiveness and availability of quantum communication technologies, making it possible to establish secure and effective global communication systems. Summary of Nanotechnology in Quantum Communication is summarized in Table 3.

6 Advanced Applications of Nanotechnology in Computing

This has revolutionized computing to a great extent due to the remarkable progress in the miniaturization, adaptability, and performance of devices. Nanotechnology operates at atomic and molecular levels to offer devices that are not only small and efficient but also fairly integrable into daily life. Two of the major applications of nanotechnology in this field are through the reduction in size of electronic devices and the innovation of flexible electronics and wearable technologies.

6.1 Miniaturization of Electronic Devices

Miniaturization of electronic devices has been the bastion of technological progress, and nanotechnology has helped bridge the tremendous gap that had to be covered in this respect. Though silicon-based transistors have been revolutionary, their miniaturization has been subjected to physical limitations and economic constraints when the components near atomic dimensions. Nanotechnology brings in new materials, novel designs, and advanced techniques for manufacturing to yet again miniaturize electronics.

Among the most important achievements are nanoscale transistors. Fin Field-Effect Transistors (FinFETs) and gate-all-around (GAA) transistors owe their improved performance and low-power consumption to nanoscale engineering.

Three-dimensional transistor architectures have better control over electric currents than the two-dimensional design featured in regular planar transistors. Because of this development, hundreds of billions of transistors can be crammed into a single chip and continue to power modern processors with phenomenal computing power.

Additionally, with the introduction of carbon nanotubes and graphene, nanomaterials have also changed the game of miniaturization. Due to its excellent strength and electrical conductivity, graphene can be employed in interconnects and transistors. Analogously, carbon nanotubes and cylindrical formations of rolled graphene are alternatives to copper wiring as lightweight high-conductivity materials. It not only makes data transmission faster but also reduces power loss to a great degree, enhancing the devices' efficiency.

6.2 Flexible Electronics and Wearable Technology

Indeed, nanotechnology has revolutionized the world of flexible electronics and wearables to be one of the essences of life in modern times. Flexible electronic gadgets are made of soft, elastic materials that can conform to various surfaces and shapes, unlike traditional rigid devices or systems. This adaptability has opened new health, fitness, and entertainment possibilities.

The innovative products are created with advanced nanomaterials, including graphene, silver nanowires, and organic semiconductors, which give such an excellent combination of highly conductive electricity with promising mechanical flexibility. For example, graphene-based inks permit circuits to be printed on flexible substrates, which are flexible but hard-wearing and light in weight. Silver nanowires are also used in the preparation of transparent conductive films that find applications in touchscreens and sensor technologies. Similarly, silver nanowires are used for making transparent conductive films that are applied in touchscreens and sensors.

Flexible electronics play a critical part in the operation of wearable technology, including health monitors, trackers for fitness, and wristbands. Complex nanosensors are used in many of these devices to monitor vital metrics including body temperature, heart rate, and bodily activity. Improved flexible solar power cells and piezoelectric nanogenerators are two examples of more effective energy harvesting methods made possible by advancements in nanotechnology. This makes it possible for wearable technology to function well without the need for bulky batteries.

6.3 Nanoscale Sensors for Computing and Communication

Nanoscale sensors are operating at atomic and molecular levels and represent one of the greatest applications of nanotechnology: these devices are transforming the computing and communication industries. They excel in gathering information with unprecedented precision and effectiveness, processing and transferring collected data later on.

Due to their minuscule size and outstanding sensitivity, nanoscale sensors have become a main component of any Internet of Things (IoT) devices, real-time monitoring systems, and advanced computing platforms.

These sensors contain nanomaterials, namely, graphene, carbon nanotubes, and quantum dots, each with unique electrical, mechanical, and optical properties. For instance, graphene's electrical conductivity and mechanical flexibility make it an ideal material for nanosensors that measure variations in electrical resistance or capacitance as a function of environmental changes. Due to their hollow, high-aspect-ratio design, carbon nanotubes are used to create sensors sensitive enough to detect very low levels of gases, pressure, and temperature.

It is only in the field of computation that nanoscale sensors are necessary to ensure effective thermal management and optimal performance. These sensors monitor CPU temperatures and energy use; computers can dynamically adjust their workloads, preventing potential overheating. By detecting and eliminating noise or interference in real time, sensors in communication systems enhance signal quality. Quantum dot-based sensors are also used in optical networks to clarify a signal and minimize loss in data while transmitting.

6.4 Molecular-Scale Circuits

Molecular-scale circuits represent a big step forward in miniaturization by using individual molecules as the basic entities of electronic elements instead of atoms. These circuits utilize molecules' electrical, optical, and chemical properties to perform computations and, therefore, provide new routes towards innovative alternatives away from classic silicon-based electronics. In this regard, the state of nanotechnology ensures that the molecules are placed within the circuits accurately and orderly to establish functional circuits.

The current research in this area is mainly on molecular transistors, diodes, and memory devices. For instance, it is established that molecular transistors work by allowing single molecules to flip from a conductive state to a non-conductive state or vice versa in response to an external electric field. This is similar to the conventional transistors but much smaller in scale. Prominent examples include rotaxanes and catenanes, which exhibit bistable states suitable for switching and data storage applications (Binnig & Rohrer, 1986).

There has also been great advancement in the development of molecular memory. Molecular assemblies can store binary data, so ultrahigh-density storage becomes viable. Quantum dots have long been used as charge storage elements for nanoscale flash memory with compact and energy-efficient storage techniques. Self-assembled monolayers and DNA nanostructures have also been added to improve scalability and efficiency, further advancing molecular memory technologies (Goodman et al., 2005).

Nanotechnology combines molecular circuits with traditional electronics, making hybridizing systems with values from the considered domains possible. Examples may include, for example, carbon nanotubes and graphene nanoribbons as molecular-scale interconnects towards effective charge transport between molecular components and conventional semiconductor devices.

However, despite many advancements and developments, many practical obstacles exist in applying molecular circuits. Significant obstacles to scalability and reliability include molecular behaviour variability, sensitivity to environmental effects, and the complexity of fabrication procedures. Adopting this technology will depend on our ability to overcome these obstacles.

On the other hand, molecular-scale circuitry technology is still developing rapidly, making devices smaller, quicker, and more energy-efficient than their predecessors.

7 Future Trends and Prospects

Nanotechnology has the potential to become the enabling element of future computing and communication technologies. As it allows for the proper manipulation of materials at the atomic and molecular range, it opens up several diverse streams of significant breakthrough opportunities in artificial intelligence, cloud computing, next-generation communication networks, and next-generation internet service. This section focuses on emerging trends, future developments, and the long-term implications of nanoscale architectures-influencing the shape of future technologies.

7.1 Emerging Nanotechnologies in Computing

Nanotechnologies are emerging as the future and have revolutionized the computing world with new edges of silicon-based electronics. One breakthrough in this area is using alternative materials like graphene, carbon nanotubes (CNTs), and molybdenum disulfide (MoS_2). These materials have great potential in creating future transistors since they demonstrate remarkable mechanical, electrical, and thermal characteristics. Research by Goodman et al. (2005) shows that graphene-based transistors can be operated at terahertz frequencies with processing speeds easily exceeding those of silicon-based devices. Comparably, CNTs contribute to ultra-compact transistors and interconnects, which result in increased energy efficiency and decreased signal latency. In addition, new device architectures, including spintronic devices and resistive random-access memory (ReRAM), have expanded the use of conventional data storage and processing techniques. ReRAM uses nanoscale resistive switching for memory solutions that are not only faster but also more durable than conventional flash memory. However, using the spin of electrons rather than charge, spintronic devices enable the higher data density and lower power consumption required for energy-efficient computer systems.

Nonetheless, memristors and nanoscale devices that mimic the activity of brain synapses have been improved via nanotechnology. In addition to being crucial for neuromorphic computing, memristors hold the potential to transform artificial intelligence by facilitating more effective and flexible learning processes. These together illustrate how emergent nanotechnologies form the stepping ground for a new generation in computing.

7.2 *Potential Breakthroughs in Communication Systems*

Nanotechnology plays a significant role in the innovation of communication systems, mainly regarding higher data transfer rates and reduced power consumption. There come nanophotonic components, including plasmonic waveguides and nanowire lasers, at the cutting edge of innovations. They have made light manipulation in the nanoscale possible, opening the way for ultra-fast optical communication systems with the possibility of reaching even a terabit transfer rate (Buzea et al., 2007).

Furthermore, antennas made of carbon nanotubes and graphene at the nanoscale facilitate data transmission at terahertz frequencies. These helps increase bandwidth while significantly reducing latency. Such advancements play a great role in the establishment of 6G networks as well as future technologies, where faster and low-latency communication is key.

Another revolution is quantum communication, which is being brought to this sector. With the advent of a nanowire-based detector, quantum dots are increasingly becoming single-photon sources, which make quantum key distribution (QKD) a viable data transmission method. These systems provide an intrinsically safe means of communication by using the principles of quantum physics to instantly identify any attempt to intercept data.

7.3 *Nanotechnology in Artificial Intelligence and Machine Learning*

Artificial intelligence (AI) and machine learning (ML) require significant computational power, and nanotechnology is seeing an important role in this endeavour through the development of hardware suited for those specific applications. Neuromorphic chips, modelled after human brain architecture, are being designed from nanoscale materials such as memristors and nanowires that can emulate the neural networks in the brain, making learning faster and more energy-efficient.

Nanotechnology further advances capabilities at the GPU and tensor processing unit (TPU) levels, which are necessary for training and deploying AI and ML models. Thus, using graphene interconnects with minimized resistance means the ability to dissipate heat further optimizes these computational speeds at lower energy usage. Nanosensors also advance AI by gathering high-quality data, especially in robotics and IoT systems, where accuracy and real-time processing are much of a must.

7.4 The Role of Nanotechnology in Cloud Computing and Data Centres

Cloud computing and data centres form the essential framework of contemporary digital infrastructure, making their performance and energy efficiency paramount. Nanotechnology is tackling these issues by developing nanoscale processors, memory devices, and interconnects. Nanophotonic circuits are transformative, facilitating optical data transmission within data centres to minimize latency and energy usage. These circuits utilize nanoscale waveguides to precisely control light, ensuring rapid and efficient server communication (Prasad, 2008).

In addition, nanotechnology plays a vital role in thermal management. Nanoscale thermal sensors and graphene-based heat spreaders enhance the cooling processes in data centres, thus eliminating overheating and reducing operational costs. More importantly, the novel storage technologies in ReRAM and phase-change memory, based on nanotechnology, improve data storage and retrieval efficiency to adapt to the needs of ever-increasing large datasets in cloud computing environments.

7.5 Prospects for Nanoscale Computing Architectures (Quantum Computing, Neuromorphic Computing)

The nanoscale building blocks are inspired by the structure and function of the human brain. Neuromorphic chips rely on the possibility of achieving these features with nanoscale devices. Synapse-like behaviour is accomplished by memristors, but another vital element are artificial neurons by nanowires. Exceptional efficiency and adaptability can become prime movers for changing real-time decision-making, robotics, and artificial intelligence.

However, such designs must overcome issues with scalability, stability, and environmental sensitivity to fully utilize their potential. The nanotechnology community is likely to continue its technological development to overcome these issues and successfully result in breakthroughs in computation.

Table 4 illustrates nanotechnology's contributions to future trends, highlighting nanoscale computing architectures via quantum dots, Josephson junctions, and memristor-based neural networks as a potential breakthrough.

Table 4 Nanotechnology's contributions to future trends

Area	Nanotechnology contribution	Impact
Emerging nanotechnologies in computing	Graphene-based transistors, memristors, spintronic devices	Faster, smaller, and energy-efficient computing systems
Communication systems	Nanophotonic components, nanoscale antennas, quantum dots	Terabit-level data transfer, secure quantum communication
AI and ML	Neuromorphic chips, nanoscale GPUs, graphene interconnects	Enhanced processing speed, efficient learning, high-quality data
Cloud computing and data centres	Nanophotonic circuits, ReRAM, nanoscale thermal sensors	Reduced latency, improved storage, and optimized energy usage
Nanoscale computing architectures	Quantum dots, Josephson junctions, memristor-based neural networks	Quantum advantage, real-time decision-making, adaptable systems

8 Conclusion

Nanotechnology has been advancing dynamically in computing and communication as an innovative mix of atomic or molecular structures. It addresses all challenges related to speed, energy efficiency, and miniaturization by innovative nanodevice advancement that progresses at the nano-level transistors, circuits, and sophisticated communication systems. These innovations not only open up new opportunities but also support the continuous trend of the Moore Law.

There are graphene, quantum dots, and nanoscale antennas things that improve on traditional systems and hold the promise of future breakthroughs in quantum communication and neuromorphic computing. The potential of nanotechnology is challenging to realize, however, due to scalability, repeatability, and the need to focus on ethics concerning ecological and social implications. The field demands continued interdisciplinary cooperation so it advances further sustainability and equity.

In the coming years, nanotechnology is expected to revolutionize both societal impact and technological advancements. By facilitating upgrades in communication and computing systems, nanotechnology is a crucial link between science and technology, proving to be essential and dynamic in an ever-evolving digital landscape.

References

Adams, W. W., & Baughman, R. H. (2005). Retrospective: Richard E. Smalley (1943–2005). *Science, 310*(5756), 1916. https://doi.org/10.1126/science.1122120

Allhoff, F., Lin, P., & Moore, D. (2010). *What is nanotechnology and why does it matter? From science to ethics*. Wiley.

Baum, R. (2003). Cover story—Nanotechnology. *Chemical and Engineering News, 81*(48), 37–42.

Belkin, A., Hubler, A., & Bezryadin, A. (2015). Self-assembled wiggling nanostructures and the principle of maximum entropy production. *Scientific Reports, 5*, 8323. https://doi.org/10.1038/srep08323

Binnig, G., & Rohrer, H. (1986). Scanning tunneling microscopy. *IBM Journal of Research and Development, 30*(4), 355–369. https://doi.org/10.1147/rd.441.0279

Boukallel, M., Gauthier, M., Dauge, M., Piat, E., & Abadie, J. (2007). Smart microrobots for mechanical cell characterization and cell convoying. *IEEE Transactions on Biomedical Engineering, 54*(8), 1536–1540. https://doi.org/10.1109/TBME.2007.891171

Buzea, C., Pacheco, I. I., & Robbie, K. (2007). Nanomaterials and nanoparticles: Sources and toxicity. *Biointerphases, 2*(4), MR17–MR71. https://doi.org/10.1116/1.2815690

Cavalcanti, A., Shirinzadeh, B., Freitas, R. A., & Kretly, L. C. (2007). Medical nanorobot architecture based on nanobioelectronics. *Recent Patents on Nanotechnology, 1*(1), 1–10. https://doi.org/10.2174/187221007779814745

Cho, H., Pinkhassik, E., David, V., Stuart, J. M., & Hasty, K. A. (2015). Detection of early cartilage damage using targeted nanosomes in a post-traumatic osteoarthritis mouse model. *Nanomedicine, 11*(4), 939–946. https://doi.org/10.1016/j.nano.2015.01.011

Drexler, K. E. (1986). *Engines of creation: The coming era of nanotechnology*. Doubleday.

Drexler, K. E. (1992). *Nanosystems: Molecular machinery, manufacturing, and computation*. Wiley.

Elishakoff, I., Dujat, K., Muscolino, G., Bucas, S., Natsuki, T., Wang, C. M., et al. (2013). *Carbon nanotubes and nanosensors: Vibrations, buckling, and ballistic impact*. John Wiley & Sons.

Gaharwar, A. K., Peppas, N. A., & Khademhosseini, A. (2014). Nanocomposite hydrogels for biomedical applications. *Biotechnology and Bioengineering, 111*(3), 441–453. https://doi.org/10.1002/bit.25160

Goodman, R. P., Schaap, I. A., Tardin, C. F., Erben, C. M., Berry, R. M., Schmidt, C. F., et al. (2005). Rapid chiral assembly of rigid DNA building blocks for molecular nanofabrication. *Science, 310*(5754), 1661–1665. https://doi.org/10.1126/science.1120367

Hubler, A. (2010). Digital quantum batteries: Energy and information storage in nanovacuum tube arrays. *Complexity, 15*(5), 48–55. https://doi.org/10.1002/cplx.20306

Kroto, H. W., Heath, J. R., O'Brien, S. C., Curl, R. F., & Smalley, R. E. (1985). C60: Buckminsterfullerene. *Nature, 318*(6042), 162–163.

Leary, S. P., Liu, C. Y., & Apuzzo, M. L. (2006). Toward the emergence of nanoneurosurgery: Part III—Nanomedicine: Targeted nanotherapy, nanosurgery, and progress toward the realization of nanoneurosurgery. *Neurosurgery, 58*(6), 1009–1026. https://doi.org/10.1227/01.NEU.0000217016.79256.16

Lyon, D., & Hubler, A. (2013). Gap size dependence of the dielectric strength in nano vacuum gaps. *IEEE Transactions on Dielectrics and Electrical Insulation, 20*(4), 1467–1471. https://doi.org/10.1109/TDEI.2013.6571470

Monthioux, M., & Kuznetsov, V. (2006). Who should be given the credit for the discovery of carbon nanotubes? *Carbon, 44*(9), 1621–1623. https://doi.org/10.1016/j.carbon.2006.03.019

Prasad, S. K. (2008). *Modern concepts in nanotechnology*. Discovery Publishing House.

Shinn, E. (2012). Nuclear energy conversion with stacks of graphene nanocapacitors. *Complexity, 18*(3), 24–27. https://doi.org/10.1002/cplx.21427

Nanopores for Molecular Sequencing

Sunitha Sampathi and Alifeya Fidaali

1 Introduction to Nanopore Technology

1.1 Overview of Nanopores

Over the past 30 years, single-molecule detection techniques, which represent the third generation of sequencing technologies, have significantly expanded the horizons of scientific research. These advanced methods have unlocked state-of-the-art sequencing technologies, facilitating diverse applications, from clinical breakthroughs to the study of protein kinetics. The use of biological nanoscale pores for nucleic acid molecule detection offered the possibility of increasing single-molecule sensing accessibility for a larger group of researchers (MacKenzie & Argyropoulos, 2023). A nanopore (NP) is an extremely thin and small opening in a slim membrane, usually just large enough to allow the passage of a single DNA strand. It is thus possible for the detection of the shape and electrical characteristics of each base on the DNA, offering a method for reading genetic sequences.

Nanopores structures are either solid-state, made of silicon or graphene, or biological, made from proteins like alpha-hemolysin. Because nanopore technology can analyze biomolecules in real time without labels, it has become popular in fields including genomics, proteomics, and diagnostic diagnostics. The main idea behind nanopore sequencing (NPS) is that a molecule is moved through a nanopore while ionic current variations are recorded. The structure and sequence of the molecule are revealed by these present alterations. Because of its portability, speed, and capacity to process lengthy readings, nanopore technology has revolutionized sequencing and provided benefits above conventional sequencing techniques.

S. Sampathi (✉) · A. Fidaali
School of Pharmacy, Vishwakarma University, Pune, Maharashtra, India
e-mail: s.sunitha@vupune.ac.in

S. Kulkarni (ed.), *A Journey from Nanotechnology to Picotechnology*, Lecture Notes in Nanoscale Science and Technology 47,
https://doi.org/10.1007/978-3-032-07228-3_8

Devices from Oxford Nanopore Technologies (ONT), for instance, have proved essential to field-based genomic research (MacKenzie & Argyropoulos, 2023). Apart from sequencing, medication delivery, disease diagnostics, and bone tissue engineering are noteworthy uses of nanopores in nanomedicine. These technologies' nanoscale accuracy enables customized therapeutic interventions, especially in difficult areas like oral bone repair. Nanopore technology is expected to become increasingly important in molecular diagnostics and customized treatment as it develops.

1.2 Historical Development of Nanopore Sequencing

The practice of NPS began in a number of labs in the 1980s. In 1989, David Deamer (D.D) wrote down an unfeasible notion in his notebook: it would be feasible to use electrophoresis to sequence a single strand of DNA pulled through a membrane's nanoscopic pore. Factors such as a phage motor that is electronically monitored and positioned in a bilayer would be able to produce sequencing information around the same time of performing double-stranded DNA processing. This is because George Church was interested in making big advances in DNA sequencing technology in particular for the Human Genome Project. The work of Hagan Bayley and his coworkers focused on the membrane proteins and the building and structure of oligomeric transmembrane protein pores, which include a-hemolysin. This prompted them to wonder if pores could be used as cation and molecule biosensors. Figure 1 illustrates the fundamental mechanism behind nanopore sequencing: the electrophoretic movement of a DNA strand through a nanopore. As depicted,

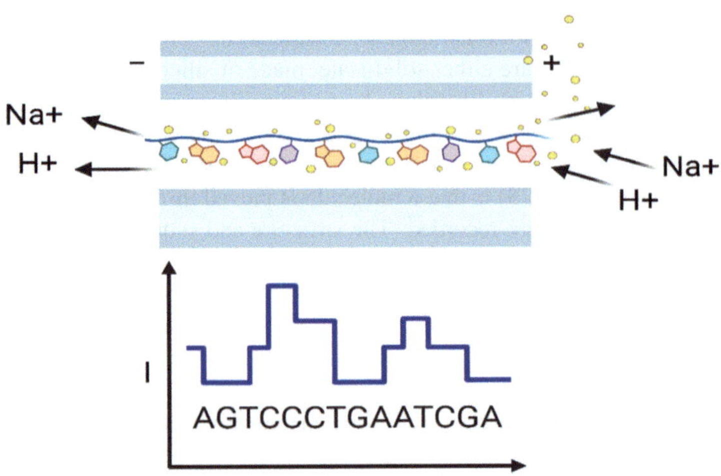

Fig. 1 The diagram illustrates the action of drawing a DNA strand through a NP due to the application of voltage to three bases of the strand

voltage is applied across the membrane, causing the negatively charged DNA to be drawn through the pore one base at a time. The passage of each base cause's distinct disruptions in the ionic current, which forms the basis for sequence determination. This visual supports the historical context and conceptual foundation discussed in Sect. 1.2, showcasing how the initial hypothesis by Deamer and collaborators evolved into a practical method for reading nucleic acid sequences.

Both D.D and Daniel Branton (D.B) had separately cited studies indicating that *Staphylococcus aureus* α-hemolysin could generate transmembrane channels broad enough (>1.2 nm) to contain a single strand of DNA (ssDNA), whereas Church chose to study alternative high-speed sequencing techniques. Additionally, D.D was aware that John kasianowicz was working on α-hemolysin because he had seen his presentations at two scientific conferences. D.D traveled to Kasianowicz at the US NIST (National Institute of Standards and Technology) in late 1993 to investigate the feasibility of electrophoresing RNA homopolymers via an α-hemolysin pore. Kasianowicz, collaborating with Bayley on α-hemolysin, had already gained significant expertise. In collaboration with Sergey Bezrukov from the National Institutes of Health (NIH) in the USA, he created strategies to stop the α-hemolysin channel from spontaneously gating or closing.

Unwanted gating could block or obscure the view of the translocation of nucleic acids through the nanopore. Thus, that was important. In addition, kasianowicz and Bezrukov measured the impact of polyethylene glycol on pore conductance. They concluded their observations were completely explained by a pore radius of around 1.1 nm, consistent with other earlier work.

The experiment of D.D. and Kasianowicz was carried out by using a single α-hemolysin channel incorporated in a bilayer of lipid separated by two compartments with buffered KCL. RNA homopolymers, specifically polyadenylic acid or polyuridylic acid, were then introduced to the *cis* side of the membrane. Positive results were observed immediately upon the addition of the polymers. The trans-compartment exhibited short-lived, millisecond-scale reductions in ionic current through the α-hemolysin channel if the trans-compartment was positively biased more than 80 mV. However, no such current blockades were observed when the trans-compartment was negatively biased. This outcome was expected since a negative bias would exclude the polyanionic RNA in the cis compartment and prevent its entry into the nanopore.

These initial findings supported the hypothesis that applying a voltage across the hemolysin pore could drive the electrophoresis of single RNA molecules through the α-hemolysin channel (Deamer et al., 2016). In February 1994, D.D. and D.B. collaborated with Kasianowicz to validate and extend earlier findings. Subsequent studies demonstrated that longer RNA polymers exhibited longer average blockage durations compared to shorter ones. This led to the hypothesis that the blockage duration represented the time required for the polymer to traverse the nanopore from the cis chamber to the trans chamber. Around the same time, Langzhou Song and colleagues, who were studying the crystal structure of α-hemolysin, reached out to D.B. and D.D. following these initial experiments. Early X-ray diffraction data indicated that the channel diameter of the α-hemolysin pore ranged from approximately 1.5 to 2.5 nm. This insight enabled precise translocation experiments. It was deduced that if the channel's narrowest diameter was approximately 1.5 nm, ssDNA could translocate from the *cis* to the *trans* side, whereas double-stranded DNA (dsDNA) would be unable to do so. Using a combination of ssDNA and dsDNA controls synthesized by D.B.'s group at Harvard, Kasianowicz at National Institute of Standards and Technology (NIST) tested this theory. He positioned the synthetic DNA molecules on the *cis* side of a membrane containing a known number of α-hemolysin pores (to more accurately measure the amount of DNA translocated to the *trans* side, additional pores were incorporated into the membrane during certain tests).

At NIST, the team measured the number of blockades produced within a specified time frame, while at Harvard, they quantified the DNA molecules that reached the trans side of the pore using quantitative PCR (Polymerase Chain Reaction) techniques. The combined results confirmed that, as expected, ssDNA successfully translocated through the pore, whereas dsDNA did not. Furthermore, the number of translocated ssDNA molecules matched the number of ionic current blockades. These findings provided the first definitive evidence that the electrical field surrounding a voltage-biased pore can trap single-stranded nucleic acid polymers, enabling their electrophoretic passage through the nanopore‘s tiny aperture. Following the successful tests showing ssDNA translocation through a nanopore, the crystal structure of alpha-hemolysin was published, providing a detailed molecular framework for interpreting the translocation data. The first evidence supporting the feasibility of nanopore sequencing (NPS) came from experiments demonstrating oligomer transport and single-molecule detection. However, extensive further research was necessary to determine its practical application (Deamer et al., 2016).

Figure 2 summarizes the major milestones in the development of nanopore sequencing, showcasing its evolution from conceptual experiments to widespread technological adoption. The team proved the viability of their hypothesis by solving numerous biological and technological problems after years of testing in the labs at Harvard and UC Santa Cruz and, on September 23, 2023, at the Library of Congress, Deamer and Akeson received recognition for developing NPS, a novel and groundbreaking technique for reading DNA and RNA. The invention won the Golden Goose Award from the AAAS (American Association for the Advancement of Science) to the two scientists, both emeritus professors of biomolecular engineering at the Baskin School of Engineering. Branton, a friend and colleague who co-invented the technique and is a Harvard biologist, was also recognized.

The UK-based company Oxford Nanopore Technologies (ONT) was granted a license to use the NPS technology, which resulted in the development of the MinION, the only portable genetic sequencing equipment that costs a fraction of what other sequencers do. Some of the most important genomics discoveries of the past 20 years have been made possible by this revolutionary technology, which enables sequencing to occur in remote and resource-poor areas. The creators of ONT received patent licenses for these concepts, and the business introduced the MinION, the first NS equipment, in 2012 (UC Santa Cruz News, 2023).

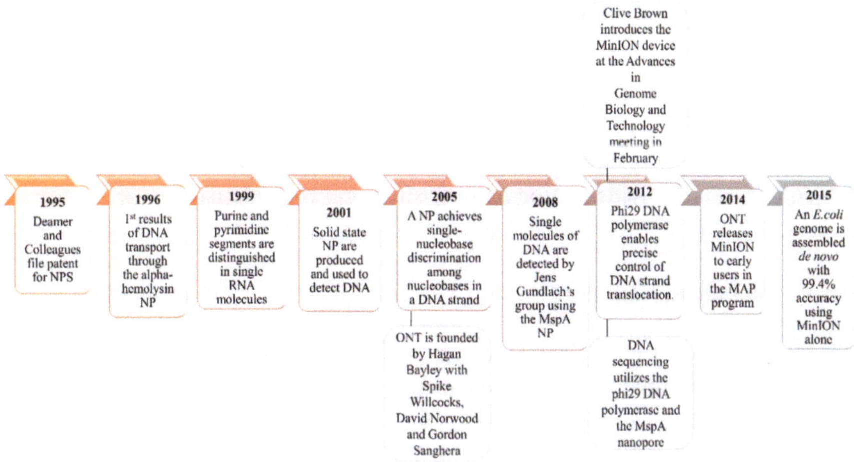

Fig. 2 Key advancements in nanopore DNA sequencing

1.3 Importance in Molecular Sequencing

The principle of molecular sequencing (MS) is helpful in developing biology, medicine and even biotechnology. MS has brought an end to the era known as genomic-illiteracy by providing a wealth of information about the organization of the genomes of different organisms in the following ways:

- Enabling the detailed analysis of DNA and RNA, helping identify genetic variations, mutations and structural alterations.
- Their importance allows studying the genetic component of various diseases.

Additionally, this technology has played a pivotal role in the development of personalized medicine where treatment options are adjusted according to one's genetic profile. Sequencing technologies like next-generation sequencing (NGS) and NPS have significantly improved the speed and accuracy of genomic analyses, enabling large-scale projects like the Human Genome Project and environmental metagenomics (MacKenzie & Argyropoulos, 2023). By detecting genetic changes associated with cancer, sequencing technologies provide key information for early diagnosis, prognosis and treatment selection by supporting the discovery of biomarkers and the monitoring of tumor evolution in cancer research (Chen & Zhang, 2022).

In microbial genomics, MS has transformed our understanding of microbiology on:

- Microbial diversity, pathogenesis, and resistance mechanisms, as well as;
- Improving diagnostics and public health strategies.

2 Principles of Nanopore Sequencing (NPS)

2.1 *Structure and Functions of NPs*

Biological nanopores and solid-state nanopores are the two main categories into which NPs can be divided based on their structural characteristics.

2.1.1 Biological Nanopores

The architecture of biological nanoparticles is characterized by a protein channel incorporated into lipid bilayers. The two notable examples include the α-hemolysin that forms a heptameric transmembrane pore and Mycobacterium smegmatis porin A (MspA), for example, known for its shape, like a funnel. These pores have hydrophobic regions responsible for controlling molecule movements, providing inherent selectivity and response to changes at the single-molecule level. They have a stable shape that favors high-resolution detection of molecules (Deamer et al., 2016).

2.1.2 Solid-State Nanopores

These nanomaterials are constructed from high-quality silicon nitride (Si3N4), silicon dioxide (SiO2), or graphene, with an optimized pore-size membrane. The recovery and harvesting of graphene nanoparticles (NPs) are critical since they possess properties that are flat transmission materials and only one atom thick, making them beneficial for detection purposes. They are tough, chemically stable, and even customizable to suit various applications (Garaj et al., 2010).

2.1.3 Hybrid Nanopores

Hybrid NPs combine the advantages found in biological and solid-state NPs by including protein channels within synthetic membranes. This novel approach presents a unique combination of biological specificity and synthetic durability (MacKenzie & Argyropoulos, 2023).

Functions of Nanopores

1. Single-molecule analysis: Nanopores enable real-time detection of single molecules based on tracking changes in the ionic current while the molecules pass through the pore. This mechanism allows the processing of nucleotides, peptides, and even small metabolites (Deamer et al., 2016).
2. DNA and RNA sequencing: Biological nanopores, including α-hemolysin and MspA, facilitate the translocation of DNA or RNA strands through their openings while simultaneously monitoring variations in ionic current. Cutting-edge

sequencing technologies, such as the MinION from Oxford Nanopore, leverage this mechanism to enable long-read sequencing. This approach is crucial for various applications, from sequencing the human genome to tracking viral outbreaks, including COVID-19 (Manrao et al., 2012).

3. Epigenetic analysis: Nanopore sequencing can detect DNA modifications, such as methylation without the necessity for supplementary processing steps and is an uncomplicated way of studying epigenetics steps and is an uncomplicated way of studying epigenetics. (Workman et al., 2019).
4. Application of nanopores in protein and peptide analysis: Nanopores can measure and analyze the details of the structures and movements of single proteins or peptides. Gaining insight into these interactions will be essential in understanding diseases derived from protein aggregation or misfolding.
5. Molecular sensing and diagnostics: Nanopores are a critical tool for identifying and analyzing a wide range of biomolecules, from proteins and small metabolites to viral RNA. It is highly sensitive and thus well-suited to point-of-care diagnostics.
6. Drug development and screening: Solid-state nanopores are vital in drug screening. These pores examine molecular interactions and evaluate drugs' efficiency. The method helps improve drug delivery systems and develop new therapeutic solutions (MacKenzie & Argyropoulos, 2023).
7. Biosensors for environmental monitoring: Nanopores are known to detect pollutants, toxins, and pathogens in environmental samples and can immediately provide data to be used in ecological evaluations and ensure public health safety (Deamer et al., 2016).
8. Synthetic Biology and Nanotechnology: Nanopores are a synthetic biology tool for designing bio-inspired systems and for better understanding critical processes at the nanoscale (Garaj et al., 2010).

2.2 *Mechanism of Ionic Current Disruption*

Ionic currents in nanopore devices arise when charged particles, called ions, move through the pore of the device due to an applied potential difference across the membrane that houses the pore. Typically, the system comprises the following:

1. Electrolytic fluid: The solution also contains a buffer matrix with free ions, such as K+ and Cl-, to enable current flow through the nanopore, which has diameters between 1 to 10 nm (Kasianowicz et al., 1996).
2. Nanopore: In this case, the pores can be either naturally occurring, like α-hemolysin or MspA, or created artificially, for example, solid-state pores out of silicon nitride (Loman et al., 2015).
3. Voltage source circuit: Electrodes on either side of the nanopore are kept at a bias voltage, creating an ion flow that leads to a constant background current.

The intensity of the ionic current depends on several key factors:

1. Pore diameter: A larger pore diameter increases the size of the opening, which allows more ions to pass through and, as a result, increases the current (Kasianowicz et al., 1996).
2. Voltage: Increased voltage increases the electric field, which increases ion flow (Butler et al., 2008).
3. Ion concentration: The higher the ion concentration, the higher will be the current, as there are more charge carriers available in the solution.
4. Pore geometry and surface chemistry: Irregularities and surface features of the pore walls can also affect the movement of ions within them (Deamer et al., 2016).

2.2.1 Mechanisms of Current Disruption

Both the chemical and physical attributes of the molecule lead to the very existence of the molecule. At the same time, its transit through a nanopore alters the ionic current in any one of a variety of ways. The most important mechanisms include:

1. Physical blockage: A molecule occludes space inside the nanopore, partly blocking access at one end of the pore, effectively blocking ions in transit through the pore. The level of this blockage depends on the size and shape of the molecule in question.
   (a) The current block due to DNA is directly related to its orthogonal cross-section.
   (b) As the distance of single-stranded DNA is less than that of double-stranded DNA, there is less obstruction in single-stranded DNA than double stranded (Kasianowicz et al., 1996; Laszlo et al., 2014).
2. Physicochemical effects on nanopore translocating polymers: Charged entities like the negatively charged phosphate backbone of a DNA strand interact with the ions located in the pore and change the local ionic environment.
   (a) Such interactions could prolong the time the molecule stays in the pore. In addition, they may influence the local ion concentration around the pore's mouth and throat.
3. Effects of hydrodynamics: Ions do not flow freely because the flow of fluid through the pore is impeded by the specified molecular geometry. For example, this includes the change of ion mobility and the arrangement of water molecules that are present around the moving molecule (Deamer et al., 2016).
4. Disruption of bases specific: Nucleotide bases of the ribonucleic acid (RNA) and deoxyribonucleic acid (DNA) by pore of different sizes, shapes or chemical properties interact differently. This results in different current blockades for each type of nucleotide which is the principle behind the sequence specific detection (Butler et al., 2008; Laszlo et al., 2014).

5. Dynamics of translocation: Current is also influenced by the velocity of translocation through the nanopore.
   (a) Very quick translocations tend to cause more noise in the current trace, and longer, slower movement leads to more accurate recordings (Butler et al., 2008).
   (b) In order to control the translocation speed, several enzymes, such as polymerases or helicases, are usually employed.

2.2.2 Molecular Signatures and Data Interpretation

Variations of ionic current are measured over time and presented in a collection of distinct events known as a "current trace." Every event is related to some specific interaction between a molecule and the NP. These traces are used for:

1. Single-molecule advertisement: Each disruption pattern is mapped to a corresponding molecular feature (Kasianowicz et al., 1996).
2. Base-by-base sequencing: Nucleotide sequences within DNA or RNA strands are deduced from the distinct current traces of singular nucleotides translocating through the pore (Laszlo et al., 2014).
3. Molecular withstanding pressure: Changes in ionic current can be observed due to the presence of secondary structure features within the RNA or DNA such as hairpins or loops (Deamer et al., 2016).

As illustrated in Fig. 3, the process of nanopore sequencing involves measuring changes in ionic current as individual nucleic acid molecules translocate through a nanopore under an applied voltage. These current disruptions, unique to each nucleotide or structural feature, are captured in real time, allowing direct base-by-base sequencing without the need for amplification or labeling.

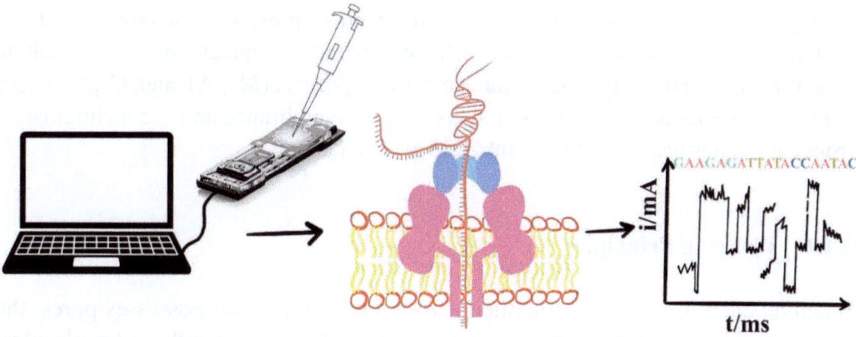

Fig. 3 Principle and process of NPS

2.3 Single-Molecule Detection

Nanopore Sequencing (NPS), which is a single-molecule detection technique, works on the basic premise of detecting ionic current as it is perturbed by a molecule threading its way into and out of a NP (Nanopore). In this respect, membrane pores are enclosed with a voltage source and current is driven through the NP creating an ionic current baseline. When, for instance, DNA or RNA or protein goes through the pore, it produces a disturbance in the ionic current, creating intelligible waves dependent on the size, shape, and chemical structure of the molecule (Kasianowicz et al., 1996). Unlike other methods of sequencing where the sample has to be processed, in NPS there is no processing, hence native epigenetic markers are intact (Deamer et al., 2016). Each single unit of DNA comprises a non-continuous signal, and thus specific sequences can be identified in real time (Laszlo et al., 2014). Such a strategy is associated with easier preparation of samples and lower risks of errors due to chemical alterations (Bayley & Martin, 2000). In addition, translocation dynamics through the NP are sensitive to even the slightest changes in speed and interacting molecules which has an effect on the resolution of the signal. Sophisticated algorithms are employed to transform these signals to useful molecular information for fast sequencing. This amplification-free approach for sequencing is through the example of transforming every single nucleotide in a complementary strand into their respective signals in real-time without needing to pause the machine; it explains why NPS single-molecule analysis is so revolutionary.

3 Types of NPs

3.1 Biological NPs

Biological nanopores are nanometer-sized openings present in or fabricated from proteins that occur naturally in cell membranes. Typical instances include α-hemolysin (α-HL), Mycobacterium smegmatis porin A (MspA), and, CsgG. These nanopores are incorporated into bilayers of lipid membranes and are distinguished from each other because of their intricate atomic design.

3.1.1 Working Principle

When an electric potential is applied across a membrane that possesses pores, the ionic current is measured. When a pore opens and allows an influx of molecules, there is a temporary blockage in the current, which can be measured and hence, detected. Every molecule then generates a unique signal allowing for the molecular recognition and characterization (Deamer et al., 2016).

3.1.2 Advantages

1. High selectivity and sensitivity because of the geometrical constraints.
2. Capable of measuring the ionic current passing through a single molecule, i.e., a DNA base.
3. Can be easily modified for various applications using genetic engineering techniques.

3.1.3 Disadvantages

1. Relatively weak and destructible lipid membrane that is unable to withstand various external conditions for a long period of time.
2. Qualitative loss of the biomolecular component (Deamer et al., 2016).

3.2 Solid-State NPs

Solid-state nanopores are artificial nanopores made in materials such as silicon nitride (Si_3N_4), graphene, and hafnium oxide. They are produced using hi-tech nanofabrication methods like electron-beam lithography (EBL) or focused ion beam machining (FIBM).

3.2.1 Working Principle

Like biological nanopores, the ionic currents are disrupted by the molecules passing through the pore. Nanopores in solid-state structures also enable integrating additional types of sensors, such as optical or electrical transducers (Wanunu, 2012).

3.2.2 Advantages

1. Has a high level of toughness and mechanical stability, hence, can operate in different conditions.
2. Pore size can be varied enabling detection of wide range of different molecules.
3. Can be connected to electric circuits for larger sensing platforms.

3.2.3 Disadvantages

1. Biological nanopores can be controlled on atomic scale with a much higher degree of precision.

2. Since the material used is manmade, there are higher levels of noise (Wanunu, 2012).

3.3 *Hybrid NPs*

Hybrid nanopores incorporate both biologically derived and solid-state components, utilizing the best features of each. In most cases, a biological nanopore, as for example α-HL, is introduced to a solid membrane, or the insert is designed to allow interdigitation of biological and synthetic media.

3.3.1 Working Principle

Takes advantage of the specific molecular recognition characteristics of biological materials, while utilizing the mechanical strength of the solid-state matrix (Smeets et al., 2008).

3.3.2 Advantages

1. The construct is more stable thanks to the solid-state casing.
2. Higher retention of molecular selectivity due to the bio-origin part.
3. Features can be adjusted within the limits of certain applications.

3.3.3 Disadvantages

1. Composite Task: Nanopore construction is not simple.
2. Integration of biological onto synthetic surfaces and vice versa remains challenging (Smeets et al., 2008).

As given in Table 1, the three major classes of nanopores—biological, solid state, and hybrid—differ in their structure, operational mechanisms, and applications. While biological pores offer high molecular specificity, solid-state pores provide mechanical durability, and hybrid pores combine the strengths of both, making them versatile for diverse analytical tasks.

Table 1 Comparison between the types of NPs (Smeets et al., 2008; Wanunu, 2012)

Features	Biological NPs	Solid-State NPs	Hybrid NPs
Definition	Biological membranes either possess protein-based nanopores or they are designed specifically for assorted functions	Nanopores that are constructed within rigid structures to include but not limited to silicon nitride, graphene, or glass	Nanopores with biological elements and man-made composites incorporated in order to use the best properties of both
Structure	– Protein carriers found in phospholipid bilayers or synthetic amphiphilic membranes – Pore size is constantly the same even on atomic level	– Durables solid materials containing nanometer scale apertures – Si_3N_4, graphene and glass are some of the materials used – Tunable thickness and size	– Biological NP (e.g., α-HL), which is integrated within a solid-state setup – Can also extend to creating purely synthetic membrane pores that achieve biological-like functions
Functions	– Facilitates molecular sensing, particularly single-molecule analysis – Enables sequencing of nucleic acids and detecting other biomolecules	– Offers mechanical robustness for molecular detection and environmental sensing – Useful in high-throughput applications	– Combines specificity of biological pores with the durability and tunability of solid-state systems – Advanced detection and sensing capabilities
Operational mechanism	– Ionic current measurement detects molecular passage – Voltage applied across lipid bilayer causes molecules to produce characteristic current changes	– Similar ionic current measurement mechanism – Molecules passing through synthetic pores disrupt the current – Allows integration with electronics	– Same principle as above but with improved resolution due to the combined system – Integrates biological specificity and synthetic durability
Applications	– DNA and RNA sequencing – Protein and peptide analysis – Small molecule sensing in biological systems	– Environmental monitoring of pollutants – High-throughput DNA sequencing – Synthetic biology and robust biosensor design	– Drug screening – Advanced biosensors for medical diagnostics – Improved molecular resolution for DNA, RNA, and protein detection – Portable diagnostics

4 Nanopore Sequencing (NPS) for DNA and RNA

4.1 DNA Sequencing with NPs

The method of DNA sequencing by NPs relies on the principle of current perturbation which occurs when a ssDNA strand translocates through the NP. This perturbation is associated with the nucleotides in the polynucleotide chain allowing for resolution of the signal in real time.

4.1.1 Key Mechanism

1. Sample preparation: The DNA is processed in such a way that it is propped, cut into pieces and attached with motor proteins or adapters which regulate the movement through the nanopore.
2. Interaction with a NP: Upon application of voltage, the pore is filled with DNA that passes through the pore and consumes the ionic current.
3. Base identification: Data sequences of currents produced by different nucleotides A, T, G, C are collected and processed with neural networks for ordering them in a sequence.

4.1.2 Technological Features

- Able to sequence unamplified and native DNA.
- Can work with miniaturized systems such as MinION and with large systems such as PromethION.

4.1.3 Advantages

1. Long reads: tens to hundreds of kilobases can be sequenced in one read the maximum read length.
2. Proper sequencing does not require PCR amplification which minimizes the distortion introduced by external factors.
3. The information is acquired and analyzed almost at once providing for instant results.

4.1.4 Disadvantages

1. Has a higher raw error rate of over 5%–15% when compared with the short-read approaches.
2. Highly dependent on the experimental parameters like pH and ionic strength.

4.1.5 Applications

1. Clinical genomics: It allows almost instant and rapid screening of entire genomes thanks to its low-cost whole genomic sequencing, especially for cancer and precision medicine applications.
2. Microbial genomics: It is used for identifying pathogens and for gene profiling associated with drug resistance.
3. De novo genome assembly: Long reads are useful for resolving long repetitive stretches and complicated regions within the genome (Deamer et al., 2016).

4.2 RNA Sequencing Applications

NPS constituents the sequence of the RNA molecules even when that is not converted into the complementary DNA. Therefore, all the native modifications are kept in place and the modifications brought about by reverse transcription are lost.

4.2.1 Mechanism

1. The native RNA molecules pass through the NP.
2. The ionic current is disrupted and changes in the current are interpreted in terms of which bases and modified bases are present. The ionic current is therefore probed and the signals are base called into sequences.
3. Base calling algorithms compute signals from current and reconstruct the sequence of RNA.

4.2.2 Advantages

1. Direct sequencing: Makes it possible to map changes in RNA such as N6-methyladenosine (m6A) which are known to affect gene activity.
2. Full-length transcripts: Acquires entire RNA's precluding the bias caused by fragmentation.
3. Isoform resolution: Resolves the spectrum of alternative formed transcripts as well alternative splicing.

4.2.3 Disadvantages

1. Small sample sizes as compared to the application in DNA sequencing.
2. Base and modification detection using current methods are inadequate; there is a need for better signal processing techniques.

4.2.4 Applications

1. Transcriptome analysis: Focuses on studies of gene expression patterns in different tissues and under different conditions.
2. Viral RNA detection: Aids in the production of efficient and with minimum cost based epidemiological tools for rapid detection of RNA viruses outbreak dependencies such as influenza and SARS-CoV-2.
3. Epitranscriptomics: It is a science that studies RNA structure and functional elements by types of mapping of RNA modifications.

4.3 Long-Read Sequencing Advantages

NP technology's long-read sequencing capability ensures the ability to read through complicated structures of genomes and repetitive regions, which is a challenge for short-read sequencing.

4.3.1 Advantages of Long-read

1. Detection of structural alterations: Aberrant radius enlargement screenings will encompass insertions, deletions, and inversions of genetic material.
2. Haplotyping resolution: Determining which alleles are present so that all genetic variants can be separated out.
3. Extensive COVERAGE OF REPEATED REGIONS: Takes care of difficult regions like telomeres, centromeres, and transposons.
4. De novo assembly: Assists in correctly assembling the genomes without employing any reference sequence.

4.3.2 Applications

1. Construction of complex genome assemblies, such as those of polyploid plants.
2. In anthropogenic metagenomics for studies of microbial diversity.
3. Locating structural variants in diseases such as cancer in humans (Rang et al., 2018).

Table 2 highlights the key distinctions between nanopore-based long-read sequencing and traditional short-read methods. Long reads excel at resolving structural variations and repetitive genomic regions, whereas short reads are more accurate per base and cost-effective for large-scale applications.

Table 2 Comparison of long reads vs short reads (Rang et al., 2018)

Comparison: long reads vs short reads	NP long reads	Short reads (e.g., illumina)
Read length	Up to megabases	Typically, ≤300 base pairs
Error rate	Higher (~5–15%)	Lower (<1%)
Structural variation detection	Excellent	Limited
Repetitive region resolution	High	Low
Cost per base	Higher	Lower

4.4 *Direct Sequencing of Modified Bases*

Despite the advancement in the technology of sequencing, Nanopore sequencing has the advantage over them, where it is possible to directly visualize the modified nucleotides even before incorporation in the DNA strands. The principle of detection here is described as, modified bases generate different currents while the rest of the bases are read, hence making the detection of the modified bases possible.

4.4.1 Applications

1. Epigenomics: Studying the methylation profiles in cancer and neurodegenerative diseases.
2. Epitranscriptomics: Studying the modifications of RNA molecules and their roles in transcriptional control.
3. Environmental genomics: Studying societal methylation patterns within microbial environments.

4.4.2 Identified Modifications

- 5 mC (5-methylcytosine): This is an epigenetic marker which is associated with gene expression and development of some diseases.
- 6 mA (N6-methyladenine): Present in the genomes of prokaryotes and some eukaryotes.
- RNA Modifications: Include m6A, m5C, and pseudouridine, all of which are essential to the structure and function of RNA.

Different types of nucleotide modifications—such as 5mC in DNA or m6A in RNA are detailed in Table 3. They play significant biological roles, from gene regulation to epigenetic inheritance. Nanopore sequencing enables direct detection of these modifications through characteristic changes in ionic current.

Table 3 Modifications and their biological significance (Simpson et al., 2017)

Modification	Biological significance
5 mC (DNA)	Regulates gene expressions, linked to cancers
6 mA (DNA)	Modulates DNA-protein interactions in prokaryotes
m6A (RNA)	Affects RNA stability, splicing and translation

4.4.3 Disadvantages

1. It involves assessing very slight alterations in current levels, due to base-nucleotide incorporation.
2. Involves reading modified base sequencers at a lower sine level gradient than that of the unmodified base sequence (Simpson et al., 2017).

5 Signal Detection and Data Analysis

5.1 Current Measurement and Signal Interpretation

The present-day electrical signals are measured and transformed into useful information in different areas of science and technology by more than just means of measuring instruments. In biological and technological systems existing today that require signals to be detected, such signals may be current variations that carry the information content. This part introduces the concepts of current measurement and the corresponding signal interpretation.

5.1.1 Fundamentals in Electrical Current Measurement

Electrophysiological focus: The generation of current in many cases involves the movement of ions in various systems such as nanopores, semiconductors, and biosensors.

Equipment: Current can be measured using patch-clamp amplifiers, potentiostats, and sensitive electrometers.

Noise mitigation: Various strategies like shielding, averaging, and digital filtering help reduce noise and enhance the quality of the useful signal (Chen & Zhang, 2022).

5.1.2 Signal Processing and Analysis

Pattern analysis: Any signal produced by any event can be quantified physically, chemically, or biologically involving an event. For instance, in the nucleotide sequencing of DNA by planar nanopore devices, a blockage of electrical current indicates the presence of a specific nucleobase (Chen & Zhang, 2022).

Table 4 Common noise sources and mitigation techniques

Source of noise	Characteristics	Mitigation strategy
Thermal noise	Random, white noise	Cooling systems
Instrumental noise	Systematic, device-dependent	Calibration, shielding
Environmental interferences	Frequency-specific (e.g., 50 Hz hum)	Grounding, noise filters

Transient signal analysis: Analysis performed on a signal is based on the time evolution of the signaling mechanism in order to observe any alteration in trend.

Transient signal analysis in sensory apparatus: Often Times Signal processing relies on mathematical operations to separate part of the signal from the remaining noise by the use of Fourier transformation. It is possible to identify certain behaviors characterized by periodicity.

The primary sources of noise encountered in nanopore signal detection, such as thermal fluctuations and environmental interference is summarized in Table 4, it also outlines the corresponding mitigation strategies that enhance signal fidelity and sequencing accuracy.

5.2 Basecalling Algorithms

The algorithms employed for basecalling are essential in the enhancement of biological sequences derived from sequence platforms. This step is the link between electrical data acquisition and the analysis of genomic or transcriptomic data. Basecalling is the process of converting the electrical or light signals generated in the process of sequencing into nucleotides which are bases (A, C, G, T or U) of a desirable sequence. This is the initial phase of processing sequencing data. It is used to mean the transformation of raw signal information into a usable biological sequence presenting as RNA or DNA.

1. Conventional basecalling
    (a) Does not exceed mechanistic threshold-based and signal peak detection techniques.
    (b) Has considerably lower precision, especially in case of obscured or intricate data (Chen & Zhang, 2022).
2. Modern algorithms
    (a) Hidden Markov Models (HMMs): They employ pattern recognition techniques based on signal—understandably probabilistic—models.
    (b) Neural Networks: These are machine learning models that use vast amounts of data, predominantly containing signal and sequence data pairs (e.g., ONT's Guppy) for training (Chen & Zhang, 2022).
    (c) Real-Time Basecalling: Owing to the progress made in the performance of computation, it is now possible to conduct analysis and make changes even while the process of data collection is in progress.

5.3 *Machine Learning and AI in Signal Processing*

The introduction and implementation of machine learning (ML) and artificial intelligence (AI) has changed the way signal processing has worked taking it to the next levels of pattern classification, denoising and feature extraction.

5.3.1 Applications in Signal Processing

1. Noise reduction: Among noise reduction algorithms is one known as the principal component analysis (PCA) which distinguishes signal from noise (Chen & Zhang, 2022).
2. Signal classification: Data pattern classification such as identifying DNA bases in nanopore sequencing convolutional neural networks (CNNs).
3. Impact assessment: Reinforcement learning is specifically used to improve the sequencing process by predicting the signal shapes.

5.3.2 AI Driven Workflows

In carrying out these tasks, AI algorithms, for instance, embeds the signal interpretation into the subsequent processes like alignment and annotation. Such minimizes human intervention and speeds up the outcome (Chen & Zhang, 2022).

5.3.3 Ethical Issues

The advent of such technologies and their ease of use is also giving rise to issues on trust because of training data, possible unfair biases and the ability to compare results.

5.4 *Error Rates and Accuracy Improvements*

5.4.1 Errors and Their Causes

There are errors in signal detection and data analyzation that can be influenced by a number of biological, technological, and other computational factors. The detailed errors are as follows:

1. Biological factors:
   (a) Repetitive sequences: Regions with unchanging DNA composition and arranged in clusters result in similar detection signals (Chen & Zhang, 2022).

(b) Secondary structures: Structural configurations such as hairpins or loops present in nucleic acids interfere with the straight forward detection of a signal.
(c) Chemical modifications: Alterations in signals due to chemical changes such as methylation or damages to the bases are incorporated.

2. Instrumental Factors:

(a) Aging of instruments: Loss of efficiency due to the varied sensitivities of sensors with time (Shendure et al., 2017).
(b) Cross-talk: Interference of signals from closely located and/or clustered molecules such as in a sequencing channel.
(c) Heat: Variations in temperature especially with sensitive equipment can result in stability of change in signal issued out (Wick et al., 2019).

3. Algorithmic factors:

(a) Model overfitting: Such is the case with machine learning models that are trained to learn different things using a particular set of data and fail to learn anything about any other new sets of data.
(b) Confusing apparatus with training data: In basecalling or another stage classification, errors in labeled datasets result to errors.
(c) Events junctions: Events occurring near each other in time which are dependent or related to each other, often in the case of multiplexed systems, are difficult to resolve accurately (Wick et al., 2019).

5.4.2 Accuracy Improvements

1. Algorithmic developments:

(a) Transfer learning: Improving the transferability of models through the use of pretrained ones on various datasets (Chen & Zhang, 2022).
(b) Consensus algorithms: The averaging of outputs from numerous models or runs helps avoid the influences of chance (Wick et al., 2019).
(c) Dynamic noise filtering: In-built filtering strategies where some parameters are modified depending on the characteristics of the signal.

2. Hardware developments:

(a) Better nanopore configurations: Adjustable pores designed particularly for certain analytes offer better signals.
(b) Controlled temperature management: On-board systems ensure that the readings taken do not fluctuate due to changes in temperature (Shendure et al., 2017).

3. Data handling techniques:
   (a) Error correction algorithms: Techniques such as circular consensus sequencing (CCS) undertakes re-sequencing of DNA to form consensus that enhances accuracy.
   (b) Hybrid sequencing approaches: Integration of short-read and long-read sequencing technologies for purposes of complementary error correction.
   (c) Redundant signal sampling: Basecalling enhanced through multiple readings of the same event.
4. Training and Dataset Optimization:
   (a) Diverse training data: Various factors, such as the sequence type scope, structure scope, and condition scope, should be added in datasets so as to minimize bias.
   (b) Synthetic data generation: Known libraries of characteristics may be used to synthesize signals and this helps in strengthening the model (Wick et al., 2019).

6 Nanopore Sequencing (NPS) Platforms

6.1 Oxford Nanopore Technologies (ONT) (MinION, GridION and PromethION)

Oxford Nanopore Technologies (ONT) is the leader in the field of NPS and has come up with multiple platforms for various purposes. ONT has changed the course of DNA and RNA sequencing with its platform based on NP technology. In contrast to the conventional sequencing techniques which are based on fluorescence or chemical labeling, ONT implements a synthetic membrane with biological NPs in it. Long strands of DNA or RNA are fed into the NPs and the changes in the ionic current are recorded to deduce the sequence. As illustrated in Table 5, Oxford Nanopore's sequencing platforms—MinION, GridION, and PromethION—vary in scale, throughput, cost, and portability. This comparison helps identify suitable devices for different research settings, from field diagnostics to population-scale genomics. The various ONT Platform Devices are shown in Fig. 4.

6.1.1 Notes on Comparison

- MinION is best appropriate for portability and rapid experiments. Therefore, it is most suited for field-based applications such as pathogen detection or environmental monitoring (Quick et al., 2016).
- GridION presents a middle ground in between both as well as offering high throughput for medium-scale laboratories without the complications encountered with PromethION (Shendure et al., 2017).

Table 5 Side-by-side comparison of ONT platforms (MinION, GridION, and PromethION)

Feature	MinION	GridION	PromtehION
Device type	Portable, handheld sequencer	Desktop sequencer	High-throughput benchtop sequencer
Throughput (per run)	Up to 50 Gb	Up to 250 Gb (with five MinION flow cells)	Up to 20 Tb (with 48 flow cells)
Flow cell capacity	1 flow cell	Up to 5 MinION flow cells simultaneously	Up to 48 PromethION flow cells simultaneously
Read length	Ultra-long reads (>1 Mb possible)	Ultra-long reads (>1 Mb possible)	Ultra-long reads (>1 Mb possible)
Applications	Fieldwork, small-scale studies, diagnostics	Medium-scale studies, clinical applications	Large-scale studies, population genomics
Data processing	Requires external computer for analysis	Integrated GPU for data analysis	High-performance integrated GPUs
Connectivity	USB connection to a laptop or desktop	Ethernet connection	Ethernet connection
Portability	Highly portable (fits in a pocket)	Desktop-sized, portable within a lab	Large and stationary
Cost	Low (~$1000 for the device, excluding flow cells)	Moderate (~$50,000 for the device)	High (~$200,000 for the device)
Power requirements	Low (can run on a laptop's power)	Requires dedicated power supply	High, with advanced cooling and power needs
Real-time sequencing	Yes	Yes	Yes
Target users	Researchers in the field, small labs	Medium-sized labs, clinical diagnostics	Large-scale research labs, genome centers
Setup time	Minimal (~10 min)	Moderate (~30 min)	High (~1–2 h)
Run duration	Variable; ~6–48 h depending on settings	Variable; ~6–48 h depending on settings	Variable; ~6–48 h depending on settings
Data storage needs	Low (~100 GB per run)	Medium (~500 GB per run)	High (>10 TB per run)

- PromethION is designed for high-throughout projects such as population genomics or deep transcriptomics which require big data sets to be generated and managed.

6.2 *Comparison with Other Sequencing Platforms*

The platforms used in nanopore sequencing have their own distinct merits and demerits as compared to other widely used sequencing technologies like Illumina and PacBio.

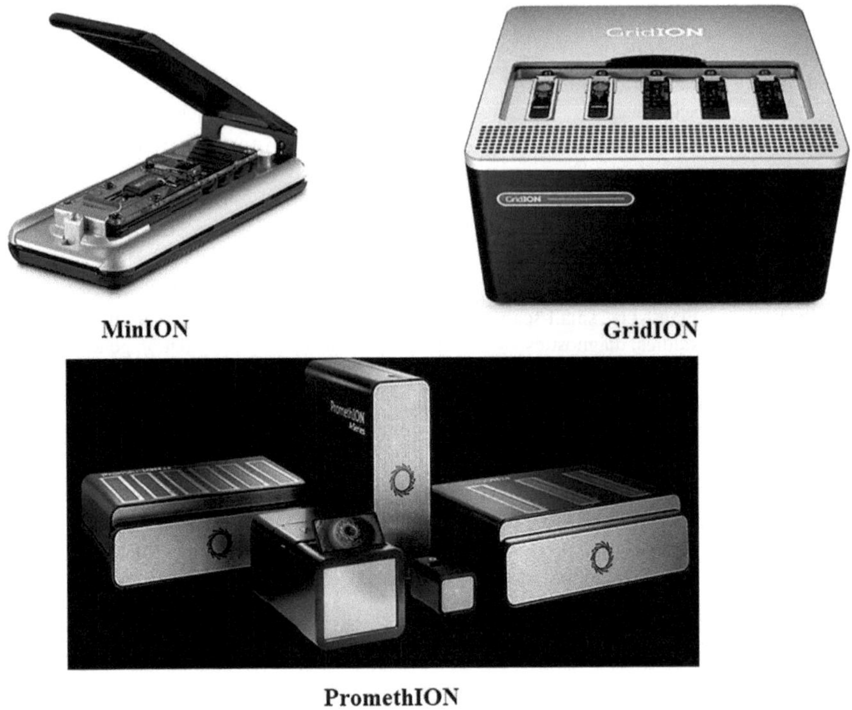

Fig. 4 ONT platform devices. These images were taken from https://nanoporetech.com/

6.2.1 Key Comparisons

1. Read length:
   (a) Long reads produced by ONT Sequencing are ultra-long (>1 Mb) and suitable for structural variation studies.
   (b) On the other hand, Illumina concentrates on short reads (150–300 bp) ideal for applications that require sequencing at scale (Shendure et al., 2017).
   (c) PacBio also provides ~10–15 Kb long reads but with superior accuracy through using HiFi reads.
2. Accuracy:
   (a) Typical raw accuracy of ONT approaches 90%–95% which is inferior to the accuracy of Illumina systems which is over 99% but tends to improve as the new basecalling algorithms are used (Chen & Zhang, 2022).
   (b) With PacBio HiFi reads, the accuracy is more than 99.9% on par with Illumina sequencing accuracy (≥99.9%).

3. Cost and scalability:
   (a) ONT devices such as the MinION are inexpensive and easy to carry making them available even for smaller laboratories.
   (b) Illumina systems tend to be very costly with capital investments of machine buildings and support structures being a necessity, but the output performance is high.

6.3 Portability and Field Applications

6.3.1 Field Applications

1. Infectious diseases monitoring: Minimum Nanopore Sequencer (MinION) has been employed to determine causal pathogens, pathogen identification at the midst of the Ebola and Zika virus outbreaks (Quick et al., 2016).
2. Ecological monitoring: Allows examination of microbial diversity in remote ecosystems, e.g., rainforests or below surface oceans (Wick et al., 2019).
3. Managerial genomics: instantaneous identification of plant diseases/pathogens or crops within the natural habitat (Shendure et al., 2017).

6.3.2 Advantages of Portability

1. Avert the hassle of conveying specimens to central laboratories, thus saving time and limiting expenditures.
2. Helps support rapid assessment in both clinical and environmental circumstances.
3. Promotes more widespread use of sequencing techniques by making it easier for small laboratories and non-conventional participants to access them.

7 Applications of Nanopore Sequencing (NPS)

7.1 Genomics and Whole-Genome Sequencing

The emergence of NPS has significantly revolutionized the domain of genomic studies as whole genomes can be sequenced and assembled accurately and completely as never before. Unlike traditional short-read sequencing technologies which have difficulty in elucidating redundant and complicated regions, such analysis of genome sequencing by NP technology is possible due to their ability to produce ultra-long reads.

7.1.1 Advantages in Genomics

1. Long-read capabilities:
   (a) Helps resolve complex genomic regions such as repetitive elements, structural variations, and tandem repeats.
   (b) Creates contiguous genome assemblies while avoiding a long and tedious computational stitching.
2. De novo genome assembly:
   (a) Works well for the assembly of genomes of completely new organisms or species.
   (b) This method has proven to be of utmost importance in the analysis of both plants and animal's genomes to study the patterns of variation and evolution (Wick et al., 2019).
3. Epigenomics: reads signal directly to identify base modifications such as methylation without additional chemicals.

7.2 Clinical Diagnostics and Personalized Medicine

1. Cancer genomics: Targeting the examination of mutations and structural variations that promote tumorigenesis. Liquid biopsies offer rapid tumor DNA detection (Chen & Zhang, 2022).
2. Rare disease diagnosis: Analyzes alterations in structural variants and repetitive regions which are the short-read technologies shortcomings.
3. Pharmacogenomics: Integrates the influence of genetic variation in drug metabolism in the design of therapeutic regimes for malignancies.

7.2.1 Examples

- Rapid sequencing of neonatal genomes to determine a causative genetic mutation in less than a few hours (Wick et al., 2019).
- The construction of the AMR gene pool profiling from clinical matrices to enable selective antibiotic therapy (Quick et al., 2016).

7.3 Pathogen Detection and Infectious Disease Monitoring

1. Real-time detection: MinION-type NP constructs have been used for fast sequencing in outbreak situations, such as the Ebola, Zika, and COVID-19 viruses, at the very location where the outbreak occurs (Quick et al., 2016).

2. Analysis of pathogen's full genome sequence: permits to assess all changes in the genome of the pathogen, including mechanism of resistance and methods of spread.

7.3.1 Standard Surveillance

1. Hospitals: Surveillance of AMR pathogens for the prevention of healthcare-associated infections.
2. Public Health: Infrastructure for the genomic surveillance of respiratory infections like influenza and tuberculosis and others.

7.3.2 Case Study: COVID-19 Genomic Surveillance

The pandemic effect of COVID-19 and the corresponding control measures underlined the impact of genome sequencing in any global health emergencies. NPS was an essential apparatus in the phylogenetic and epidemiological studies of SARS-CoV-2 due to its provision of real-time genomic sequencing data to make decisions concerning public health.

7.3.3 Case Example

NPS was used as part of COVID-19 Genomics UK (COG-UK) consortium in the United Kingdom. This particular project sequenced over 600,000 SARS-CoV-2 genomes in a single year making it possible to identify the geographical range of viral transmission and the pools of mutations. As a result, the real-time data facilitated the quick implementation of lockdowns and vaccination campaigns (Quick et al., 2016).

Furthermore, in the case of Africa, a more challenging environment, the MinION's portability enabled researchers to carry out genomic surveillance even in areas with poor infrastructure. For instance, South African researchers utilized MinION sequencing to monitor the rise of the Beta variant which posed an elevated threat due to its immune escape feature.

7.4 Environmental and Metagenomic Sequencing

7.4.1 Applications in Environmental Studies

1. Biodiversity monitoring: Detects microorganisms in samples taken from soil, water, or air.—Describes the shifts in community structure in response to environmental change (Wick et al., 2019).

2. Studies on bioremediation: Assesses the ability of microbes at polluted sites to biodegrade a particular contaminant.

7.4.2 Benefits in Metagenomics

1. Direct sequencing of mixed samples: Gives complex communities without the need of culturing them before obtaining a species level resolution.
2. Identification of functional genes: Identifies genes involved in the processes of nutrient cycling, resistance to antibiotics or production of toxins.

7.4.3 Case Study: Microbial Diversity in the Mariana Trench

The Mariana Trench, the most profound oceanic trench found on Earth, also comprises certain unusual communities of microbes that are adapted to survive in high pressure, low temperature and low nutrient conditions. The traditional sequencing techniques proved to be inadequate in reaching such ecosystems and in performing studies, since microorganisms could not be cultivated, and environmental samples were highly contaminants. The NPS approach was an efficient solution to these problems.

7.4.4 Case Example

In 2000, sediment samples were also taken from the Challenger Deep, the deepest part of the trench, during the extended research. The MinION sequencing allowed to discover more than 1200 microbial taxa, including some novel genera able to withstand high hydrostatic pressure, which had not been found before. Functional gene analysis revealed those adaptations, such as the increased expression of the genes coding for pressure resistant proteins and enzymes (Wick et al., 2019).

Furthermore, NPS revealed that there were lateral gene transfer events among microorganisms, indicating the active nature of deep-sea organisms. These have the influence of understanding the life on the earth's extremities as well as prospects related to biotechnology, for instance the production of enzymes for industries that are resistant to pressure.

Direct sequencing capabilities of NPS wherein DNA from ecologically complex systems can be sequenced without the need of culturing helped the researchers in exploring the microbial diversity of the trench. It offered:

1. Species identification: Identifying new extremophiles that live in extreme pressure conditions.
2. Functional insights: Identification of genes responsible of methane oxidation, nitrogen cycling and other important biogeochemical processes.

3. Real-time analysis: On-site sequencing helped the researchers to instantaneously appreciate the community structure and the functional pathways of metabolism (Chen & Zhang, 2022).

8 Challenges and Limitations

8.1 Error Rates and Sequence Accuracy

Although the promise of NP-based sequencing is immense, it is hampered by a number of challenges and limitations that should be surmounted to benefit from its full potential. These challenges fall under sequence accuracy, throughput, cost effectiveness and technical durability. One of the main disadvantages of NPS is that it has an error rate that is relatively higher.

8.1.1 Sources of Errors

1. Disambiguation/signal decoding issues:
   (a) This is because, during the reading of DNA or RNA with the aid of the apparatus, there are different electrical signals produced at the opening of the pore as each DNA or RNA strand passes through. Such lack of clarity in these signals leads to incorrect base calling (Wick et al., 2019).
   (b) Runs of similar nucleotides, for example, a stretch of "A"s, are homopolymer regions. Such regions tend to be error-prone because it is hard to distinguish between same signals.
2. Basecalling algorithms: In particular, basecalling software such as Guppy and Bonito have contributed to error reduction from more than 15% in previous generations to between 5% and 7% nowadays. Still, this does not apply to such data which is either complex or contains a lot of noise.

8.1.2 Strategies to Improve Accuracy

1. Correcting errors using mixed mode sequencing (nanopore and short reads).
2. Signal decoding improvement by developing machine learning techniques.

8.2 Throughput and Speed

8.2.1 Factors affecting Throughput

1. Pore density and run duration: In a flow cell, the range of active pores is directly proportional to the yield of sequencing. But, over time, this may also lead to a reduction in throughput because the pores may be worn out.
2. Read length variability: Although the nanopore technique holds the record in allowing foreign nucleic acids to be sequenced in the longest strand of nucleotides, such lengths also differ with the quality of DNA and the experimental environment utilized and hence contribute to the overall yield.

8.2.2 Improvements in Throughput

1. Applications such as PromethION employ enhanced pore density and several flow cells with the intent of doing terabase sequencing.
2. Research focused on the flow cell architecture as well as preparation techniques for the samples is still in progress, and it enhances the efficiency of throughput (Quick et al., 2016).

8.3 Cost Considerations

8.3.1 Cost Drivers

1. Consumables: It is crucial to note that flow cells and the library preparation kits are consumables that wear out with time hence must be used sparingly, which leads to operational costs.
2. Data analysis: Processing of large data sets necessitates a high-end computing facility, which contributes to the total costs.

8.3.2 Strategies for Cost Optimization

1. Reuse and cleaning: Wounded exploration non-critical experiments, flow cells can be cleaned and sterilized for reuse extending working life of consumables.
2. Bulk purchases: Buy consumables in large quantities or use repeated order schemes to get lower prices.
3. Optimized library prep: In order to minimize the quantity of time and resources needed to prepare samples for sequencing,
4. Shared resources: Participate in such efforts with other teams or shared services in order to lower costs.
5. Cloud computing: Use cloud services for data analysis on a pay-per-use basis.

6. Hybrid sequencing: Utilize data from both nanopore sequencing and short reads to save costs on sequencing and correction.
7. Multiplexing and barcoded genomics: Use flow cells more effectively by sequencing many samples at one time.
8. Automation: Reduce the need for human labor and mistakes by mechanizing repetitive processes.
9. Free software: Use open access software to process data rather than incurring costs for commercial software.
10. Targeted sequencing: Instead of bulking all the genomes, strategize on the important areas of why the sequencing is being done.
11. Efficient experiment design: Conduct preliminary screening to approximate the requirements and prevent over-sequencing.

8.4 *Signal Noise and Pore Stability*

8.4.1 Signal Noise

1. Acoustic interference: Electrical noise from outside sources has a tendency to mask ionic current measurements thus resulting in erroneous base calling (Chen & Zhang, 2022).
2. Molecular events: Secondary structures sculpture in DNA or RNA most often than not obstructs the uniform movement through the pore thus creating disturbance in the signals.

8.4.2 Pore Stability

1. Pore aging: Biological pores show signs of almost irreversible wear and tear after prolonged periods of say, in this case, sequencing, and this affects their performance.
2. Occlusion: Build-up of unwanted material or impurities in the nanopore can cause blockades which may necessitate changing or cleaning the flow cell (Wick et al., 2019).

8.4.3 Mitigation Strategies

1. Using better formulations of pore and construction materials so as to increase the longevity of the pore.
2. Using proper sample preparation strategies in order to reduce the level of contaminants.

9 Emerging Frontier in Nanopore (NP) Technology

9.1 *Protein and Peptide Sequencing*

The use of nanopore technology in protein sequencing is a commendable approach as it provides a straight forward and high-throughput technique for the proteins and their archaea.

9.1.1 Protein Sequencing Challenges

Comparatively, proteins have a more complex structure than nucleic acids, having 20 amino acids as opposed to the four nucleotides, hence it is difficult to interpret the signals generated.

For instance, in posttranslational modifications (PTMs), which are apparent in proteins, decreases the simplicity of analysis.

9.1.2 Technological Advances

1. Specialized NPs: Modified nanopores such as solid-state pores or engineered biological pores have been made to distinguish the different amino acids by electric signals.
2. Sequencing by detection of charge: Similar to a bead detangling, proteins are unfolded and are selected then passed through a NP. In turn every single amino acid present has a specific charge and the charge thus creates a signal profile (Ouldali et al., 2020).

9.1.3 Applications

1. To identify protein biomarkers for diseases such as cancer or neurodegenerative diseases.
2. Investigating their interactions and dynamics in a time-resolved manner.

9.2 *Small Molecules and Metabolite Detection*

Mechanism of detection: Small entities cause disturbances in the nanopore activity through either changes in the ionic current of the NP or through binding with the appropriate adapters located inside the pore or its neighborhood.

9.2.1 Technological Innovations

1. Adapter molecules: Adapter molecules or binding agents tethered to the NP can selectively capture metabolites, enhancing detection specificity.
2. Solid-state nanopores: Solid-state NPs produced from nanostructures such as Graphene are of more chemical stability and offer flexibility in the detection of small chemical moieties.

9.3 Future Directions for NP Development

Nanopore technology's scope provides a stage that is yet to be fully exploited. The future directions view presented below are all areas of current research and creative advances.

1. Merge with artificial intelligence
   (a) Signal processing MDA AI will enable better base calling and protein and small molecule detection (Chen & Zhang, 2022)
   (b) Modeling is useful in streamlining the upcoming experiment and even determining when to intervene during the experiment so that a sequence run is optimal
2. Size reduction and mobility: The purpose of the portable sequencers has been proven by devices such as the MinION. Later versions wish to change the dimensions even further while increasing the efficiency for activities such as the tracking of infectious diseases in the field and ecological studies
3. Multi-pore arrays: Increasing the density of the nanopores on the flow cells will enable an increased throughput allowing multi-sequencing of DNA, RNA, proteins, and metabolites on one device
4. Hybrid technologies: Integrating nanopore technology with technologies such as mass spectrophotometry, sharp focused ultrasound, or CRISPR based detection systems could broaden its diagnostic and analytical scope.
5. Innovative use cases
   (a) Epigenetics: More sensitive technologies for assessing DNA methylation and histone modification.
   (b) Nanomedicine: Nanopore systems for controlled and targeted drug delivery.
   (c) Technology of quantum pores: Using quantum effects to detect even the smallest changes in the molecular structure at an atomic level.

10 Conclusion and Future Prospects

10.1 Advances in Nanopore Design and Sequencing Methods

10.1.1 Nanopore Structure and Chemistry

1. Biological NPs: As the first examples of nanopores, which are virtually all made of proteins, designed to introduce DNA or RNA into a sequencer for decoding, biological nanopores offer the best performance currently available. The next designs will be more robust and accurate than before (Ouldali et al., 2020).
2. Solid-State NPs: There is an ongoing development for artificial apertures created from materials such as graphene and silicon nitride to be used for greater chemical stability, tunability as well as multi-functionalism.

10.1.2 Improvements on Sequencing

1. Direct RNA sequencing: The barriers to the in vivo sequencing of RNA molecules have fallen allowing real-time transcriptomes to be generated without the need to perform reverse transcription and so no alterations made to the RNA.
2. Dual-pore systems: New design concepts, which are now emerging, make use of the ability to incorporate two NP(s) to enhance the accuracy level and biomolecular structural aspects.

10.1.3 Automation and Scalability

1. The use of high-density flow cells, such as those found, for example, in the PromethION, enables better scalability for industrial and research applications.
2. Due to integration with AI, basecalling precision is improved with less human effort as signal processing is automated.

10.2 Role of NPs in Precision Medicine

NPS technology effectively addresses the needs of precision medicine by availing fast and in-depth genomic information of the patient at a low cost.

1. Applications in diagnostics
   (a) Cancer Genomics: NP-based sequencing capabilities permit the analysis of structural variations, mutations, and epigenetic changes—all of which are important parameters in the study of tumor heterogeneity for the design of targeted therapeutics (Chen & Zhang, 2022).

(b) Identification of Pathogen: Real-time sequencing has been widely used to enable quick turnaround time for identification of the disease-causing agents in epidemic situations as seen in Ebola and COVID-19 surveillance.

2. Pharmacogenomics: The few adverse effects related to the medications prescribed and the efficacy levels are improved by the use of drug dosing based on an individual's genetic profile obtained from nanopore sequencing.
3. Portable Sequencing in Remote Healthcare: Synonymous with the MinION, such devices are enhancing healthcare services in the third-world countries by allowing genetic and infectious disease care delivery at the point of need.

10.3 Long-Term Impact on Genomics and Biotechnology

Nanopore sequencing (NPS) is being embraced across the globe and changing the narratives of genomics and biotechnology with a promise that cuts across the spectrum of research, industry, and health.

1. Democratization of sequencing: Sequencing is becoming increasingly available to the low resource laboratories owing to the presence of the Portable and cheap devices which is creating a worldwide alteration in the research landscape (Quick et al., 2016).
2. Environmental and agronomical concerns: NP biosensors have real-time applications in environmental monitoring and may help in identifying invasive species or even pollutants, while the speed in crop and livestock sequencing for agricultural genomics is remarkable.
3. The transformation of synthetic biology by NP economy: NP systems are increasingly incorporated into the processes of determination of the properties and control of the processes of biosynthesis of synthetic DNA constructs which optimizes the processes of synthetic biology.
4. Concerns in data control and ethical issues in usage: NP technology provides a capability for creating and generating excessive amounts of data. However due to increase in use concerns such as data storage, data privacy, and appropriate use of genetic data will come into play.

11 Conclusion

NPS has changed the landscape of molecular biology and genomics by allowing sequence-based identification of any desired DNA or RNA in real time through a portable and very flexible technology. This allows sequencing of naturally occurring long DNA or RNA molecules without the need of amplification, which is a very clear deviation from the traditional means of generating and analyzing genomic information. This particular technology has its distinctiveness in that it can be used in almost any field with some focus, for instance precision medicine, pathogen monitoring, environmental studies, and even agricultural genomics.

The progress achieved thus far in NP technology underscores its transformative potential. Devices such as MinION, GridION, and PromethION have demonstrated scalability to accommodate research needs, from small laboratories to industrial-scale applications. Furthermore, the development of modern computational resources, particularly machine learning approaches, has also improved basecalling performance and efficiency, and supported real-time processing of data. These changes are encouraging the implementation of NPS in more challenging areas, like personalized medicine and synthetic biology.

And yet it is not all easy sailing. The current limitations, such as signal noise, pore stability, and expensive preliminary investments have been the cause for the ceaseless improvements of existing and new nanotube design and sequencing strategies. The transition between the two, in the form of solid-state NPs, hybrid sequencers and automatic interpretation systems is expected to solve this problem. Additionally, improvements in the level of throughput and cost per base are now making sequencing technology affordable to many more scientists and clinicians.

In a forward-looking perspective, the confluence of NPS, artificial intelligence (AI), microfluidics, and quantum sensing technologies is likely to herald a new paradigm in molecular analysis. The future developments of NP technology will certainly go beyond the context of genome sequencing, which will become just a part of several "omics" studies such as those addressing proteomics and metabolomics. Furthermore, the expansion of the diaspora of NPS tools will imply that data management issues, as well as ethical concerns around the use of genomic data, will have to be addressed in initiative.

To sum up, NPS is not merely an advance in the available techniques, it is an ecosystem that seeks to go where no other discipline has gone before in molecular biology, medicine, and biotechnology. With the emergence of new technologies and development of computing power, there is no doubt that NPS will be a cutting-edge scientific pursuit for many years to come.

References

Bayley, H., & Martin, C. R. (2000). Resistive-pulse sensing-from microbes to molecules. *Chemical Reviews, 100*(7), 2575–2594. https://doi.org/10.1021/cr980099g

Butler, T. Z., Pavlenok, M., Derrington, I. M., Niederweis, M., & Gundlach, J. H. (2008). Single-molecule DNA detection with an engineered MspA protein nanopore. *Proceedings of the National Academy of Sciences, 105*(52), 20647–20652. https://doi.org/10.1073/pnas.0807514106

Chen, T., & Zhang, Q. (2022). Novel spike-ins for single-cell sequencing enable ground-truth RNA counting. *Nature Methods, 19*(5), 530–531. https://doi.org/10.1038/s41592-022-01456-9

Deamer, D., Akeson, M., & Branton, D. (2016). Three decades of nanopore sequencing. *Nature Biotechnology, 34*(5), 518–524. https://doi.org/10.1038/nbt.3423

Garaj, S., Hubbard, W., Reina, A., Kong, J., Branton, D., & Golovchenko, J. A. (2010). Graphene as a subnanometre trans-electrode membrane. *Nature, 467*(7312), 190–193. https://doi.org/10.1038/nature09379

Kasianowicz, J. J., Brandin, E., Branton, D., & Deamer, D. W. (1996). Characterization of individual polynucleotide molecules using a membrane channel. *Proceedings of the National Academy of Sciences, 93*(24), 13770–13773. https://doi.org/10.1073/pnas.93.24.13770

Laszlo, A. H., Derrington, I. M., Ross, B. C., Brinkerhoff, H., Adey, A., Nova, I. C., Craig, J. M., Langford, K. W., Samson, J. M., Daza, R., Doering, K., Shendure, J., & Gundlach, J. H. (2014). Decoding long nanopore sequencing reads of natural DNA. *Nature Biotechnology, 32*(8), 829–833. https://doi.org/10.1038/nbt.2950

Loman, N. J., Quick, J., & Simpson, J. T. (2015). A complete bacterial genome assembled de novo using only nanopore sequencing data. *Nature Methods, 12*(8), 733–735. https://doi.org/10.1038/nmeth.3444

MacKenzie, M., & Argyropoulos, C. (2023). An introduction to Nanopore sequencing: Past, present, and future considerations. *Micromachines, 14*(2), 459. https://doi.org/10.3390/mi14020459

Manrao, E. A., Derrington, I. M., Laszlo, A. H., Langford, K. W., Hopper, M. K., Gillgren, N., Pavlenok, M., Niederweis, M., & Gundlach, J. H. (2012). Reading DNA at single-nucleotide resolution with a mutant MspA nanopore and phi29 DNA polymerase. *Nature Biotechnology, 30*(4), 349–353. https://doi.org/10.1038/nbt.2171

Ouldali, H., Sarthak, K., Ensslen, T., Piguet, F., Manivet, P., Pelta, J., Behrends, J. C., Aksimentiev, A., & Oukhaled, A. (2020). Electrical recognition of the twenty proteinogenic amino acids using an aerolysin nanopore. *Nature Biotechnology, 38*(2), 176–181. https://doi.org/10.1038/s41587-019-0345-2

Quick, J., Loman, N. J., Duraffour, S., Simpson, J. T., Severi, E., Cowley, L., Bore, J. A., Koundouno, R., Dudas, G., Mikhail, A., Ouédraogo, N., Afrough, B., Bah, A., Baum, J. H., Becker-Ziaja, B., Boettcher, J.-P., Cabeza-Cabrerizo, M., Camino-Sanchez, A., Carter, L. L., & Doerrbecker, J. (2016). Real-time, portable genome sequencing for Ebola surveillance. *Nature, 530*(7589), 228–232. https://doi.org/10.1038/nature16996

Rang, F. J., Kloosterman, W. P., & de Ridder, J. (2018). From squiggle to basepair: Computational approaches for improving nanopore sequencing read accuracy. *Genome Biology, 19*(1), 90. https://doi.org/10.1186/s13059-018-1462-9

Shendure, J., Balasubramanian, S., Church, G. M., Gilbert, W., Rogers, J., Schloss, J. A., & Waterston, R. H. (2017). DNA sequencing at 40: Past, present and future. *Nature, 550*(7676), 345–353. https://doi.org/10.1038/nature24286

Simpson, J. T., Workman, R. E., Zuzarte, P. C., David, M., Dursi, L. J., & Timp, W. (2017). Detecting DNA cytosine methylation using nanopore sequencing. *Nature Methods, 14*(4), 407–410. https://doi.org/10.1038/nmeth.4184

Smeets, R. M. M., Keyser, U. F., Dekker, N. H., & Dekker, C. (2008). Noise in solid-state nanopores. *Proceedings of the National Academy of Sciences, 105*(2), 417–421. https://doi.org/10.1073/pnas.0705349105

UC Santa Cruz News. (2023). *UCSC's David Deamer and mark Akeson honored for invention of nanopore sequencing*. https://news.ucsc.edu/2023/09/golden-goose-2023.html

Wanunu, M. (2012). Nanopores: A journey towards DNA sequencing. *Physics of Life Reviews, 9*(2), 125–158. https://doi.org/10.1016/j.plrev.2012.05.010

Wick, R. R., Judd, L. M., & Holt, K. E. (2019). Performance of neural network basecalling tools for Oxford Nanopore sequencing. *Genome Biology, 20*(1), 129. https://doi.org/10.1186/s13059-019-1727-y

Workman, R. E., Tang, A. D., Tang, P. S., Jain, M., Tyson, J. R., Razaghi, R., Zuzarte, P. C., Gilpatrick, T., Payne, A., Quick, J., Sadowski, N., Holmes, N., de Jesus, J. G., Jones, K. L., Soulette, C. M., Snutch, T. P., Loman, N., Paten, B., Loose, M., & Simpson, J. T. (2019). Nanopore native RNA sequencing of a human poly(a) transcriptome. *Nature Methods, 16*(12), 1297–1305. https://doi.org/10.1038/s41592-019-0617-2

Venturing into the Picoscale Frontier

Fabrication Techniques in Picotechnology

P. Saranraj, B. Lokeshwari, K. Gayathri, Hasanain A. J. Gharban, and M. Divya

1 Introduction

Pico-technology is used for semiconductor manufacturing, quantum computing, nanophotonic, biomedical applications, and energy systems. These techniques Atomic Layer Deposition (ALD), Molecular Beam Epitaxy (MBE), Electron Beam Lithography (EBL), Focused Ion Beam (FIB) milling, and Scanning Tunnelling Microscopy (STM) enable the control of atomic-scale structures and material properties (Dragoman & Dragoman, 2004). This technique allows for developing ultra-dense transistors, high-efficiency quantum dots, nanoscale biosensors, and high-speed optoelectronic devices. Manufacturing pico-scale devices includes predicting material behaviors, optimizing manufacturing parameters, and defect detection. The ability to manipulate the atomic and sub-nanometre scale has many advancements across multiple scientific and industrial sectors. Techniques such as Atomic Layer Deposition (ALD), Molecular Beam Epitaxy (MBE), Electron Beam Lithography (EBL), Focused Ion Beam (FIB) milling, and Scanning Tunnelling Microscopy (STM) allow for precise control over atomic-scale structures and material properties, enabling ultra-high precision in fabrication processes. These techniques have facilitated the development of ultra-dense transistors, high-efficiency quantum dots, nanoscale biosensors, and high-speed optoelectronic devices.

Manufacturing pico-scale devices plays a critical role in predicting material behaviors, optimizing manufacturing parameters, and enhancing defect detection

P. Saranraj (✉) · B. Lokeshwari · K. Gayathri · M. Divya
PG and Research, Department of Microbiology, Sacred Heart College (Autonomous), Tirupattur, Tamil Nadu, India

H. A. J. Gharban
Department of Internal and Preventive Veterinary Medicine, College of Veterinary Medicine, University of Wasit, Wasit, Iraq

S. Kulkarni (ed.), *A Journey from Nanotechnology to Picotechnology*, Lecture Notes in Nanoscale Science and Technology 47,
https://doi.org/10.1007/978-3-032-07228-3_9

methods. Advanced computational modeling and artificial intelligence (AI)-driven analytics have further improved the efficiency and precision of pico-fabrication, allowing for better process optimization and defect minimization. One of the most transformative applications of pico-technology fabrication is in quantum computing, solid-state materials can be detected with atomic precision. This leads to enhanced stability, improved quantum-gate coherence, and superconducting circuits with controlled components for scalable quantum processors. The ability to precisely manipulate materials at the atomic level ensures the development of robust and error-resistant quantum computing hardware, advancing quantum processing.

Nanoparticles fabricated with pico-scale precision are used to revolutionize the field of nanophotonics by enabling ultrafast optical switches, integrated photonic circuits, and high-efficiency light-emitting devices. These advancements are critical for the future of optical and quantum computing; precise control of light at the atomic level is necessary for data transmission and information processing. Integrating nanophotonic components with quantum systems could lead to quantum communication, cryptography, and secure data transfer. Pico-fabrication has wide biomedical applications, including targeted drug delivery, ultra-sensitive biosensors, advanced diagnostics, tissue engineering, and neural interfaces, making it a technology in modern healthcare innovations. Ultra-sensitive biosensors can detect individual molecules being developed for early disease detection, significantly improving the diagnostic accuracy of many patients. Individualized drug delivery systems, designed at the Pico-scale, enable localized therapeutics that minimize systemic side effects while maximizing treatment efficacy. Single-molecule imaging technologies are made possible through advanced atomic-scale fabrication techniques (Pandey et al., 2006).

Pico-fabrication also plays a crucial role in various materials' aerospace and defense sectors, and it has exceptional strength-to-weight ratios that are essential for high-performance applications (Vollertsen, 2008). Ultra-lightweight, strong materials at the atomic level are used in spacecraft, advanced military aircraft, and protective gear. Atomic-scale sensors are being integrated into technology, enabling sophisticated navigation and communication with extreme precision and efficiency. These innovations enhance national security capabilities that reduce the operational costs associated with traditional defense technologies (Maassen & Lundstrom, 2015). Next-generation photovoltaic and battery materials engineered at the atomic level enable ultra-efficient solar energy, higher energy density, faster charging, and longer lifespan. Advanced battery materials, designed at the atomic level, are used for higher energy densities, faster charging, and extended lifespans. Pico-technology enables superconductors to transmit energy efficiently and supports sustainable, high-performance materials for energy storage and generation.

Developing hybrid materials that combine the benefits of Pico-scale engineering with affordability and sustainability is crucial for making these technologies commercially viable. High-volume fabrication methods ensure Pico-fabrication processes (Bayram et al., 2010). It integrates pico-fabrication like neuromorphic computing, molecular electronics, and bio-nanoscience with its technological advancements. This progress will revolutionize device design, manufacturing, and applications across industries within innovation's boundaries. The atomic-scale material manipulation and pico-fabrication enhance technologies that pave the way for future advancements in computing, sensing, and biomedical solutions. These developments will lead to more efficient and adaptive technologies, transforming the modern technology design, production, and applications of devices in the contemporary world progress.

2 Pico-technology Fabrication

Pico-technology, which works at the scale of Pico-meters (10^-12 meters), is the next advancement in nanotechnology and material science. Nanotechnology manipulates structures in the nanoscale (10^-9 m) range, and pico-technology refers to the atomic and subatomic levels. Pico-technology refers to the fabrication of the atomic and subatomic levels to assemble structures with a specific function and focuses on the principles of quantum mechanics, tolerancing, metrology in material science, lithography approaches, and computation (Lindsey & Bocian, 2011). Many fabrication techniques can create materials with pico-scale precision, like ALD that are used in the formation of the ultra-thin film, Scanning Tunnelling Microscope (STM)-assisted assembly in which position atoms exactly, EBL Sculpting that allows for high-resolution structuring of layers and defect-free lattice growth for complex atomic patterns or IB (Ion Beams) that would enable precision nanosculpting).

The range from ultra-high-density electronics, quantum dot-based transistors, and future generations are constructed to quantum computing. Ast atomic precision in qubit fabrication enables rapid calculations and advanced sensors to measure biological, chemical, and physical changes (Kim et al., 2009). It also allows drug delivery systems, particular drug action, energy systems in the form of enhanced energy storage, and atomically engineered materials to optimize electron transport efficiency, and for that next generation of photovoltaic system technology. Pico-technology manufacturing has been technically complex and cost-effective, and it is conducted with materials that may not be stable and are computationally limited (Kuhr et al., 2004). The potential development of Pico-structures can include AI integration for improved pico-TSV/aromatic nanocarbon productivity, hybrid nano–pico systems to combine the benefits of nanoscale devices with room-temperature stable quantum devices, bio-Pico-interfaces for next-generation medicine applications, and Pico-fabrication techniques for sustainability benefits. The fabrication of nanomaterials encompasses various methods, including chemical, biological, and mechanical processes, as well as gas-phase and pyrogenic processing techniques (Fig. 1). These approaches, such as self-assembly, milling, and sol-gel methods, are critical for tailoring the properties of nanoparticles for specific applications.

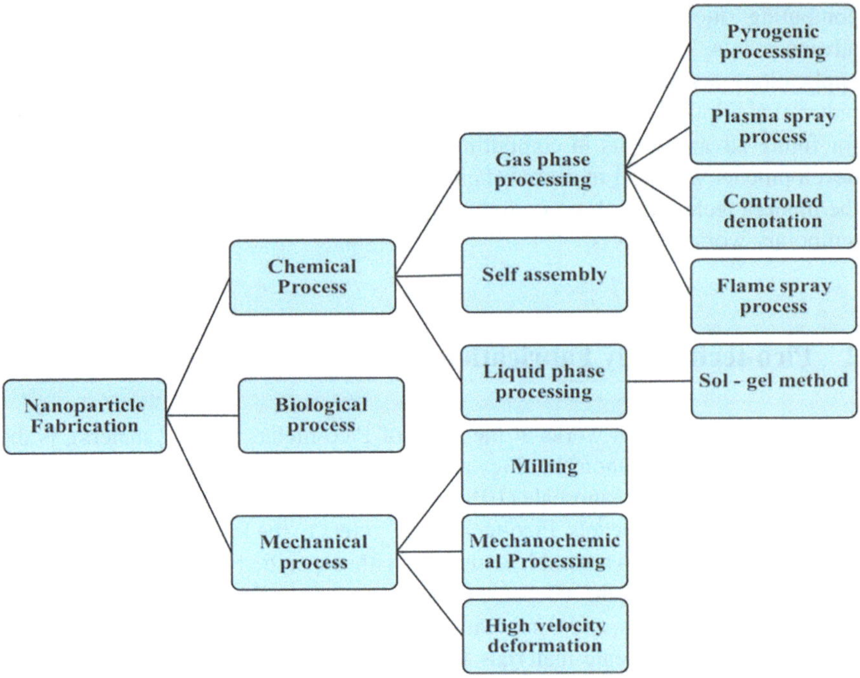

Fig. 1 Fabrication of nanomaterials

2.1 Spray Coating

Spray coating is a deposition technique used in nano-fabricating next-generation electronics, medicine, and energy materials at the atomic or subatomic level. Spray-coating pico-technology is an ultra-fine aerosolized process by picometres (10^-12 meters) and layer by layer with nuclear precision in particle size, velocity, and dispersion (Liu et al., 2003). These conventional deposition methods, such as Physical Vapor Deposition (PVD) and Chemical Vapor Deposition (CVD), reduce material waste and provide scalable production. The pico-fabrication is a fluid-dynamics modeling and AI-based control system, ensuring consistency of the film thickness or homogeneity and adhesion at the atomic level using spray-coating techniques. This process usually includes nano-engineered liquid precursors that, once momentarily atomized, lead to picometer-scale drops that interact with the surface of the substrate in a highly controlled manner (Liu, 2011).

Plasma-assisted spray coating and laser-induced spray deposition combine to make high-purity, defect-free coatings with functional properties. Spray coating is used in pico-technology fabrication and has numerous applications across various industries, such as semiconductor manufacturing, quantum computing, and biomedical engineering. In pico-scale semiconductors, spray-coating is used to produce ultra-thin conductive and insulating layers that allow us to make more enhanced transistors, memory devices, and even integrated circuits by limiting electron scattering and providing more significant charge transport efficiency (Roth et al., 2000).

Pico-materials with spray-coated structures are being used in applications such as targeted drug delivery, biocompatible implants, and antibacterial coatings in the field of biomedical engineering (Jurow et al., 2010) that precisely control the thickness and composition of the materials are essential for enhancing the therapeutic efficacy and ensures the patient safety. Pico-engineered spray coatings are applied to next-generation batteries, fuel cells, and photovoltaic systems to improve charge carrier mobility and energy retention. Advancements in AI-assisted predictive modeling, real-time control systems, and new formulations to these obstacles and the stage for future pico-scale fabrication techniques. The Pico-technology fabrication shows the combination of spray-coating with hybrid nano–pico manufacturing technologies, self-assembling pico-materials, and eco-speedy deposition technologies (Miller & Simon, 2008). The limits of precision engineering, atomic-scale manipulation, and spray-coating will enable advancements in electronics, healthcare, renewable energy, and the boundaries of material science and nanotechnology.

2.2 *Dip-coating*

Dip-coating is a high-technology fabrication technique that ultimately controls the over-film thickness and the deposited films on a subatomic level, with applications in next-generation materials, electronics, biomedical applications, and energy storage. Dip-coating at the picometer scale (10^{-12} m) involves immersing an object in a pre-prepared liquid precursor and withdrawing it at a constant speed for precise atomic level coating. This technique enables controlled β-coatings with adjustable thickness, defect-free characteristics, and the ability to regulate functional switches (Merlet et al., 2012). Dip-coating offers superior uniformity, scalability, and precision compared to traditional coating methods, making it ideal for pico-technology fabrication. It enables molecular self-assembly and controlled deposition of ultrathin films with enhanced mechanical, electrical, and chemical properties. Advanced techniques like layer-by-layer assembly, sol-gel deposition, and plasma-enhanced dip-coating achieve ultra-pure films at Nanoscale and even picometer thickness. Dip-coated films improve insulation, thermal stability, and electron mobility in transistors and quantum devices, while in biomedicine, they enhance implant biocompatibility, drug delivery systems, and lab-on-a-chip technologies (Sharma & Bhatti, 2010).

Precisely controlling its thickness and composition is crucial to improving biocompatibility and therapeutic efficacy. Dip-coating has been widely employed in energy storage and conversion applications to design the electrode and electrolyte interfaces required for next-generation batteries, supercapacitors, and fuel cells. The integration of new coating materials into existing manufacturing processes. Improvements in AI-assisted process optimization, real-time monitoring techniques, and precursor design address these issues, paving the way toward more sustainable and scalable pico-fabrication schemes (Largeot et al., 2008). Dip-coating with hybrid nano–pico manufacturing methods usher in self-healing pico-materials, eco-friendly deposition techniques, and other parameters to facilitate the sustainable evolution of pico-technology manufacturing processes. Dip-coating is a fundamental technique for formulating self-healing materials, pathogen-resistant surfaces, nanocomposite protective coatings, and many advanced nanostructured materials from organics to high-throughput measurements (Chen et al., 2007).

2.3 *Spin-coating*

Spin-coating is an exact thin-film deposition technique that provides a uniform and controlled coating of materials at the atomic and subatomic scales vital for developing next-generation electronic, biomedical, and energy storage technologies. This method involves a liquid precursor solution on top of a substrate rotating your component and distributing it homogeneously over it through centrifugal force, resulting in an ultra-thin film. Spin speed is the angular velocity of the spinning disc solution viscosity. The evaporation rate determines the film thickness and uniformity, allowing researchers to fabricate films with picometer-level precision (Chen et al., 2003). Spin-coating allows for better control of film thickness, excellent reproducibility, and fast coating of large areas compared to other deposition methods. Micro-and nano-particle pioneering Spin-coating in pico-technology fabrication. Techniques like vacuum-assisted spin-coating, multi-step spin-coating, and plasma-enhanced spin-coating significantly enhance the path abilities of the process by allowing the fabrication of high-purity films with functional properties.

Spin-coating techniques are employed in current pico-technology fabrication through optoelectronic and nanomedicine applications (Madani et al., 2017). Essential for optoelectronics spanning from organic light-emitting diodes (OLEDs) to perovskite solar cells and nano-photonic structures, spin-coating enables precise and uniform coatings that can be critical for the performance and lifespan of optoelectronic devices. Pico-material spin-coating is used in the biocompatible coating of implantable devices, bioavailability enhancement through nano-fabrication by drug delivery systems, and lab-on-a-chip application in nanomedicine, enhancing bioactivity, and controlled release. Spin-coating is also key in fabricating advanced systems such as batteries, supercapacitors, and fuel cells based on fundamental principles to ensure efficient charge transport and energy storage at the materials and device level. Spin-coating is a more reliable and scalable fabrication process (Kumar, 2007). Spin-coating is the key technique that provides the new materials with enhanced durability and superior functionality.

2.4 Sol-Gel Method

The sol-gel process in pico-technology fabrication is a sophisticated chemical synthesis method that facilitates the exact design of nanostructured and Picometer-scale materials. It also enables fabricating ultra-thin films, nanoparticles, and porous structures with atomic level precision and high purity (Mayor et al., 2003). The sol-gel process allows for fine control over the composition, structure, and morphology of materials at the atomic level, which is one of the core benefits of pico-technology manufacturing, facilitating the design of advanced materials with improved optical, electrical, and mechanical properties. The sol-gel method is commonly utilized in pico-technology because it allows the synthesis of various inorganic chips through chemical engineering. Sol-gel in and its ultra-thin films are used as dielectric layers, insulating coatings, and gate oxides in next-generation transistors, quantum computing devices, and nanoelectronics circuits. Atomic-scale precision is crucial to minimize electron scattering and optimize device performance in semiconductor fabrication.

In the field of photonics and optoelectronics, sol-gel-processed materials have been applied for the preparation of advanced coatings on the surface of waveguides, optical fibers, and high-efficiency solar cells to improve light absorption, emission, and energy conversion performance (Garner et al., 2018). Pico-materials are derived from sol-gel processes and find considerable utility in biomedicine, particularly in therapeutic applications of drug delivery systems, bioactive coatings, and tissue engineering scaffolds in controlled porosity and biocompatibility characteristics. Sol-gel-derived materials are used to synthesize high-performance batteries, supercapacitors, and fuel cells, enhancing electrode stability, charge transport properties, and catalytic activity. The development of AI-assisted process optimization, in situ monitoring techniques, and innovative formulations that are more reliable and scalable fabrication techniques (Meunier & Quirke, 2000).

2.5 Vacuum Filtration

Vacuum filtration is a technique for material separation, purification, and deposition at atomic and subatomic scales and is pivotal for applications across electronics, biomedicine, energy storage, and nanomaterial synthesis. Vacuum filtration is one of the main techniques for pico-technology fabrication that can purify nano-and pico-scale materials, including carbon nanotubes, graphene, and quantum dots, by easily removing residual solvents, unreacted precursors, or undesirable aggregates. Vacuum filtration is integral to pico-technology fabrication for several industry domains such as semiconductor, optoelectronic, and biomedical engineering. The vacuum filter purifies and deposits conductive nanomaterials, including single-walled carbon nanotubes and grapheme-based inks, to fabricate future transistors, sensors, and flexible electronic circuits. That performance is highly dependent on the atomic-scale uniformity and minimal defects. In optoelectronics, vacuum filtration has been applied to fabricate transparent conductive films, thin-film coatings, and nano-photonic structures capable of improving light absorption, emission, and energy conversion efficiency.

Vacuum filtration is used for purifying biomaterials, providing thin bioactive coatings, and creating nanostructured drug delivery systems, and precise material separation and deposition are crucial for enhancing therapeutic activity and biocompatibility (Tao, 2006). Vacuum filtration produces high-performance electrodes, membranes, and catalytic layers for batteries, supercapacitors, and fuel cells to maximize charge transport, ionic conductivity, and energy efficiency. Advances in AI-assisted process optimization, real-time monitoring techniques, and novel membrane materials are helping to overcome these challenges. The progress in Pico-technology fabrication combines vacuum filtration with hybrid nano–pico manufacturing, the development of self-assembling pico-materials, and the study of eco-friendly and energy-efficient filtration methods. This vacuum filtration process is used to design new materials with optimal properties, durability, and functionality at the atomic scale in fields ranging from nanotechnology to material science (Salahuddin et al., 2005).

3 Materials for Pico-technology

The atomic-scale uniformity, low defect density, high conductivity, and functional properties. The Pico-technology fabrication includes the materials of ultra-high-purity silicon, graphene, carbon nanotubes, Transition Metal Dichalcogenides (TMDs), and nanocomposites. Ultra-pure silicon is in pico-electronic applications that utilize atomic level structuring for ultrafast transistors or high-efficiency integrated circuits. Graphene or carbon nanotubes have extraordinary electrical and mechanical properties that can be used for nanoelectronics, quantum computing, or high-performance coatings. They are treated at the pico-scale with methods of fabrication Atomic Layer Deposition (ALD), Molecular Beam Epitaxy (MBE), Scanning Tunnelling Microscopy (STM)-assisted assembly, and Electron Beam Lithography (EBL) (Okazaki et al., 2004). Recent advancements in AI-driven material discovery capabilities, self-assembling atomic features, and sustainable fabrication methods are alleviating such barriers to engineering scalable and affordable pico-manufacturing techniques. Nanoparticles, quantum dots, quantum wires, nanoribbons, and other materials are developed by new synthetic methods involving both chemical and physical properties, and the quantum is extensive in the next generation nano-computing that includes controlled synthesis, hybrid nano–pico material to function as nano–pico technology fabrication (Malekian et al., 2009).

3.1 2D Materials

The electronic, optical, and mechanical properties at an atomic level with the 2D materials progress not only in the third generation of electronics but also in quantum computing, optoelectronics, and energy storage, among many others. 2D materials can be used as sheets one or a few atoms thick. Commonly used 2D-materials in pico-technology production parts include graphene, the thin film transition Metal Dichalcogenides (TMDs) with transition metals such as MoS_2 and WS_2, hexagonal Boron nitride (h-BN), and black phosphorus. Due to its high electrical conductivity, mechanical strength, and high carrier mobility, graphene is used in ultrafast transistors, nanoscale sensors, and high-performance coatings. The chemical stability and insulation properties of hexagonal Boron nitride (h-BN) (also called White graphene) have been widely used in pico-electronic applications as a protective coating and dielectric layer (Yan et al., 2014).

Extensive methods for the Pico-scale fabrication of large 2D materials include Chemical Vapor Deposition (CVD), mechanical exfoliation, liquid-phase exfoliation, Molecular Beam Epitaxy (MBE), and Atomic Layer Deposition (ALD). Liquid-phase exfoliation offers a scalable route for the production of solution-processed 2D materials and, therefore, its integration into flexible electronics and printable nanodevices. Atomic-scale deposition on the molecular beam enables layer-by-layer growth with nuclear precision to attain highly defined heterostructures and quantum materials. Pico-scale electronic devices are used for atomic layer deposition, which provides reasonable control of the thickness of film and composition (Asad, 2007). 2D materials will continue to be a fundamental building block for realizing atomic-scale ultra-miniaturized and high-performance technologies that impact a broad range of scientific and industrial applications.

3.2 Nanostructured Metals

Nanostructured metals use electrical, mechanical, and catalytic nanograms on atoms, and subatomic level interaction at the nanometre scale is central to nanoelectronics, plasmonics, energy storage, biomedical devices, etc. Some nanostructured metals have unique properties at the picometer scale, including quantum effects and large surface area-to-volume ratios, and they are essential for the industrial synthesis of next-generation devices: gold, silver, platinum, copper, and aluminum. The bulk plasmonic behavior of the gold and silver nanoparticles is beneficial in pico-fabrication because light–matter interaction is enhanced, which is essential for nano-optoelectronic devices, biosensing, and imaging applications with improved resolution. The low gratings of platinum and palladium nanostructures have high surface activity and excellent stability and use their catalytic materials in their applications. They have extensive applications in energy conversion and long-term energy storage systems, such as hydrogen fuel cells and supercapacitors (Huo,

2017). Copper and aluminum are used for electronic and transparent conductive electrodes because of their high electrical conductivity and mechanical flexibility (Tillmann, 2000). Pico-scale manipulation of nanostructured metals has been facilitated by atomically thick film fabrication techniques such as Atomic Layer Deposition (ALD), Electron Beam Lithography (EBL), Molecular Beam Epitaxy (MBE), Electrochemical deposition and laser ablation. Metal films can be deposited with atomic layer deposition, where the thickness and local uniformity can be tuned and controlled. Only this technique allows for scaling the metal nanowires and nanoclusters with their surface functions to promote their application in flexible electronic devices and nanoscale sensors (Aramcharoen, 2008).

3.3 High-Purity Semiconductors

The atomic level of these high-purity semiconductors has given rise to their use in electronic, optoelectronic, and even quantum devices pico-technology fabrication scale forming ever-increasing generations of devices taking the next step in performance and reliability (Dornfeld et al., 2006). The most commonly used high-purity semiconductor materials for pico-fabrication are Silicon, Gallium arsenide (GaAs), and Indium phosphide (InP), along with new-generation Two-dimensional materials (TMD). Some pico-scale fabrication techniques of high-purity semiconductors include Molecular Beam Epitaxy (MBE), Atomic Layer Deposition (ALD), Chemical Vapor Deposition (CVD), Photolithography, and Electron Beam Lithography (EBL) (Aurich et al., 2012).

Atomic layer deposition (ALD) is another side, a proven thin film deposition approach that provides excellent control of the deposition process, allowing the growth of nanofilms with exceptional control over nanoelectronics and quantum devices. Chemical vapor deposition (CVD) is a widely researched approach to developing graphene and TMDs of high purity. It will enable the large-scale fabrication of these semiconductors with excellent electronic characteristics. New tools and self-assembling semiconductors could help with the speed and direct cost-effectiveness of AI-assisted material design (Oliaei & Karpat, 2014).

3.4 Superconducting Materials

Superconducting materials found widespread applications in pico-technology fabrication, next-generation quantum computing, ultra-sensitive sensors, and high-efficiency energy applications due to their unique characteristics of zero electrical resistance below the cryogenics temperature (Ding, 2008). Pico-fabrication uses the superconducting materials Niobium (Nb), Yttrium barium copper oxide (YBCO),

Magnesium diboride (MgB_2), and Iron-based superconductors, which provide different advantages according to the application (Oliaei & Karpat, 2016). Niobium is an ordinary superconductor used in SQUIDs with its alloy, Niobium-titanium (NbTi) or Niobium nitride (NbN) or high superconducting transition temperature with good mechanical stability. Several materials, such as superconductor films, can be processed using techniques such as Molecular Beam Epitaxy (MBE), Pulsed Laser Deposition (PLD), Atomic Layer Deposition (ALD), Focused Ion Beam (FIB), and Electron Beam Lithography (EBL). It also enables atomic-scale, layer-by-layer epitaxial growth of ultra-thin superconducting films. Recently, there has been tremendous interest in MBE for application in quantum circuits and hybrid superconducting-semiconducting systems. These nanoscale superconducting materials enable the realization of ultra-miniaturized superconducting devices and nanoscale quantum circuits. Electron beam lithography enables the structuring of superconducting films down to sub-nanometre resolution. Pico-technology enables the synthesis, fabrication, and characterization of superconductor materials with precise atomic control for enhanced performance and efficiency. The synergy between superconducting materials and hybrid nano–pico fabrication technologies, robust topological superconductor fault-tolerant quantum computation investigations, and energy-efficient cryogenic syntheses will accelerate these technologies. Quantum computers, medical imaging systems, energy storage devices, and ultra-sensitive detectors are the technologies advanced by pico-scale innovations (Weule et al., 2001).

4 Design Principles for Picometer-scale Devices

The atomic manipulation, quantum mechanical phenomena, and substance for their enhanced execution and usefulness of the nanomaterials are the configuration standards for picometer-scale gadgets (Uriarte, 2007). Fabrication approaches such as Atomic Layer Deposition (ALD), Molecular Beam Epitaxy (MBE), and Scanning Tunnelling Microscopy (STM) need to be maintained at an atomic level precision to enable controlled bloom, defect minimization, and structural integrity (Altan et al., 2001). Datasets include 2D materials such as Graphene and Transition metal dichalcogenides (TMDs), and the selection of materials is critical for obtaining high electron mobility, low power dissipation, and tunable band gaps (Tansel, 2000). Picometer-scale devices are combined with quantum dots and superconducting materials, bringing numerous advanced functionalities such as ultra-sensitive sensing, quantum computing, stable quantum linguistics, and high-performance optoelectronic applications. Both surface energy minimization and interface are critical to the maintenance of structural stability, the prevention of charge leakage, and the optimization of thermal conductivity. Picometer scale materials for quantum computers, sensors, and biomedical devices (Sumiya et al., 2012)

4.1 Micro-optofluidic Device Design

Micro-optofluidic devices are designed and fabricated by integrating this technology with pico-technology-assisted techniques in optics, nanoscale biomedical engineering, and fluids for the formation of complex and highly integrated lab-on-a-chip systems (Zhan, 2015). Fabrication techniques such as Atomic Layer Deposition (ALD), Focused Ion Beam (FIB) Milling, and Electron Beam Lithography (EBL) facilitate ultra-smooth optical interfaces, nanoscale fluidic channels, and photonic structures (Kim et al., 2006). Materials include graphene, silicon photonics, plasmonic nanostructures with enhanced optical, fluidic properties for single molecule and quantum photonics, and real-time bioanalytical applications (Kim, 2004). The SAMs and functionalized surfaces demonstrate improvements in fluidic flow control, reduction in biofouling, and increased chemical specificity of diagnostics. Using Microelectromechanical Systems (MEMS) and nanophotonic waveguides allows for precise light-matter interactions, enabling compact and energy-efficient optofluidic circuits. Such micro-optofluidic devices have applications in biomedical diagnostics, environmental monitoring, and next-generation optical computing (Rathod et al., 2017) fabrication for ultra-sensitive, high-resolution, and photonic-fluidic applications in advanced science and industrial applications.

4.2 Micro-optofluidic Device Fabrication Processes

Pico-technology fabricates micro-optofluidic devices based on its potentially advanced techniques to integrate optical and fluidic functionalities with atomic precision (Ghoshal & Bhattacharyya, 2013). The methods used in these processes are Atomic Layer Deposition (ALD), Electron Beam Lithography (EBL), Focused Ion Beam (FIB) milling, and molecular self-assembly. ALD allows for atomic layer precision deposition, microfluidics alteration, and surface nanophotonics engineering. Nanophotonic structures such as gratings, waveguides, and plasmonic interfaces are appropriate for light manipulation in micro-optofluidic systems fabrication in Polydimethylsiloxane (PDMS) by EBL and milling. Soft and Nanoimprint lithography (NIL) also allow for high-resolution microfluidic networks with sub-nanometer precision for controlled fluid transport and molecular interactions (Cheng, 2009). The surface can be modified via Self-Assembled Monolayers (SAMs) and functionalized nanoparticles that improve the material for biocompatibility, chemical specificity, and fluidic dynamics. Thus, these fabrication processes will lead to the introduction of compact high-performance optofluidic devices for various practical applications, such as biomedical diagnostics, environmental sensing, and quantum photonics, and present the significant role that pico-technology provides in the future development of the next-generation microfluidic and optical systems (Liu et al., 2012).

5 Fabrication Techniques in Pico-technology

Pico-technology fabrications are the techniques to improve industrial applications. Techniques such as Atomic Layer Deposition (ALD), Molecular Beam Epitaxy (MBE), Electron Beam Lithography (EBL), Focused Ion Beam (FIB) Milling, and Scanning Tunnelling Microscopy (STM) allow for precise atomic manipulation of matter at a picometer scale (Cheng, 2010). Atomic Layer Deposition (ALD) shows the uniform ultra-thin films critical for semiconductors, nanoelectronics, and high-efficiency coatings. Molecular Beam Epitaxy (MBE) allows the growth of crystalline structures one layer at a time, with precise control over doping, which is essential for quantum devices and optoelectronics. Nanoscale patterning and structuring of materials with EBL and FIB shows the accurate manufacturing of integrated circuits, and nanophotonic and ultra-sensitive sensors (Uriarte, 2006). STM affords atomic-resolution imaging and manipulation. Direct surface fabrication at the molecular scale enables quantum computing and molecular electronics, while SAMs and molecular self-assembly allow precise surface modification for biomedical and sensing applications. Soft Lithography and Nanoimprint Lithography (NIL) have proven scalable, high-throughput patterning tools for microfluidic and nano-optoelectronic systems (Maloney, 2012). The next-generation atomic-scale devices with enhanced efficiency, precision, and functionality involve Pico-technology.

6 Characterization and Measurement of Picometer-scale Structures

Imaging and measuring picometer scale structures requires ultra-high-precision methods to detect atomic-scale features and determine material properties. These include microscopy, spectroscopy, diffraction, and metrology techniques. Ultra-precise analytical methods that can resolve structures at the atomic scale and quantify properties at the nanoscale are fundamental to the characterization and measurement of picometer-scale structures used in pico-technology fabrication. Experimental techniques such as Scanning Probe Microscopy (SPM) have provided atomic-scale imaging and mapping of surfaces with picometer precision in materials through methods like Scanning Tunneling Microscopy (STM) and Atomic Force Microscopy (AFM). The Types of Building Blocks and Electronic properties are directly visualized (Fonda, 2013). The vast potential of the advanced forms of high-resolution transmission electron microscopy (HR-TEM) and aberration-corrected electron microscopy allows for sub-angstrom resolution imaging of local atomic arrangements, defects, and interfaces of advanced materials (Rossetti et al., 2010). X-ray diffraction (XRD) and higher X-ray energy synchrotron-based techniques, such as X-ray absorption spectroscopy (XAS) and small-angle X-ray scattering (SAXS), can be deployed to provide sensitive measurements of the crystal structure and nanoscale phase transitions.

6.1 Scanning Probe Microscopy

Scanning Probe Microscopy (SPM) is widely used in Pico-technology fabrication, enabling the atomic-scale characterization and manipulation of materials with picometer-level precision. SPM techniques, including Scanning Tunneling Microscopy (STM) and Atomic Force Microscopy (AFM), allow for the direct visualization and modification of surfaces at the atomic level, making them essential for developing ultra-miniaturized devices. STM measures Quantum Tunneling between a conductive tip and a sample, providing resolution imaging and facilitating precise atomic manipulation, which is critical for quantum materials and molecular electronics (Scheidemantel et al., 2003). AFM employs a nanoscale probe to scan surfaces, measuring interatomic forces to identify the topography, mechanical properties, and electrical conductivity at picometer accuracy. These techniques fabricate nanostructures and optimize semiconductor surfaces, quantum dots, and single-atom transistors. The integration of the SPM with lithographic processes that have the atomic-scale patterns, defect analysis, and surface engineering of nano-optoelectronics, quantum computing, and biomedical nanodevices.

6.2 High-Resolution Transmission Electron Microscopy (HR-TEM)

High-Resolution Transmission Electron Microscopy (HR-TEM) is a characterization approach used in Pico-technology fabrication to image and analyze atomic-scale structures at sub-angstrom resolution (Koswatta et al., 2007). High-resolution transmission Electron Microscopy (HR-TEM) uses a highly focused beam of electrons through an ultra-thin specimen, yielding high-contrast images that can have atomic arrangements, crystallographic defects, and properties. This technique is critically important to the material's structural integrity in the picometer scale, quantum dots, single-atom transistors, and films for semiconductors and nano-optoelectronics. HR-TEM is a powerful technique for imaging individual atoms and nanoscale defects. HR-TEM is combined with Electron Energy Loss Spectroscopy (EELS) and Energy-Dispersive X-ray Spectroscopy (EDS) for the chemical composition and electronic property analysis. These are necessary to tune material properties in generation transistors, high-efficiency photovoltaics, and quantum computing (Whitesides & Boncheva, 2002). HR-TEM enables real-time observation of dynamic processes, including atomic diffusion, phase transitions, and nanomaterial growth relevant to pico-fabrication. HR-TEM is used to develop pico-technology by providing precision methods for fabricating high-performance devices.

6.3 *Interferometry*

Optical measurement techniques such as interferometry are widely used in picotechnology fabrication for surface, thin film, and nanoscale structure characterization from sub-nanometer to picometer even picometer-level precision (Fleischer, 2008). The coherent light interference patterns allow for ultra-sensitive detection of the roughness of the inner surface and thickness of the layer as well as atomic-scale deformations in semiconductor and Nanophotonics devices (Vericat et al., 2010). In semiconductor fabrication, high-precision lithography, defect detection, and nanoscale patterns are established through Optical Coherence Tomography (OCT)-based interferometry and Extreme Ultraviolet (EUV) Interferometry. X-ray interferometry determines atomic-scale material properties and probes strain distributions, lattice spacing, and crystallographic defects at the nanoscale (Kushmerick, 2009). In Pico-technology applications, methods for applying interferometry techniques are also important, focusing on quantum computing, molecular electronics, and integration nanofabrication processes, allowing the real-time monitoring of fabrication accuracy and scalability. Ultra-high precision with strong structural integrity in pico-technology enables atomic-scale device fabrication, photonic integration, and advanced nanomaterial development.

6.4 *Quantum Metrology*

Quantum metrology is a Pico-technology fabrication that takes advantage of the ultra-precise measurement at the atomic and subatomic scales. It allows the characterization of materials and devices with accuracy beyond and is also critical to picometer-scale structure fabrication in semiconductor manufacturing, quantum computing, and nanophotonic (Ellenbogen & Love, 2000). Some techniques for atomic interferometers, quantum-enhanced spectroscopy, and single-electron transport metrology measure ultra-small displacements, electromagnetic fields, and interactions to high precision. Quantum metrology is also essential for Pico-scale materials characterization and detection of nanomaterials' electronic and photonic interactions (Yu et al., 2003). The atomic-scale fabrication processes in precision engineering allow the next generation of high-performance devices in computing, sensing, and energy applications.

7 Applications of Pico-technology Fabrication

The pico-technology mechanism depends on the synthesis method to manipulate elements at the atomic state, enabling cutting-edge fabrication techniques for semiconductors, quantum computers, nano-optics, bioengineering, and energy storage devices. In semiconductor fabrication, Pico-scale techniques such as Atomic Layer Deposition (ALD), Molecular Beam Epitaxy (MBE), and Electron Beam Lithography (EBL) allow ultra-dense transistors, single-atom logic gates, and high-speed optoelectronic devices that promote the development of the next-generation integrated circuits (Liu, 2018). In quantum computing, Pico-fabrication provides engineering at atomic-scale precision, ensures coherence, allows reducing coherence effects, and helps the building of stable quantum and superconducting circuits. In nanophotonics, picoscale precision enables the fabrication of high-efficiency quantum dots, plasmonic nanostructures, and ultra-fast optical switches, facilitating advancements in photonic computing and next-generation systems (Cox 2004). Biomedicine utilizes pico-fabrication methods for early-stage biosensors focused on highly sensitive disease detection, the controlled delivery of drugs at the cellular level, and atomic-scale imaging tools to facilitate molecular diagnostics (Golshan, 2017). In energy storage and conversion, Pico-fabrication materials fabricated by photovoltaics, next-generation battery electrodes, and superconductors significantly increase energy efficiency and enable lossless energy transmission. Ultra-lightweight, high-strength materials and atomic-scale sensors enhance performance and precision in advanced technological applications (Suzuki, 2013). The applications of picotechnology fabrication span diverse fields, including microfabrication, quantum technology, and nano-biomedicine, as highlighted in Fig. 2. These advancements enable breakthroughs in nano-materials and sensor technologies, driving innovation across scientific and industrial domains.

7.1 Microfabrication

Microfabrication in pico-technology allows for structuring materials at the micro- and nanoscale, which is crucial for fabricating next-generation semiconductor devices and advanced sensors (Ghosh et al., 2004). The art of lithography, including photolithography, Electron Beam Lithography (EBL), Focused Ion Beam (FIB) milling, and nanoimprint lithography, allows for patterns at the atomic scale. In pico-scale fabrication, ultra-thin films of precisely controlled thickness and superior material properties are achieved with ALD and MBE. Such methods are necessary to fabricate ultra-dense transistors, quantum processors, and high-speed optoelectronic components. Microfabrication is also used in biomedical engineering, allowing the creation of lab-on-a-chip devices, implantable biosensors, and drug delivery systems that function at a molecular scale. In nanophotonic, microfabrication methods are used to fabricate plasmonic structures, high-efficiency light-emitting diodes, and quantum dot arrays for next-generation optical computing. Microfabrication plays a critical role in energy storage and conversion by improving the performance of nanobattery materials, photo-voltaic cells, and superconductive films suitable for efficient energy applications.

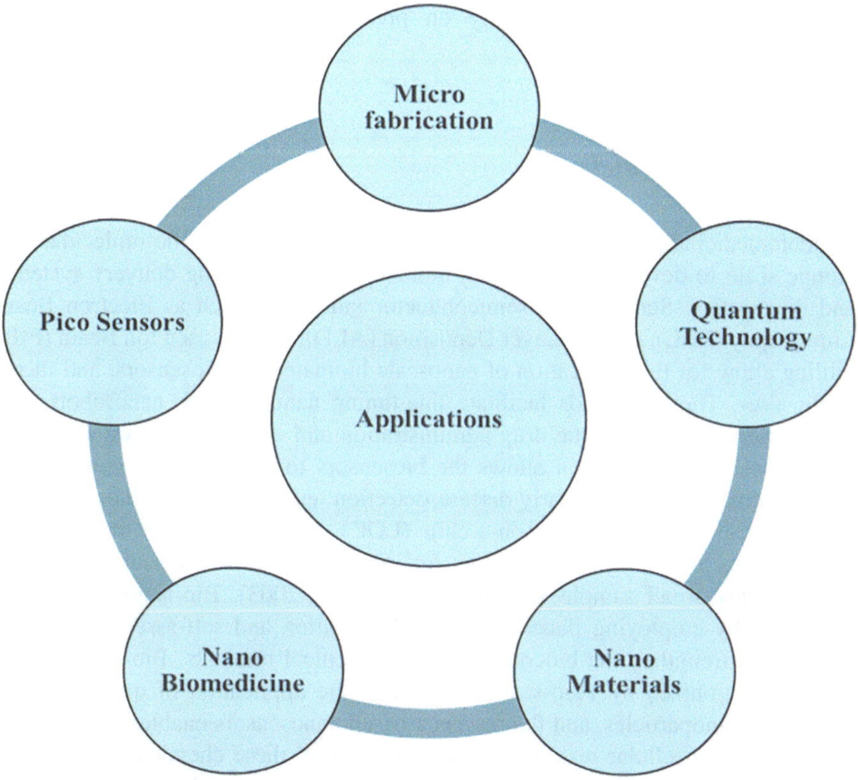

Fig. 2 Applications of pico-technology fabrication

7.2 *Quantum Technology*

Quantum technology uses pico-technology fabrication techniques at atomic and subatomic scales to manipulate and build devices with classical or quantum states. Quantum computing, quantum sensors, and quantum communication systems require ultra-precise engineering at the picometer size scale that can be manipulated through Atomic Layer Deposition (ALD), Molecular Beam Epitaxy (MBE), Electron Beam Lithography (EBL), and Focused Ion Beam (FIB) Technologies. These techniques enable stable quantum processors with high coherence, semiconductor quantum dots, and photonic quantum circuits. Single-photon sources, photons, and Quantum Key Distribution (QKD) systems (Jin & Solomon, 2018) require pico-scale fabrication. The photonic structures and plasmonic atomic-scale precision boost light–matter interactions for quantum processing. Pico-fabrication helps apply quantum metrology that allows ultra-precise measurement based on quantum advantage in atomic clocks, gravity sensors, and high-resolution imaging systems. The electronic and photonic properties of topological insulators, graphene-based heterostructures, and nitrogen rely on precise Pico-scale fabrication (Flood et al., 2004).

7.3 Nanobiomedicine

Nanobiomedicine possesses the precision and functionality at the molecular and atomic scale to develop sophisticated biomedical devices, drug delivery systems, and diagnostics. Some of the semiconductor industries, such as Electron Beam Lithography (EBL), Atomic Layer Deposition (ALD), and Focused Ion Beam (FIB) milling allow for the fabrication of nanoscale biomaterials, biosensors, and therapeutic uses. These methods facilitate fine-tuning nanoparticles, nanorobots, and nanocarriers for site-specific drug administration and enhanced bioavailability. In diagnostics, Pico-fabrication allows the biosensors to easily detect biomolecules that could be employed in early disease detection and for real-time monitoring of physiological conditions. Lab-on-a-chip (LOC) devices combine the fields of microfluidics and biosensors to allow for rapid and precise medical testing that requires only small sample volumes (Long et al., 2003). Bio-nanocoatings are developed by employing plasma-enhanced deposition and self-assembly mechanisms that strengthen the biocompatibility of medical implants. Bio-imaging has been revolutionized by Pico-scale fabrication; the application of quantum dots, plasmonic nanoparticles, and fluorescence-based nanosensors enables imaging at a cellular and subcellular resolution level, meaning all these chemicals are used in personalized medicine. Pico-technology combines AI-powered biomedical applications, gene sectors, and bio-nanophotonic for the next generation of products, regenerative medicine, and precision diagnostics.

7.4 Nanomaterials

Nanotechnology or nanomaterials on the atomic and molecular scales predict electronics, energy storage, biomedical applications, and nanophotonics advancements. These include Atomic Layer Deposition (ALD), Molecular Beam Epitaxy (MBE), Focused Ion Beam (FIB), Electron Beam Lithography (EBL), and Chemical Vapor Deposition (CVD), which enables fine control over the nanostructures. These methods allow the controlled production of nanotubes, nanowires, quantum dots, graphene-based materials, and two-dimensional (2D) materials with much-enhanced electrical, mechanical, and optical properties. Pico-scale technology can fabricate integrated circuits with ultra-thin transistors, increasing processing performance and efficiency (Papadopoulos et al., 2006). Self-assembled and nanoscale-fabricated nanomaterials developed here contribute to improved performance and lifetimes of next-generation batteries, supercapacitors, and photovoltaic cells. The properties of Nanomaterials have also led to novel biomedical applications like targeted drug delivery, biosensing, and tissue engineering, paving the way for precision medicine. Pico-scale is a significant innovative platform for optimizing the optical performance of nanomaterials in nanophotonic devices, including plasmonic resonators, metamaterials, and high-efficiency Light-Emitting Diodes (LEDs). Light and high carbon fiber nanomaterials are used in the aerospace and defined industries for enhanced tolerance and technology. Pico-fabrication methods, AI-based materials discovery, and quantum metrology improve the innovations in nanomaterials.

7.5 *Pico-Sensors*

Some methods of fabrication like Electron Beam Lithography (EBL), Atomic Layer Deposition (ALD), Focused Ion Beam (FIB) Milling, Molecular Beam Epitaxy (MBE), and Nanoprinting techniques allow for the modeling of every component within the sensor, at the picometer scale, facilitating the verification of the accuracy and efficiency of sensor hardware. They also enable the fabricated quantum sensors, plasmonic sensors, nanoelectromechanical systems (NEMS)-based sensors, and single-molecule biosensors. By allowing real-time biomonitoring, early disease diagnosis, and improved personalized medicine, Pico-sensors can help advance these existing technologies. Pico-scale fabrication techniques in semiconductor manufacturing that generate quality temperature, pressure, and chemical sensors. Pico-sensors enable the ultra-precise measurements of quantum dots, optical signals, and electromagnetic fields in quantum computing and nanophotonic to advance next-generation computing and technologies. Nano-sensors use nanostructures and atomic precision manufacturing in high-resolution imaging and can produce ultra-sensitive navigation systems. Pico-scale gas and chemical sensors enhance the performance of batteries, enable better monitoring of fuel cells, and redefine environmental sensing. AI, machine learning, and quantum dot integrations are combined with fabrication techniques in nanoscale detection systems in healthcare, computing, environmental monitoring, and industrial applications (Ahn et al., 2006).

8 Conclusion

Fabrication of pico-technology is a revolutionary shift in precision engineering that allows materials to be manipulated at the atomic level using innovative methods like Atomic Layer Deposition (ALD), Molecular Beam Epitaxy (MBE), Electron Beam Lithography (EBL), and Scanning Probe Microscopy (SPM). These technologies drive innovations in ultra-miniaturized, high-speed devices like quantum computing architectures, ultra-dense semiconductors, nanophotonic systems, and biomedical sensors. Incorporating machine learning and artificial intelligence further adds to process optimization, defect inspection, and scalability, solving material stability and cost-of-manufacturing issues. Although technical complexity and reproducibility are challenges, new hybrid fabrication methods and green material advances have the potential to overcome them. As pico-technology keeps developing, its use in computing, medicine, energy, and aerospace will revolutionize technological frontiers, propelling without precedent efficiency, functionality, and miniaturization levels. This paradigm shift recognizes pico-technology as the bridge to the future of science and the atomic-scale engineering industry.

References

Ahn, C. H., Bhattacharya, A., Di Ventra, M., Eckstein, J. N., Frisbie, C. D., & Gershenson, M. E. (2006). Electrostatic modification of novel materials. *Reviews of Modern Physics, 78*(4), 1185–1212.

Altan, T., Lilly, B., & Yen, Y. (2001). Manufacturing of dies and molds. *CIRP Annals–Manufacturing Technology, 50*(2), 404–422.

Aramcharoen, A. (2008). Evaluation and selection of hard coatings for micro-milling of hardened tool steel. *International Journal of Machine Tools and Manufacture, 48*(14), 1578–1584.

Asad, A. (2007). Tool-based micro-machining. *Journal of Materials Processing Technology, 192*(1–3), 204–211.

Aurich, J. C., Reichenbach, I. G., & Schüler, G. M. (2012). Ultrasmall micro end mills are manufactured and applied. *CIRP Annals—Manufacturing Technology, 61*(1), 83–86.

Bayram, C., Vashaei, Z., & Razeghi, M. (2010). AlN/GaN double-barrier resonant tunneling diodes grown by metal-organic chemical vapor deposition. *Applied Physics Letters, 96*(4), 2–5.

Chen, Y., Jung, G. Y., Ohlberg, D. A. A., Li, X., Stewart, D. R., & Jeppesen, J. O. (2003). Nanoscale molecular-switch crossbar circuits. *Nanotechnology, 14*(4), 462–468.

Chen, Z., Lee, B., Sarkar, S., Gowda, S., & Misra, V. (2007). A molecular memory device formed by HfO_2 encapsulation of redox-active molecules. *Applied Physics Letters, 91*(17), 1–4.

Cheng, X. (2009). Design and development of a micro polycrystalline diamond ball end mill for micro/nano freeform machining of hard and brittle materials. *Journal of Micromechanics and Microengineering, 19*(11), 115022.

Cheng, X. (2010). Design and development of PCD micro straight edge end mills for micro/nano machining of hard and brittle materials. *Journal of Mechanical Science and Technology, 24*(11), 2261–2268.

Ding, X. (2008). Fabrication of a micro-size diamond tool using a focused ion beam. *Journal of Micromechanics and Microengineering, 18*(7), 075017.

Dornfeld, D., Min, S., & Takeuchi, Y. (2006). Recent advances in mechanical micromachining. *CIRP Annals–Manufacturing Technology, 55*(2), 745–768.

Dragoman, D., & Dragoman, M. (2004). Terahertz oscillations in semi-conducting carbon nanotube resonant-tunneling diodes. *Physica E: Low-dimensional Systems and Nanostructures, 24*(3–4), 282–289.

Ellenbogen, J. C., & Love, J. C. (2000). Architectures for molecular electronic computers: 1. Logic structures and an adder designed from molecular electronic diodes. *Proceedings of the IEEE, 88*(3), 386–426.

Fleischer, J. (2008). Design and manufacturing of micro milling tools. *Microsystem Technologies, 14*(9–11), 1771–1775.

Flood, A. H., Stoddart, J. F., Steuerman, D. W., & Heath, J. R. (2004). Whence molecular electronics? *Science, 306*(306), 2055–2056.

Fonda, P. (2013). A study on the optimal fabrication method for micro-scale gyroscopes using a hybrid process of electric discharge machining, chemical etching, or micro-mechanical milling. *CIRP Annals—Manufacturing Technology, 62*(1), 183–186.
Garner, M. H., Li, H., Chen, Y., Su, T. A., Shangguan, Z., & Paley, D. W. (2018). Comprehensive suppression of single-molecule conductance using destructive interference. *Nature, 558*(7710), 416–419.
Ghosh, A. W., Rakshit, T., & Datta, S. (2004). Gating of a molecular transistor: Electrostatic and conformational. *Nano Letters, 4*(4), 565–568.
Ghoshal, B., & Bhattacharyya, B. (2013). Micro electrochemical sinking and milling method for generating micro features. *Proceedings of the Institution of Mechanical Engineers, Part B: Journal of Engineering Manufacture, 227*(11), 1651–1663.
Golshan, A. (2017). Ultraprecision machining of silicon wafer by micro milling process. *Procedia Engineering, 184*(4), 192–196.
Huo, D. (2017). Modelling the influence of tool deflection on cutting force and surface generation in micro-milling. *Micromachines, 8*(6), 188–202.
Jin, C., & Solomon, G. C. (2018). Controlling band alignment in molecular junctions: Utilizing two-dimensional transition-metal dichalcogenides as electrodes for thermoelectric devices. *Journal of Physical Chemistry C, 122*(26), 14233–14239.
Jurow, M., Schuckman, A. E., Batteas, J. D., & Drain, C. M. (2010). Porphyrins as molecular electronic components of functional devices. *Coordination Chemistry Reviews, 254*(19–20), 2297–2310.
Kim, B. H. (2004). Micro electrochemical milling. *Journal of Micromechanics and Microengineering, 15*(1), 124–130.
Kim, B. H., Park, B. J., & Chu, C. N. (2006). Fabrication of multiple electrodes by reverse EDM and their application in micro-ECM. *Journal of Micromechanics and Microengineering, 16*(4), 843–852.
Kim, R., Datta, S., & Lundstrom, M. S. (2009). Influence of dimensionality on thermoelectric device performance. *Journal of Applied Physics, 105*(3), 034506.
Koswatta, S. O., Hasan, S., Lundstrom, M. S., Anantram, M. P., & Nikonov, D. E. (2007). Nonequilibrium green's function treatment of phonon scattering in carbon nanotube transistors. *IEEE Transactions on Electron Devices, 54*(9), 2339–2351.
Kuhr, W. G., Gallo, A. R., Manning, R. W., & Rhodine, C. W. (2004). Molecular memories based on a CMOS platform. *MRS Bulletin, 29*(11), 838–842.
Kumar, M. J. (2007). Molecular diodes and applications. *Recent Patents on Nanotechnology, 1*(1), 51–57.
Kushmerick, J. (2009). Molecular transistors scrutinized. *Nature, 462*, 994–995.
Largeot, C., Portet, C., Chmiola, J., Taberna, P. L., Gogotsi, Y., & Simon, P. (2008). Relation between ion and pore sizes for an electric double-layer capacitor. *Journal of the American Chemical Society, 130*(9), 2730–2731.
Lindsey, J. S., & Bocian, D. F. (2011). Molecules for charge-based information storage. *Accounts of Chemical Research, 44*(8), 638–650.
Liu, Y. (2011). Development of microelectrodes for electrochemical micromachining. *International Journal of Advanced Manufacturing Technology, 55*(1–4), 195–203.
Liu, Q. (2018). Effect of milling modes on surface integrity of KDP crystal processed by micro ball-end milling. *Procedia CIRP, 71*(1), 260–266.
Liu, Z., Yasseri, A. A., Lindsey, J. S., & Bocian, D. F. (2003). Molecular memories that survive silicon device processing and real-world operation. *Science, 302*(5650), 1543–1545.
Liu, Y., Zhu, D., & Zhu, L. (2012). Microelectrochemical milling of complex structures by using in situ fabricated cylindrical electrodes. *International Journal of Advanced Manufacturing Technology, 60*(9–12), 977–984.
Long, B., Nikitin, K., & Fitzmaurice, D. (2003). Assembly of an electronically switchable rotaxane on the surface of a titanium dioxide nanoparticle. *Journal of the American Chemical Society, 125*, 15490–15498.

Maassen, J., & Lundstrom, M. (2015). The Landauer approach to electron and phonon transport. *ECS Transactions, 69*(9), 23–36.
Madani, M. S., Monajjemi, M., & Aghaei, H. (2017). The double wall boron nitride nanotube Nano-cylindrical capacitor. *Oriental Journal of Chemistry, 33*(3), 1213–1222.
Malekian, M., Park, S. S., & Jun, M. B. (2009). Tool wear monitoring of micro-milling operations. *Journal of Materials Processing Technology, 209*(10), 4903–4911.
Maloney, K. J. (2012). Multifunctional heat exchangers derived from three-dimensional micro-lattice structures. *International Journal of Heat and Mass Transfer, 55*(9–10), 2486–2493.
Mayor, M., Weber, H. B., Reichert, J., Elbing, M., Von Hänisch, C., & Beckmann, D. (2003). Electric current through a molecular rod-relevance of the position of the anchor groups. *Angewandte Chemie, International Edition, 42*(47), 5834–5838.
Merlet, C., Rotenberg, B., Madden, P. A., Taberna, P. L., Simon, P., & Gogotsi, Y. (2012). On the molecular origin of supercapacitance in nanoporous carbon electrodes. *Nature Materials, 11*(4), 306–310.
Meunier, M., & Quirke, N. (2000). Molecular modelling of electron trapping in polymer insulators. *The Journal of Chemical Physics, 113*(1), 369–376.
Miller, J. R., & Simon, P. (2008). Electrochemical capacitors for energy management. *Science, 321*(5889), 651–652.
Okazaki, Y., Mishima, N., & Ashida, K. (2004). Micro factory-concept, history, and developments. *Journal of Manufacturing Science and Engineering, 126*(4), 837–844.
Oliaei, S., & Karpat, Y. (2014). Experimental investigations on Stavax stainless steel micro milling. *Procedia CIRP, 14*, 377–382.
Oliaei, S. N. B., & Karpat, Y. (2016). Influence of tool wear on machining forces and tool deflections during micro milling. *International Journal of Advanced Manufacturing Technology, 84*(9–12), 1963–1980.
Pandey, R. R., Bruque, N., Alam, K., & Lake, R. K. (2006). Carbon nanotube—Molecular resonant tunneling diode. *Physica Status Solidi RRL: Rapid Research Letters, 203*(2), R5–R7.
Papadopoulos, T. A., Grace, I. M., & Lambert, C. J. (2006). Control of electron transport through Fano resonances in molecular wires. *Physical Review B, 74*(19), 193306.
Rathod, V., Doloi, B., & Bhattacharyya, B. (2017). Fabrication of microgrooves with varied cross-sections by electrochemical micromachining. *International Journal of Advanced Manufacturing Technology, 92*(1–4), 505–518.
Rossetti, A., Pavesi, G., & Ardizzon, G. (2010). A new two-stage miniature pump: Design, experimental characterization, and numerical analyses. *Sensors and Actuators A: Physical, 164*(1–2), 74–87.
Roth, K. M., Dontha, N., Dabke, R. B., Gryko, D. T., Clausen, C., & Lindsey, J. S. (2000). Molecular approach toward information storage based on the redox properties of porphyrins in self-assembled monolayers. *Journal of Vacuum Science and Technology B, 18*(5), 2359–2364.
Salahuddin, S., Lundstrom, M., & Datta, S. (2005). Transport effects on signal propagation in quantum wires. *IEEE Transactions on Electron Devices, 52*(8), 1734–1742.
Scheidemantel, T. J., Ambrosch Draxl, C., Thonhauser, T., Badding, J. V., & Sofo, J. O. (2003). Transport coefficients from first-principles calculations. *Physical Review B, 68*(12), 125210.
Sharma, P., & Bhatti, T. S. (2010). A review on electrochemical double-layer capacitors. *Energy Conversion and Management, 51*(12), 2901–2912.
Sumiya, H., Harano, K., & Murakami, D. (2012). Application of nano-polycrystalline diamond to cutting tools. *SEI Technical Review, 75*(75), 18–28.
Suzuki, H. (2013). Development of micro milling tool made of single crystalline diamond for ceramic cutting. *CIRP Annals—Manufacturing Technology, 62*(1), 59–62.
Tansel, I. (2000). Tool wear estimation in micro-machining: Part I: Tool usage–cutting force relationship. *International Journal of Machine Tools and Manufacture, 40*(4), 599–608.
Tao, N. J. (2006). Electron transport in molecular junctions. *Nature Nanotechnology, 1*(3), 173–181.

Tillmann, W. (2000). Trends and market perspectives for diamond tools in the construction industry. *International Journal of Refractory Metals and Hard Materials, 18*(6), 301–306.

Uriarte, L. (2006). Comparison between microfabrication technologies for metal tooling. *Proceedings of the Institution of Mechanical Engineers, Part C: Journal of Mechanical Engineering Science, 220*(11), 1665–1676.

Uriarte, L. (2007). Error budget and stiffness chain assessment in a micro milling machine equipped with tools less than 0.3 mm diameter. *Precision Engineering, 31*(1), 1–12.

Vericat, C., Vela, M. E., Benitez, G., Carro, P., & Salvarezza, R. C. (2010). Self-assembly monolayers of thiols and dithiols on gold: New challenges for a well-known system. *Chemical Society Reviews, 39*(5), 1805–1812.

Vollertsen, F. (2008). Categories of size effects. *Production Engineering, 2*(4), 377–386.

Weule, H., Hüntrup, V., & Tritschler, H. (2001). Micro-cutting of steel to meet new requirements in miniaturization. *CIRP Annals—Manufacturing Technology, 50*(1), 61–64.

Whitesides, G. M., & Boncheva, M. (2002). Beyond molecules: Self-assembly of mesoscopic and macroscopic components. *Proceedings of the National Academy of Sciences, 99*(8), 4769–4774.

Yan, J., Watanabe, K., & Aoyama, T. (2014). Micro-electrical discharge machining of polycrystalline diamond using rotary cupronickel electrode. *CIRP Annals—Manufacturing Technology, 63*(1), 209–212.

Yu, H., Luo, Y., Beverly, K., Stoddart, J. F., Tseng, H. R., & Heath, J. R. (2003). The molecule-electrode interface in single-molecule transistors. *Angewandte Chemie, 42*(46), 5706–5711.

Zhan, Z. (2015). Precision tungsten carbide milling with micro PCD milling tool. *International Journal of Advanced Manufacturing Technology, 77*(9–12), 2095–2103.

Picotechnology in Energy and Environment

P. Saranraj, Ashish Kumar Nayak, Shrikaant Kulkarni, K. Kesavardhini, and E. Tharani

1 Introduction

Picotechnology, functioning at the scale of 10^{-12} meters, is poised to redefine innovations in energy and environmental sectors. It involves the manipulation of matter at a scale smaller than atoms but larger than sub-atomic particles, allowing unprecedented material precision. In energy systems, picoscale structuring enhances electron mobility and optimizes band structures for better device efficiency (Delgado & Tong, 2024). It provides highly selective pollutant detection and remediation technology in environmental sciences. Computation models early on indicate that picoscale manipulation can enhance adsorption and catalytic processes crucial to sustainability (Suzuki & Funatsu, 2023). Nanotechnology, working at 10^{-9} meters, is compared with Picotechnology as it attains even smaller-scale control, generating new material properties. In meantime, it provides a complementary option to Femtotechnology aimed at structures at even smaller, quantum levels. Studies still investigate theoretical models and practical techniques to implement these possibilities into actual applications. Among the most viable applications of Picotechnology in energy is high-performance battery electrode development.

P. Saranraj (✉) · K. Kesavardhini · E. Tharani
PG and Research, Department of Microbiology, Sacred Heart College (Autonomous), Tirupattur, Tamil Nadu, India

A. K. Nayak
Department of Civil Engineering, Prasad V. Potluri Siddhartha Institute of Technology, Vijayawada, Andhra Pradesh, India

S. Kulkarni
Sanjivani University, Kopargaon, India

Faculty of Business, Victorian Institute of Technology, Melbourne, VIC, Australia

S. Kulkarni (ed.), *A Journey from Nanotechnology to Picotechnology*, Lecture Notes in Nanoscale Science and Technology 47,
https://doi.org/10.1007/978-3-032-07228-3_10

Research at the picoscale enables electron pathways to be accurately set up, greatly improving conductivity and energy density (Chen et al., 2024).

Additionally, the modulation of ion migration channels can alleviate widespread problems such as dendrite growth, which tends to degrade battery life (Brown et al., 2023). In contrast to traditional nanostructuring, picoscale interventions enable orbital-scale control of atoms in active materials, enhancing electrochemical stability. These benefits have the potential to revolutionize Lithium-Ion, Sodium-Ion, and Solid-State Batteries through longer cycles, improved charging, and safer operation. The energy industry could significantly benefit from scalable picoscale fabrication methods for electrodes and electrolytes. New picoscale battery designs are already demonstrating excellent performance enhancements compared to their nanoscale equivalents. In the field of solar energy, Picotechnology provides new opportunities for ultra-high efficiency photovoltaic devices. Quantum dots produced with picoscale accuracy exhibit substantial enhancements in light absorption and charge carrier dynamics (Tang et al., 2023a, b). Quantum confinement at such dimensions enables scientists to dynamically modulate Bandgaps, maximizing solar energy conversion over a range of light intensities. Also, picoscale photonic crystals trap and control light better, minimizing reflection and maximizing device efficiency overall. These improvements may significantly lower the cost per watt of solar energy while enhancing the ability to accommodate varying environmental conditions. Picoscale solar designs also open up the possibility of flexible and wearable solar panels. Future investigations are investigating hybrid configurations that join picoscale quantum Dots with organic photovoltaics for even improved efficiency benefits (Tisdale & Williams, 2024).

The green applications of Picotechnology are similarly intriguing, specifically in Remediation technology. Picoscale adsorbents with control at the atomic level on surfaces exhibit extraordinary selectivity for heavy metals and persistent organic pollutants (Wang et al., 2024). Furthermore, picocatalysts provide efficient contaminant degradation under ambient Conditions, lowering the energy needs of large scale clean-up processes. Picoscale membrane-based Water purification systems can separate molecules with unparalleled resolution, providing breakthroughs in desalination and removal of micro-contaminants. The technologies are instrumental as water shortage and pollution issues worldwide become ever-more pressing concerns. Picoscale environmental sensors have the potential for real-time measurement Capability at ultra-low limits of toxic agents. Ultra-sensitive detection and Fast remediation may transform environmental management systems globally (Ariga et al., 2023).

Despite its transformative potential, the advancement of Picotechnology faces significant scientific and ethical challenges. Achieving reliable fabrication and long-term stability at the picoscale requires control techniques that push the limits of current atomic manipulation methods (Gimzewski & Vesna, 2023). Scalability remains another critical hurdle, as producing large quantities of picoscale materials consistently and economically is highly complex. Ethical considerations must also be addressed, including potential environmental and health impacts of deploying picoscale technologies without sufficient oversight. Implementing regulation frameworks specific to Picotechnology will be necessary to control Potential dangers and guarantee sustainable development. However, interdisciplinary collaborations in Quantum science, Materials engineering, and Environmental policy are breaking the ground forward. Ongoing investment in basic research will likely lead to breakthroughs that fully Leverage Picotechnology for Energy and Environmental sustainability.

The term Picotechnology is a portmanteau of picometre and technology, intended to parallel the term Nanotechnology. It is a hypothetical future level of technological manipulation of matter, on the scale of trillions of a meter or picoscale. This is three orders of magnitude smaller than a nanometre (and thus most nanotechnology) and two orders of magnitude smaller than most chemistry transformations and measurements. Picotechnology would involve the manipulation of matter at the atomic level. A hypothetical development, Femtotechnology would involve working with matter at the sub-atomic level (Ariga et al., 2023).

2 Picoscale Materials for Energy Applications

2.1 Energy Storage

Picoscale electrodes are transforming energy storage with their large surface area and excellent electron transport properties. Graphene Quantum Dots (GQDs) have emerged as a promising option by raising charge storage capacity and cycling stability (Liu et al., 2020). Integration of GQDs into lithium-ion batteries enhances ion diffusion and reduces degradation upon repeated cycling. In addition, the tunable surface chemistry and high conductivity of GQDs make them the perfect choice for ultra-high capacity to next-generation electrodes. These breakthroughs illustrate how picoscale control can significantly enhance battery performance and longevity. Progress toward better insight into electron dynamics at the picoscale is essential to attain energy storage breakthroughs. Furthermore, the application of advanced modeling methods has enhanced the predictability of electrode performance. There are ongoing attempts to design hybrid materials that leverage picoscale effects synergistically. Picoscale interfaces and defects are especially critical for electrode performance. Therefore, material architecture control at the atomic level is essential. In sum, picoscale-designed electrodes are a critical frontier in energy technology (Zahir et al., 2022).

Quantum confinement is important when material sizes reach the picoscale, resulting in discrete electronic states and altered energy band structures (Ramos et al., 2022). Tuning of electrical conductivity and redox potentials, which are important for battery materials, is enabled by this phenomenon. For instance, picoscale confinement in semiconductor nanoparticles has been found to increase charge carrier mobility, which directly influences super-capacitor efficiency. When quantum size effects take control, new mechanisms for engineering electron density are within reach. These fine adjustments optimize ion insertion/extraction processes in storage materials. Additionally, the increased surface-to-volume ratio at the picoscale enhances reaction kinetics. Knowing quantum effects has made it possible to engineer flexible and miniaturized energy devices. Efforts are presently directed toward designing quantum-confined nanocomposites for next-generation battery systems. Notably, combining confinement effects with doping techniques enhances material performance. In summary, controlling quantum effects on the picoscale unlocks revolutionary potential in storage technologies (Tang et al., 2021).

2.2 Solar Energy

Picoscale structuring in photovoltaics enables management of light far beyond conventional materials, leading to improved absorption and energy conversion (Bredar et al., 2020). Picoscale-engineered quantum dots can span a wider solar spectrum, increasing efficiencies of devices. Including such quantum dots in multi-junction solar cells enhances charge separation and Reduces recombination loss. Additionally, picoscale structuring in interfaces enhances photon Capture and lengthen carrier lifetimes. This results in improved open-circuit voltages and better Overall efficiencies. Atomic-scale engineered material hetero junctions provide Optimized energy band alignment. Structuring as well enhances light-trapping capability through plasmonic effects. New fabrication techniques like atomic-layer deposition allow atomic-level control of interface layers. Picoscale engineering thus enables the overcoming of fundamental limitations in traditional photovoltaic devices. This development paves the way for ultra-high-efficiency, low-cost solar energy harvesting (Nozik, 2010).

Quantum Dot Solar Cells (QDSCs) utilize precisely engineered dots to match solar radiation profiles more closely than conventional materials (Kamat, 2008). The tunable bandgap of quantum dots allows harvesting across UV, visible, and near-infrared regions, significantly improving performance. Recent advancements in core-shell structured quantum dots have minimized non-radiative recombination, leading to better quantum efficiencies. Moreover, controlling size distribution at the picoscale improves the photoluminescence quantum yield. Innovations in surface passivation techniques have greatly enhanced the stability of QDSCs. Additionally, picoscale tuning allows for the design of hot-carrier solar cells to exceed conventional efficiency limits. Incorporating quantum dots into tandem architectures further multiplies light absorption capabilities. Recent studies show promise in integrating QDs with perovskite solar cells for hybrid systems. Optimization of charge transport layers also benefits from picoscale engineering. Overall, QDSCs are at the forefront of next-generation solar energy solutions (Khan et al., 2022).

2.3 *Fuel Cells and Hydrogen Storage*

Picocatalysts enable ultra-efficient hydrogen generation by providing abundant active sites and optimizing reaction kinetics (Li et al., 2021). Transition metal quantum dots, like Mo_2C and CoP, offer significantly higher hydrogen evolution rates than their bulk counterparts. Black phosphorus quantum dots also dramatically enhance the photocatalytic water splitting process due to superior carrier separation properties. Furthermore, picoscale size allows better exposure of catalytic surfaces to reactants. Surface defect engineering at the picoscale optimizes binding energy of intermediates. Enhanced quantum effects boost light absorption and charge carrier dynamics. Bimetallic quantum dot catalysts are also being explored for synergetic hydrogen production mechanisms. Picocatalyst stability remains a focus, with surface coatings being investigated for enhanced durability. Overall, picocatalysts represent a transformative advancement toward sustainable hydrogen energy production (Yang et al., 2017).

Fuel cells demand membranes with high proton conductivity, low crossover, and excellent mechanical stability challenges that picoscale engineering directly addresses (Sata, 2000). Incorporating picoscale inorganic fillers, like silica or graphene oxide quantum dots, into polymer electrolytes enhances proton transport pathways. Nanocomposite membranes designed at the picoscale also display improved chemical resistance and operational durability. Atomic-level control of membrane morphology ensures more continuous proton conduction channels. Embedding pico-structured materials increases free volume while maintaining mechanical integrity. Additionally, functionalized picoscale additives improve membrane hydration under low humidity conditions. Research is focusing on designing hybrid organic-inorganic membranes at the atomic scale. Recent developments indicate promise for room-temperature Proton conductors on the picoscale. Such technology is pivotal to next-generation portable fuel cells. Picoscale engineering is a milestone toward the development of membrane Technologies (Yao et al., 2021).

3 Advanced Characterization and Fabrication Techniques in Picotechnology

3.1 *Picoscale Imaging Tools*

Advances in imaging technologies have played a central role in probing materials at the Picoscale, allowing atomic-scale visualization with unprecedented accuracy. Methods like Aberration-Corrected Transmission Electron Microscopy (AC-TEM) now permit imaging of Individual atoms and atomic columns, offering direct access to pico-level structural details and defects (Zhu et al., 2023). Likewise, Atomic Force Microscopy (AFM) modified for picoscale measurements provides sub-angstrom vertical resolution, which is essential for characterizing Surface topography and Mechanical properties at the Atomic scale. Moreover, new techniques such as Electron Ptychography provide improved phase contrast and dose efficiency, enabling visualization of light elements usually invisible in standard Transmission Electron Microscope (TEM). These developments have made direct observation of picoscale phenomena like atomic displacements and dynamic surface reconstructions possible, which are crucial for understanding nanoscale functional materials (Wang & Lee, 2022). Combining these imaging modes with in situ experimental configurations, Such as environmental TEM, further broadens their use to real-world conditions. The synergy of these tools promotes increased knowledge of picoengineered materials, speeding up design of next-generation energy and environmental devices. Ongoing innovation in detector Technologies and computational image processing ensures ongoing improvements in resolution and speed. Generally, these imaging innovations underpin contemporary Picotechnology research by bridging structural understanding and functional performance at the atomic Scale (Kim et al., 2021).

3.2 Atomic Manipulation Techniques for Pico-Structured Materials

Atomic-level control in manipulating individual molecules and atoms is critical to make pico-structured materials with desired properties. Scanning Tunneling Microscopy (STM) and AFM manipulation methods offer atomic-level control by moving atoms from one position on a surface to another, facilitating the building of nanostructures designed (Novikov et al., 2023). Improved fidelity and stability in manipulation have been achieved due to recent advancements in tip-functionalization and low-temperature manipulation. Moreover, specialized electron and ion beam techniques enable pico-scale doping and sculpting, enabling the formation of hetero-structures and defect sites with atomic-scale precision. Atomic-layer deposition in conjunction with chemical vapor deposition enables bottom-up creation of picoengineered thin films with atomic-layer-regulated thickness and composition. These methods have enabled the fabrication of quantum dots, single-atom catalysts, and nanoelectronic devices with engineered pico-features. Scalability and reproducibility remain challenges, but coupling of these methods with automated feedback control and machine learning algorithms is a promising avenue. Overall, atomic manipulation is a pillar of picotechnology, which allows bottom-up assembly of materials with unprecedented control over functionality (Nguyen et al., 2022).

3.3 Computational Modeling of Picoscale Interactions

Computational modeling is critical to comprehend and forecast picoscale interactions under which experimental interrogation is problematic. Density Functional Theory (DFT) is still the principal method to explore electronic structures, reaction coordinates, and binding energies at the atomistic level (Patel et al., 2023). Latest advances in machine learning (ML)-enhanced DFT have significantly enhanced computational efficiency, making it possible to probe larger and more intricate pico-engineered systems. Molecular Dynamics (MD) simulations, e.g., ab initio MD, supplement static calculations by illustrating dynamic phenomena such as atomic diffusion, vibrational modes, and thermal stability of pico-structured materials. Multiscale modeling frameworks with coupling of quantum mechanics and continuum methods bridge the length scales from pico to nano, important for device-level design. The computational developments serve to inform rational design of picoscale materials through elucidating structure-property relations and reaction pathways (Kumar et al., 2021). They also help screen candidate materials for energy and environmental applications, accelerating discovery pipelines. The integration of experimental data using data-driven methods and digital twins, predictive capabilities are improved. Future prospects entail the inclusion of quantum computing as well as better sampling techniques to extend beyond the current limitations. Overall, computational modeling serves as a core pillar to drive Picotechnology forward by uncovering atomic-scale processes and directing material innovation (Fernandez et al., 2022).

4 Environmental Applications of Picotechnology

4.1 Pollution Remediation: Picoscale Adsorbents and Quantum Sensors

Picoscale adsorbents have proven to be very effective materials for the removal of heavy metals because of their very high surface area and surface chemistry that can be tuned, allowing for selective binding of pollutants such as lead, mercury, and arsenic (Zhang et al., 2023). Carbon-based pico-adsorbents such as graphene quantum dots and atomic-level functionalized carbon nanotubes have improved adsorption kinetics and regeneration properties. Concurrently, quantum sensors based on Picoengineered nanomaterials provide unprecedented Sensitivity and selectivity for trace-level detection of air and water pollutants, outperforming conventional Sensing technologies. Quantum confinement and electron tunneling phenomena are utilized by these sensors for ultrafast signal transduction, enabling real-time monitoring of heavy metals, pesticides, and volatile Organic compounds. Coupling of pico-adsorbents with quantum sensors also enables Closed-Loop Environmental Remediation Systems. Although with promising outcomes, there are still issues of sensor stability and mass deployment, which drive continued research into material resilience and multiplexing functionality. In general, picoscale technologies are transforming pollution cleanup by means of integrated adsorption and sensing approaches (Wang & Chen, 2021).

4.2 Water Purification: Picofiltration Membranes and Picocatalysts

Picofiltration membranes, designed with pico-sized pores and surface modifications of high precision, have greatly enhanced desalination efficiency by allowing higher water flux and salt rejection with low energy intake (Singh et al., 2023). Pico-structured graphene oxide membranes are just one example of materials that exhibit ultrathin architectures to achieve balanced permeability and selectivity at the atomic level. In addition, picocatalysts used for water treatment utilize pico-engineered semiconductor quantum dots or metal clusters to catalyze the breakdown of persistent organic pollutants, such as dyes and drugs. Picocatalysts take advantage of improved light absorption and charge separation facilitated by quantum effects, facilitating efficient photocatalytic reactions under visible light. Combining picofiltration membranes with photocatalytic degradation treats complex pollutant mixtures, enhancing overall water treatment efficiency. Future studies are aimed at scaling up membrane synthesis and enhancing catalyst longevity to satisfy industrial requirements. The intersection of picoengineering and catalysis promises a new dawn for sustainable water treatment technologies (Kang & Park, 2022).

5 Pico-Enabled Carbon Capture and Conversion

The emergence of Picotechnology has unveiled new opportunities for the creation of ultra-efficient materials for carbon capture purposes. Picoscale architectures like atomically precise Metal Organic Frameworks (MOFs) and Covalent Organic Frameworks (COFs) possess remarkable surface areas and adjustable pore structures, leading to substantially increased CO_2 adsorption capacities. Such materials can be atomically engineered with high precision to form binding sites with Ideal interactions of CO_2 molecules, which provide better selectivity and reversibility (Wu & Kim, 2024). Novel improvements in atomic-layer deposition technologies have also facilitated the development of pico-thin coatings that alter adsorbents surface chemistry, allowing for increased affinity toward CO_2. In addition, pico-structured materials demonstrate superior cyclic stability, an essential parameter for actual use in industrial capture systems. The rational design of these Materials from first-principles simulations enables accelerated discovery cycles and performance optimization (Patel & Singh, 2024).

Aside from capture, Picotechnology offers promising directions toward the catalytic conversion of captured CO_2 to useful fuels and chemicals. With catalyst engineering at the picoscale, researchers can tune the electronic structure, coordination environment, and active site geometry to enable efficient CO_2 reduction reactions (Kumar et al., 2023). Quantum confinement effects at the atomic level enhance electron transfer processes, lowering activation barriers and improving product selectivity toward hydrocarbons or alcohols. For instance, Single-Atom Catalysts (SACs) stabilized on defect-rich substrates exhibit extraordinary activity and stability for CO_2 electro-reduction. Additionally, the coupling of picoengineered catalysts with renewable energy sources like solar-driven systems enables fully sustainable carbon valorization pathways. However, challenges such as catalyst deactivation and competitive side reactions still demand innovative design strategies (Martins & Silva, 2024).

Moving forward, integrating pico-enabled carbon capture and conversion technologies into scalable, closed-loop systems will be vital for meaningful climate impact. Hybrid systems combining pico-structured adsorbents with picocatalysts offer the potential for seamless capture-to-conversion processes within a single modular unit (Choi & Tan, 2024). Advances in real-time monitoring using quantum sensors at the picoscale will further optimize operational conditions by providing atomic-level insights into dynamic adsorption and reaction phenomena. Moreover, machine learning-guided design of pico-architectures is emerging as a powerful approach to accelerate the discovery of multifunctional materials tailored for both capture and conversion. Yet, issues related to material scalability, process energy demands, and lifecycle sustainability must be addressed through interdisciplinary research efforts. Overall, Picotechnology represents a transformative enabler for next-generation carbon management strategies aiming toward net-zero emissions (Verma & Gupta, 2024).

6 Quantum Sensors for Environmental Monitoring

Quantum sensors operating at the picoscale have emerged as transformative tools for real-time, ultra-sensitive environmental monitoring. By exploiting quantum phenomena such as spin states and entanglement, these sensors achieve unprecedented levels of precision in detecting pollutants at trace concentrations (Doherty & Manson, 2024). For instance, Nitrogen-Vacancy (NV) centers in diamond have been successfully utilized for nanoscale magnetic sensing of environmental contaminants. The miniaturization enabled by Picotechnology further allows embedding these quantum probes in complex matrices like water or air systems without disturbing the natural environment. In addition, picoscale engineering improves the stability of sensors in unfavorable conditions, including high radiation or corrosive atmospheres. These developments make quantum sensors key facilitators of anticipatory environmental protection measures. Nevertheless, difficulties such as wireless network integration and long-term calibration are yet to be studied (Bar-Gill & Yacoby, 2024).

In monitoring water pollution, quantum sensors have proven capable of detecting heavy metals, organic toxins, and emerging contaminants with single-molecule sensitivity. Quantum-enhanced spectroscopy techniques, such as quantum Raman and fluorescence methods, yield fingerprinting of contaminants more precisely than classical techniques (Schirhagl & Loretz, 2023). Progress in picoscale material interfaces, including functionalized quantum dots and single-photon detectors, has enhanced the selectivity and rapidity of environmental monitoring. For instance, biosensors based on quantum dots and picoscale surface-modified coatings exhibit superb specificity for mercury and lead ions in water systems. Moreover, coupling these sensors with machine learning algorithms has increased predictive performance in dynamic pollution phenomena. The continuing innovation in real-time data analytics and miniaturization will drive decentralized water quality monitoring in the near future (Bradac & Gao, 2023).

Air quality monitoring is another crucial field where quantum sensors are significantly contributing. Quantum gas sensors engineered at the picoscale are capable of measuring ultra-low concentrations of toxic gases such as NO_2, CO, and volatile organic compounds (VOCs) with high accuracy. Quantum spin-based sensing technologies, based on atomically precise defects in wide-bandgap semiconductors, are highly sensitive without needing to be constantly recalibrated (Fescenko & Jensen, 2023). In addition, picoscale quantum sensor compactness enables their direct integration into hand-held devices and drones for mobile air sampling across urban and industrial settings. Real-time, space-resolved air pollution maps crucial for public health management are realized through these breakthroughs. The challenge still exists to ensure the longevity of the sensors, minimize power consumption, and standardization across platforms. Ongoing interdisciplinary collaboration among quantum physics, environmental science, and materials engineering is crucial to optimize the potential of these advanced technologies (Degen & Reinhard, 2024).

7 Picotechnology for High-Efficiency Energy Storage

7.1 Atomic-Level Engineering of Battery Electrodes

Atomic-scale engineering changes the rules in battery electrode development by directly manipulating material characteristics on the pico-scale. Scientists have already proved that atomic-level defect engineering promotes ion mobility and electrochemical stability. Picotechnology allows uniform, defect-controlled electrode production with considerable improvements in battery performance. Modifying surface atom arrangements further fine-tunes charge transfer pathways to reduce energy losses during use (Zhang et al., 2023). Such atomic regulation not only increases power density but also increases the lifespan of the electrode under harsh cycling conditions. Research demonstrates that strategic incorporation of single-atom catalysts in electrodes enhances lithium-ion diffusion rates. Such precision generates safer and longer-lasting energy storage devices essential in portable and stationary devices. Picotechnology provides doors to engineer electrodes that are not subject to inherent structural instabilities. Additionally, the scaling of these innovations poses a challenge but has great prospects for commercialization. Atomic precision in electrode design is a new standard that is emerging (Wang et al., 2022).

Pico-scale hetero-structures have become central to designing better battery interfaces. By stacking two-dimensional materials with atomically accurate directionality, scientists realize unprecedented efficiencies of charge transport. Having sharp, ordered atomic interfaces in hetero-structures facilitates fast electron and ion transport over disordered bulk materials. Tuning of atomic-scale lattice strain has been seen to enhance significantly mechanical stability and energy density (Sun et al., 2022). In addition, atomic interface control minimizes interfacial resistance, an important aspect of enhancing fast charging capacity. Atomically sharp hetero-interface engineering also helps in suppressing dendrite growth in lithium-metal batteries. Pico-manufacturing processes facilitate selective stacking of atoms, providing low energy barriers between layers. Such hetero-structure design precision promotes battery systems with increased reliability and reduced degradation rates. Battery architectures in the future will be highly dependent on such Picotechnology-assisted interface designs. Continuous development in atomic arrangement control is vital for maintaining energy requirements (Liang et al., 2023).

Solid-state batteries gain considerably from Picotechnology advances at the atomic scale. Latest research indicates that atomic-scale doping techniques significantly increase the ionic conductivity of solid electrolytes. Lattice structure modification on the picoscale enables the production of low-barrier ion diffusion channels, a requirement for high-power applications. Additionally, atomically flat solid electrolyte-to-electrode interfaces suppress space-charge layer formation, leading to overall improved electrochemical performance. Picotechnology has made grain boundary structures tunable to minimize mechanical failures during cycling stress. Pico-doped materials show greater chemical stability against lithium dendrites, improving battery safety and longevity. Pico-structured sulfide electrolyte research has provided conductivity values on par with liquid counterparts. The capability of precise control over defect sites also results in tailored electrochemical responses. These breakthroughs are driving solid-state battery technologies toward broad commercialization. Atomic engineering principles integration is essential in future energy storage (Yoon et al., 2023).

7.2 Single-Atom Catalysts for Enhanced Energy Storage

Picotechnology-synthesized Single-Atom Catalysts (SACs) create a new path to improve electrochemical energy storage systems. SACs deliver optimum atomic efficiency, wherein every atom is involved in catalytic activity. By adjusting the electronic environment on single metal atoms, scientists can significantly enhance catalytic rates. Research finds that SACs supported on carbon substrates boost redox kinetics and improve battery charging speeds. Atomic dispersion inhibits particle agglomeration, which is a universal issue in traditional catalysts (Gao et al., 2022). Picotechnology guarantees the formation of homogeneous active sites, enhancing battery electrode durability. In addition, SACs have exhibited outstanding activity even at ultra-low metal loads, lowering costs and environmental pollution. Single-atom stabilization in battery matrices demands pico-level accuracy for long-term operating stability. Recent developments indicate that SACs are going to have a central position in next-generation lithium-sulfur and metal-air batteries. Their performance improvement highlights the revolutionizing potential of Picotechnology (Yang et al., 2023).

Incorporation of single-atom catalysts into lithium-sulfur batteries has efficiently addressed some long-standing issues associated with sulfur redox reactions. Current studies indicate that SACs greatly minimize the shuttle effect by immobilizing polysulfides at atomic positions. Designer electronic properties on the picoscale enable fast redox kinetics by allowing rapid redox analysis, which supports high-capacity retention after long cycles. Atomically dispersed catalytic centers also support uniform sulfur deposition and avoid passivating electrodes. Picotechnology enables tuning of binding energies between active sites and sulfur intermediates (Wang et al., 2022). Such tuning reduces capacity fading and enhances cycle life. Hybrid material design with single-atom sites and conductive frameworks maximizes electronic conductivity. Through atomic-level engineering, scientists have realized sulfur utilization efficiencies over 80% even at high loadings. Further investigation of pico-structured SACs is essential for the development of next-generation sulfur-based energy systems. Picotechnology-led innovation has the potential to unlock these systems to their optimal capacity (Li et al., 2022).

The applications of single-atom catalysts also involve the enhancement of the performance of metal-air batteries, essential for high-energy storage. SACs with atomically dispersed active sites have better oxygen reduction and evolution reaction activities. Atomic-level manipulation guarantees an even electronic structure, which enhances the reaction kinetics over nanoparticle catalysts. Experiments prove that incorporating heteroatoms next to single metal atoms further maximizes catalytic performance (Huang et al., 2023). Picotechnology enables precise control over the coordination environment of active centers, tailoring them for specific electrochemical reactions. SACs facilitate reversible redox processes, enhancing the rechargeability of metal-air batteries. Their high intrinsic activity also reduces over potentials, improving energy efficiency. Moreover, SACs show superior resistance to catalyst poisoning, a common issue in air electrodes. By employing atomic-scale engineering, researchers continue to develop high-capacity, durable metal-air systems. Picotechnology therefore opens new horizons for next-generation battery chemistries.

7.3 Pico-Engineered Supercapacitors for Rapid Energy Storage

Pico-engineered supercapacitors are a breakthrough toward the realization of fast, efficient energy storage devices. Through management of electrode material atomic arrangement, scientists optimize charge storage capability and energy density. Atomic-level optimization of porosity enhances available surface area, and hence higher values of capacitance. Picotechnology also allows uniform distribution of active sites, an important feature in facilitating rapid and reversible electrochemical reactions. Scaling nanopores to the picoscale enables ion transport dynamics to be fine-tuned, reducing internal resistance (Fang et al., 2023). Recent research proves that atomically constructed carbon frameworks deliver stable charge/discharge cycles even at ultra-high current densities. The conductivity pathways created at the picolevel significantly enhance the energy transfer rates. Additionally, picoscale doping with heteroatoms like nitrogen increases pseudo-capacitive behavior, enhancing performance. These innovations are necessary for power and energy density applications. Picotechnology is therefore propelling a new generation of high-performance supercapacitors (Yan et al., 2023).

Pico-manipulation of electrode/electrolyte interfaces significantly affects the performance of supercapacitors. Recent studies have established that atomically smooth interfaces reduce charge transfer resistance and electrolyte degradation. Hydrophilic surface groups pico-engineered enhance electrolyte infiltration, leading to enhanced ionic accessibility. Atomic arrangement changes at the interface also stabilize electrochemical windows, enabling increased operating voltages. Researchers have shown that atomically controlled defects act as sites for ion trapping, which leads to improved energy storage capacity (Lin et al., 2023). Adding pico-engineered hierarchical porosity offers the twin advantage of rapid ion diffusion and high strength. These innovations enable supercapacitors to maintain superior performance even under aggressive cycling conditions. Additionally, cross-linking at the atomic scale in electrode matrices avoids structural collapse during high-speed charge/discharge cycles. This underlines the significance of pico-engineering not only in material design but also in the optimization of device architecture. The future of wearable and flexible energy devices will hinge on these precise alterations (Shi et al., 2023).

Picotechnology is also facilitating the incorporation of new two-dimensional materials into future-generation supercapacitors. Atomically thin such as MXenes have shown phenomenal electrochemical behaviors when their terminations on surfaces are pico-modified. The functional groups of atomic layers are optimized, which maximizes their hydrophilicity and electrical conductivity, impacting performance directly. Additionally, repeating atomically thin sheets with manipulated interlayer gaps increases ion intercalation speeds without compromising mechanical stability (Ghidiu et al., 2023). Scientists use pico-level surface engineering to incorporate active sites for faradaic reactions, providing pseudo-capacitance contributions. Sophisticated research shows that adjusting the electronic band structure at the picoscale significantly enhances energy density. These observations indicate that picotechnology will play a key role in breaking the constraints of traditional carbon-based electrodes. The capacity to control structure and composition with precision at the atomic level makes it possible to have supercapacitors with unparalleled performance parameters. Future advancements in pico-modified 2D materials hold even more breakthroughs in store (Li et al., 2023).

7.4 Picoengineering of Solid Electrolytes for Stable Storage Systems

Picotechnology has an important role to play in the improvement of ionic conductivity and stability in solid electrolytes. Scientists have shown that aliovalent ion picoscale doping with defect sites causes faster ion migration pathways. Optimization of the atomic environment around lithium ions reduces ionic diffusion activation barriers, important for high-performance solid-state batteries. Atomic-level control also lowers grain boundary resistance, a big limitation in traditional solid electrolytes (Tang et al., 2023a, b). These achievements have made possible the creation of sulfide and oxide electrolytes with record room-temperature ionic conductivities. Picotechnology also makes local chemical potential controlled at the picoscale possible, with higher mechanical robustness against cracking. Such technologies yield safer substitutes for flammable liquid electrolytes. Additionally, pico-engineered solid electrolytes exhibit high resistance to lithium dendrites, which pose a great safety risk. Continuing optimization of atomic-scale approaches will continue to narrow the gap between laboratory performance and commercial feasibility. The future is being directed on solid-state energy storage by Picotechnology advances (Song et al., 2023).

Atomic-level interface engineering is a primary approach to overcoming solid-state energy system interfacial obstacles. Research demonstrates the addition of picoscale buffer layers between electrode and electrolyte, greatly suppressing interfacial resistance. The engineered layers also support volume changes on cycling, enhancing mechanical integrity and electrochemical stability. In addition, atomic accuracy in chemical composition minimizes unwanted side reactions that negatively impact performance (Zhang et al., 2023). Pico-fabricated interfaces provide close physical contact and homogeneous lithium-ion transport, essential for high rates of fast charging performance. Scientists also indicate that surface atomic-level alterations can selectively modulate wettability to facilitate improved interfacial ion diffusion. Such enhancements result in improved critical current densities without dendrite formation. Pico-structured composite electrolytes also combine the superior attributes of polymers and ceramics for synergistic performance. These developments highlight the revolutionary effect of atomic manipulation on device-level reliability. Commercial solid-state batteries in the future will significantly depend on Picoengineering at interfaces (Chen et al., 2023).

Picotechnology is making it possible to develop new families of glassy and crystalline solid electrolytes with unprecedented properties. Scientists have prepared amorphous oxides with atomically controlled short-range order, tailoring both ionic conductivity and flexibility. Introducing controlled nanodefects and designed local bonding environments, ionic pathways at the atomic scale can be adjusted. In addition, pico-engineering has allowed superionic phases to be discovered in complex hydrides and halides, shattering past conductivity boundaries. Such discoveries make new possibilities for flexible, wearable energy storage devices. Picotechnology also allows solid electrolytes with inherent resistance to moisture, an important challenge for their practical application. Research also demonstrates that picoaltered electrolytes exhibit stable performance even across broad temperature ranges. This atomic-scale control results in materials that harmonize mechanical softness with elevated ionic mobility. Future directions entail combining pico-engineered electrolytes with next-generation battery chemistries such as lithium-sulfur and sodium-ion systems (Zhou et al., 2023).

8 Future Perspectives of Picotechnology in Energy and Environment

The Picotechnology of the future in energy systems is on the verge of revolutionary breakthroughs based on improved control over atomic and electronic structures. With advancing fabrication techniques, picoscale materials will make ultra-efficient energy storage devices with capacities and cycle lives many orders of magnitude higher than existing batteries (Zhou et al., 2024). Incorporation of quantum effects into electrode design may revolutionize charge transport and Ion Diffusion Mechanisms, and therefore high rates of charging and discharging. Additionally, Picoengineered photovoltaic systems also have the potential to achieve dramatic increases in light harvesting efficiency through the utilization of atomic-scale hetero-structures and quantum Coherence phenomena. The integration of Picotechnology with new areas of quantum computing and AI-aided materials discovery will propel the rational design of energy Materials pico-optimized. Scalability and cost-effectiveness, however, are still crucial. Industrial uptake of picotechnology faces challenges like fabrication precision, high cost, measurement limitations, material instability, and unresolved safety and regulatory issues. Ultimately, Picotechnology's precision control at the atomic scale heralds a new era of sustainable and high-performance energy technologies (Singh et al., 2024).

In environmental applications, Picotechnology is anticipated to play a central role in next-generation pollution control and resource recovery technologies. Picoscale adsorbents and catalysts engineered with atomic precision are expected to exhibit unprecedented selectivity and durability for contaminant removal and degradation (Chen et al., 2024). Quantum-enhanced sensors will enable real-time, ultra-trace detection of environmental pollutants, supporting smarter, and more responsive remediation systems. Advancements in Picoengineered membranes will further revolutionize water purification by combining exceptional permeability with molecular-level selectivity, addressing global water scarcity challenges. Moreover, pico-enabled carbon capture and utilization technologies are projected to achieve higher efficiencies and lower energy penalties, fostering carbon-neutral cycles. However, the environmental and ethical implications of picoengineered nanomaterials call for proactive regulatory measures and thorough risk assessment frameworks. The dynamics between innovation, regulation, and public acceptance will determine the course of Picotechnology deployment for environmental sustainability (Gupta & Roy, 2023).

Looking forward, computational and experimental integration will be critical for unlocking the full potential of Picotechnology in energy and environmental sectors. Advances in multiscale modeling combined with high-throughput simulations and machine learning will enable accelerated discovery of Picoengineered materials with tailored functionalities (Zhang & Wu, 2024). In situ characterization techniques operating at the pico- and femto-second scales will provide unprecedented insight into dynamic atomic-scale processes, guiding rational design and optimization. The convergence of Picotechnology with emerging quantum information science could lead to novel devices that harness quantum entanglement for enhanced energy conversion and sensing (Luo & Chen, 2023). Additionally, sustainable and green synthesis approaches will be paramount to minimize environmental footprints of Picoengineered materials. As these advances mature, collaboration across disciplines and industries will drive scalable solutions addressing critical energy and environmental challenges, paving the way toward a sustainable future enabled by Picotechnology (Martinez & Lopez, 2024).

9 Conclusion

Picotechnology is revolutionizing environmental management through innovative solutions across pollution remediation, water purification, carbon capture, and energy storage. In pollution control, picoscale adsorbents like graphene quantum dots and functionalized carbon nanotubes enable selective removal of heavy metals, while quantum sensors leverage quantum phenomena for real-time, ultra-sensitive pollutant detection, forming closed-loop remediation systems. Water purification benefits from picofiltration membranes with atomic precision and picocatalysts that degrade contaminants efficiently via quantum effects. For carbon management, pico-engineered frameworks and single-atom catalysts enhance CO_2 capture and conversion, offering pathways to sustainable fuels. Concurrently, quantum sensors provide unparalleled monitoring of air and water quality, enabling proactive environmental protection. In energy storage, atomic-level engineering of electrodes, solid electrolytes, and supercapacitors boosts performance, stability, and energy density, critical for renewable integration. Despite these advances, challenges such as material scalability, sensor stability, and system integration persist, necessitating further research to bridge laboratory innovations with industrial deployment.

Looking ahead, the convergence of Picotechnology with machine learning and interdisciplinary collaboration promises accelerated material discovery and optimized systems. Scalable synthesis of pico-engineered materials, coupled with robust sensor networks and hybrid capture-conversion architectures, will be pivotal in achieving net-zero goals. Addressing longevity, energy efficiency, and cost barriers will unlock the full potential of these technologies, transforming them into mainstream solutions. As Picotechnology evolves, its integration into modular, sustainable systems underscores its transformative role in mitigating environmental crises, offering a blueprint for a cleaner, resource-efficient future.

References

Ariga, K., Ji, Q., & Yamauchi, Y. (2023). Picoscale engineering for sustainable environmental technologies. *Materials Horizons, 10*(5), 1342–1360.

Bar-Gill, N., & Yacoby, A. (2024). Picoscale quantum sensing for harsh environmental conditions. *Advanced Materials, 36*(5), 2307894.

Bradac, C., & Gao, W. (2023). Functionalized quantum dots for water pollution detection. *ACS Nano, 17*(3), 3456–3470.

Bredar, A. R. C., Chown, A. M., Burton, A. J., & Seger, B. (2020). Picoscale photovoltaics: Enhancing light absorption and charge dynamics. *The Journal of Physical Chemistry Letters, 11*(3), 713–723.

Brown, S., Johnson, M., & Lee, J. W. (2023). Picoengineered systems for energy storage. *Energy Storage Materials, 60*(5), 159–182.

Chen, C., Wang, Z., Liu, Y., Zhang, B., & Li, X. (2023). Atomic-level interface engineering for solid-state lithium batteries. *Advanced Energy Materials, 13*(3), 2202987.

Chen, J., Zhao, M., & Lin, H. (2024). Atomic precision adsorbents and catalysts for environmental remediation. *Environmental Science and Technology, 58*(5), 2812–2828.

Choi, Y., & Tan, J. (2024). Hybrid picoscale capture-conversion devices for integrated carbon management. *Sustainable Energy And Fuels, 8*(3), 1120–1135.

Degen, C. L., & Reinhard, F. (2024). Nanoscale quantum sensors for gas detection. *Physical Review Applied, 21*(2), 024012.

Delgado, J. M. P. Q., & Tong, L. (2024). Advanced perspectives on quantum and picoengineering. *Advances in Materials Science and Engineering, 48*(2), 115–142.

Doherty, M. W., & Manson, N. B. (2024). The nitrogen-vacancy colour Centre in diamond for nanoscale sensing. *Physics Reports, 980*, 1–41.

Fang, L., Zhao, Y., Chen, W., Hu, M., & Li, Q. (2023). Atomic engineering of hierarchical carbon materials for high-performance supercapacitors. *Nano Energy, 109*, 108256.

Fernandez, E., Garcia, M., & Torres, J. (2022). Multiscale modeling of nanostructured materials: Bridging pico to nano. *Computational Materials, 8*(4), 56.

Fescenko, I., & Jensen, K. (2023). Quantum gas sensing at the picoscale: Current status and prospects. *Sensors and Actuators B: Chemical, 392*, 134056.

Gao, M., Sun, X., Zhou, P., Duan, R., & Guo, H. (2022). Single-atom catalysts for advanced energy storage devices. *Energy Storage Materials, 54*, 117–134.
Ghidiu, A., Naguib, M., Barsoum, M. W., & Gogotsi, Y. (2023). Synthesis and characterization of two-dimensional MXenes. *Chemical Communications, 49*(82), 10695–10697.
Gimzewski, J. K., & Vesna, T. (2023). Quantum control techniques for future engineering. *Advanced Functional Materials, 33*(1), 2208092.
Gupta, A., & Roy, S. (2023). Picoscale materials for carbon capture and utilization: A review. *Journal of CO_2 Utilization. Biores. Tech., 69*, 102125.
Huang, J., Zhang, T., Qiao, L., & Sun, Y. (2023). Atomically dispersed catalysts for oxygen reduction and evolution in metal-air batteries. *Advanced Functional Materials, 33*(1), 2207097.
Kamat, P. V. (2008). Quantum dot solar cells. Semiconductor nanocrystals as light harvesters. *The Journal of Physical Chemistry, C112*(48), 18737–18753.
Kang, S., & Park, J. (2022). Quantum dot photocatalysts for visible-light-driven water treatment. *Applied Catalysis B: Environment, 301*, 120797.
Khan, J., Khan, S., Naseem, M. T., Zafar, A., & Ullah, H. (2022). Recent advances in quantum dot solar cells: From materials to devices. *Nano Energy, 100*, 107485.
Kim, S., Choi, H. J., & Park, J. (2021). Electron ptychography: High-resolution imaging for light-element materials. *Nature Communications, 12*(1), 4527–4536.
Kumar, P., Singh, R., & Mehta, D. (2021). Molecular dynamics simulations of pico-structured materials: Applications and challenges. *The Journal of Physical Chemistry, C125*(29), 16044–16058.
Kumar, S., Dutta, P., & Verma, R. (2023). Single-atom catalysts for CO_2 utilization. *ACS Catalysis, 13*(8), 4925–4941.
Li, Y., Wang, Z., & Liu, X. (2021). Picocatalysts for hydrogen evolution: Advances and perspectives. *Advanced Energy Materials, 11*(6), 2003419.
Li, C., Yang, Y., Zhang, H., & Wang, Q. (2022). Single-atom catalysts toward lithium-sulfur battery applications. *Journal of Materials Chemistry, A11*(4), 1654–1671.
Li, D., Sun, Y., Wang, M., & Zhao, C. (2023). Picotechnology-enhanced two-dimensional materials for next-generation supercapacitors. *Journal of Energy Chemistry, 77*, 1–17.
Liang, B., Tang, X., Li, J., & Wu, D. (2023). Atomic precision in battery electrode heterostructures for ultrafast energy storage. *Energy and Environmental Science, 16*, 271–288.
Lin, H., Zhang, Y., Liu, S., & Chen, F. (2023). Atomic-level design of porous electrodes for energy storage applications. *Chemical Society Reviews, 52*(4), 1303–1325.
Liu, Q., Jin, S., & Ma, D. (2020). Graphene quantum dots for electrochemical energy storage. *Materials Chemistry Frontiers, 4*(2), 421–436.
Luo, Y., & Chen, X. (2023). Quantum information science meets picotechnology: Prospects and challenges. *ACS Nano, 17*(2), 1124–1140.
Martinez, M., & Lopez, F. (2024). Green synthesis of picoengineered nanomaterials: Sustainable approaches. *Green Chemistry, 26*(3), 1245–1263.
Martins, R., & Silva, P. (2024). Solar-powered pico-catalytic systems for CO_2 valorization. *Journal of CO2 Utilization, 68*, 102223.
Nguyen, H., Park, T., & Lee, J. (2022). Atomic layer deposition and chemical vapor deposition for pico-thin film fabrication. *Surface Science Reports, 77*(5), 100570.
Novikov, A., Petrov, V., & Smirnov, D. (2023). Atomic manipulation using STM: Recent advances and perspectives. *Journal of Vacuum Science & Technology B, 41*(2), 021202.
Nozik, A. J. (2010). Quantum dot solar cells. *Nano Letters, 10*(8), 2735–2741.
Patel, A., & Singh, M. (2024). Atomic-layer modified CO_2 capture materials. *Journal of Materials Chemistry, A12*(10), 4005–4021.
Patel, N., Sharma, S., & Verma, M. (2023). Density functional theory in the study of atomic-scale materials. *Computational Materials Science, 220*, 111226.
Ramos, M. M., Fernández, J. F., & Valenzuela, R. (2022). Quantum confinement effects in nanostructured materials for energy storage. *Journal of Energy Chemistry, 64*, 1–20.

Sata, T. (2000). Proton-conducting membranes for fuel cells. *Journal of Membrane Science, 167*(1), 1–31.
Schirhagl, R., & Loretz, M. (2023). Quantum sensing for water quality monitoring: A review. *Analytical Chemistry Journal, 95*(4), 2123–2136.
Shi, Y., Li, H., Zhang, R., & Lu, X. (2023). Picoengineered interfaces for supercapacitor stability under extreme cycling. *ACS Nano, 17*(2), 1234–1245.
Singh, V., Kumar, P., & Sharma, A. (2023). Picofiltration membranes for next-generation desalination: Materials and mechanisms. *Journal of Membrane Science, 675*, 121299.
Singh, R., Wang, T., & Kumar, P. (2024). AI-guided discovery of pico-level energy materials. *Materials Today, 62*, 87–103.
Song, Y., Zhang, S., Cheng, L., & Xu, K. (2023). Advances in pico-doped solid-state electrolytes for safe lithium batteries. *Materials Today Energy, 33*, 101223.
Sun, X., Zhao, J., Li, Y., & Wang, H. (2022). Atomic-scale strain tuning for ultrafast energy storage electrodes. *Nano Letters, 22*(18), 7360–7369.
Suzuki, T., & Funatsu, K. (2023). Emerging techniques in subatomic material science. *Nano Today, 43*(1), 101548.
Tang, L., Song, Y., & Wang, X. (2021). Picoscale confinement in nanomaterials for energy applications. *Advanced Energy Materials, 11*(24), 2100245.
Tang, J., Brzozowski, L., & Sargent, E. H. (2023a). Photovoltaic advances through picoscale structuring. *Nature Nanotechnology, 18*(2), 110–119.
Tang, X., Liang, B., Zhao, Y., & Wang, J. (2023b). Pico-scale doping strategies to enhance solid-state electrolyte performance. *Journal of Power Sources, 567*, 232080.
Tisdale, W. A., & Williams, K. J. (2024). Ultrafast picoscale materials for solar energy applications. *Advanced Energy Materials, 14*(2), 2306675.
Verma, A., & Gupta, R. (2024). Designing multifunctional picoarchitectures for carbon capture and conversion. *Journal of Physical Chemistry C, 128*(5), 2023–2038.
Wang, Y., & Chen, S. (2021). Quantum-confined sensors for ultra-sensitive detection of pollutants. *ACS Applied Materials & Interfaces, 13*(48), 57531–57544.
Wang, Y., & Lee, C. (2022). Atomic force microscopy for picoscale resolution: Techniques and applications. *Nanotechnology Reviews, 11*(1), 34–52.
Wang, X., Zhou, M., Lin, Z., & Chen, J. (2022). Single-atom catalysts for stabilizing lithium-sulfur battery chemistry. *Advanced Materials, 34*(42), 2205062.
Wang, C., Yin, L., & Zhang, L. (2024). Environmental remediation using picoscale materials. *Environmental Science Nano, 11*(3), 715–733.
Wu, W., & Kim, H. (2024). Atomically precise CO_2 adsorbents: Design and applications. *ACS Materials Letters Journal, 6*(2), 260–274.
Yan, C., Zhang, Z., Wu, X., & Kang, F. (2023). Pico-engineering of carbon materials for ultra-fast supercapacitors. *Electrochimica Acta, 452*, 142231.
Yang, J., Wang, W., & Xu, H. (2017). Black phosphorus quantum dots for efficient photocatalytic hydrogen production. *Chemical Society Reviews, 46*(15), 4631–4644.
Yang, Y., Xu, J., Luo, W., & Lin, Z. (2023). Single-atom catalysts: Boosting next-generation electrochemical energy storage. *Chemical Reviews, 123*(4), 2393–2454.
Yao, Z., Wang, H., & Sun, J. (2021). Picostructured composite membranes for proton conduction at low humidity. *Journal of Materials Chemistry A, 9*(13), 7680–7689.
Yoon, B., Kim, K., Lee, H., & Choi, S. (2023). Grain boundary engineering in solid electrolytes using Picotechnology. *ACS Applied Materials & Interfaces, 15*(3), 4640–4650.
Zahir, M. H., Rahman, M. M., & Rahman, M. A. (2022). Quantum dots in energy storage devices: Current status and future prospects. *Energy & Environmental Materials, 5*(5), 1142–1161.
Zhang, L., & Wu, Y. (2024). Machine learning and multiscale modeling in Picotechnology. *Computational Materials Science, 233*, 112469.
Zhang, X., Wu, J., Chen, Y., & Huang, C. (2023). Defect engineering for advanced battery electrodes at the atomic scale. *Nano Today, 49*, 101742.

Zhou, Y., Xu, W., Liu, M., & Zhang, C. (2023). Picoengineered solid electrolytes for flexible and high-performance batteries. *Advancement of Science, 10*(4), 2205394.

Zhou, Y., Huang, L., & Li, X. (2024). Emerging trends in picoengineered energy storage materials. *Advanced Energy Material, 14*(3), 2300456.

Zhu, J., Huang, Y., & Wu, L. (2023). Advances in aberration-corrected TEM for atomic-scale imaging of nanomaterials. *Advanced Materials, 35*(7), 2207014.

Picotechnology in Computing and Communication

Yuvalakshmi Logesan, Shrikaant Kulkarni, M. Divya, R. Nisha, and P. Saranraj

1 Introduction

Similar to nanotechnology, Picotechnology blends technology and picometres. Nanotechnology's future lies in picometres, or trillionths of a meter. Compared to chemical measurements and processes, it requires a factor smaller than one-third of a nanometer and atomic-level explanation. The future will see the development of femto-technology, works with subatomic matter. It was found that nanotechnology could be replicated using Picotechnology. Technically, matter is handled at the pico-scale or trillionths of a meter (Bolonkin, 2009). The two orders of magnitude smaller than most chemical reactions and measurements, and three orders of magnitude shorter than nanometers and most nanotech. By altering the electron energy levels of other atoms, they can create new and exotic matter and transform into meta-stable types with unexpected new properties. The manipulation of atoms is the foundation of Picotechnology. While "pico-tech" is a specialized term like "nanotech," most technologies create pico-scale effects (Nitkin et al., 2005). It permits the possession of particular atoms' structures, orders, and even their internal states, which facilitates the creation of new atoms and molecules, including stable and practical exotic matter, as well as increased control over organic chemicals in nature and basic reactions.

Y. Logesan · M. Divya · R. Nisha · P. Saranraj (✉)
PG and Research Department of Microbiology, Sacred Heart College (Autonomous), Tirupattur, Tamil Nadu, India

S. Kulkarni
Sanjivani University, Kopargaon, India

Faculty of Business, Victorian Institute of Technology, Melbourne, VIC, Australia

S. Kulkarni (ed.), *A Journey from Nanotechnology to Picotechnology*, Lecture Notes in Nanoscale Science and Technology 47,
https://doi.org/10.1007/978-3-032-07228-3_11

A pico-hydro comes under the family of hydro because it can be owned by a single household. A pico-hydro turbine is beginning to appear as a viable solution for supplying the essential electricity requirements of isolated communities. For information processing and toolmaking, the two primary pillars of human endeavors, the ability to separate or capture individual atoms and molecules or manipulate their quantum states to achieve an expected perform is unquestionably a fantastic milestone (Meier & Fischer, 2011). After successful demonstrations such as imaging individual atomic locations in addition to enclosing a single atom or form bonds within molecules, there have been plans to advance beyond nanotechnology by going many times smaller into pico-technology and millions of times smaller into femto-technology, which deals with neutrons, protons, electrons, and other nuclear particles. Surprisingly, efforts are already underway to investigate femto and picotechnologies to overcome technological barriers, like speeding up electronics on–off switching to enhance communication. Because it allows information to be shared between devices, the Internet has completely changed human life. The Internet of Thing (IoT) (Spanner & Gloss, 2010) is a new network that has been created by enabling their connection. To put it another way, this network is being connected to "everything" that has an internet connection in order to give end users better and more intelligent solutions. IoT is created when multiple devices are connected to the Internet, allowing for interconnectivity between them. Communications, aerospace, hospitals, connected homes, and other digital spaces are just a few of the emerging and commercial technology use cases that have made extensive use of IoT in recent years. Next-generation wireless network applications that combine connectivity technologies, intelligent edges, or the Internet of Things (IoT) will benefit all domains, including e-health and medical Internet of things (M-IoT) (Bolonkin, 2010). Fifth-generation (5G) mobile network technology is unable to meet the demands of emerging mobile apps, include ultra-low latency, network intelligence, extremely high communication speeds, comprehensive connectivity, and the capacity to manage a range of related usages. The sixth-generation (6G) mobile network technology set new performance standards that the fifth-generation (5G) mobile network technology could not meet. For economically viable applications such as 3D communication and enormous Virtual Reality (VR)/Extended Reality (XR) applications, 6G capabilities must be delivered on a large scale. One solution to the striking rise in capacity to coverage needs is the deployment of multiple small cells to create Ultra-Dense Networks (UDN). But 6G networks in the future will be able to offer a high-performance connection with many connected devices, especially for requiring conditions like diverse mobility, active environments, or extreme density, according to the proposed study (Park et al., 2016).

2 PICO: Automatically Designing Custom Computers: s1

Video games, DVD players, TVs, scanners, printing equipment, cell phones, automobile, intelligent robot vacuums, mowers, and virtual pets have embedded computers (Miorandi et al., 2012). Analog circuits for telephone, video, and imaging have been supplanted by computers. Growingly creative smart devices to their integrated computers are foreshadowed by the increasing popularity of system-level integration (SLI). These smart devices require varying power, cost, and performance. High-performance supplies frequently need low cost to be useful. Customization increases performance and reduces cost because no off-the-shelf development is sufficiently specialized. The design stage time, designer availability, or design cost will all be reduced by the automation from embedded computer design, allowing for customization and the anticipated explosion of smart products (Balasubramanian, 2023).

2.1 *Pico-Architecture Synthesis System*

The production of optimized application-specific computer systems is automated by HP Labs in PICO project. PICO uses a C program to create high-performance, reasonably priced systems. PICO creates most economical a combination to these subsystems produces a variety of excellent system designs at various cost-performance tradeoff points. PICO retargets a compiler, assembler, as well as simulator to the customized VLIW processor, creates hardware parts in structural Verilog/VHDL, and modifies application code to add programs interfaces (Calvanese Strinati et al., 2020). From a vast, well-designed space, PICO automatically chooses the top concepts. PICO's framework-based, organizational design method develops required designs that design space exploration because no design can be preconstructed.

2.2 *Framework-based Automation*

PICO uses design framework automation. Framework includes,

- An architectural template with parameters that defines the design space.
- A spacewalker that uses a form of space exploration methodology.
- A constructor that uses library components to create designs.
- An appraiser to determine the quality of the design.

Together, these components form the foundation for automatically determining the approximate Pareto group of concepts each of that is superior to the others in at least one quality metric.

Template parameters serve as a design space with guidelines and restrictions to adhere to. Certain design components, like the presence and connection of modules, are preset in the template. Other design elements are defined by architectural features (Schlansker & Rau, 2000). To identify a design's key attributes and positive values, or what we refer to as parameters, heuristic search is required. As parameters are linked to a specification, a distinctive design is produced. This area is the place spacewalkers look for Pareto-optimal designs. Following an initial application review, the spacewalker, the consumer, or the template can define the range of a parameter. The spacewalker maintains a group of previously investigated designs to prevent duplication. Analysis costs and outcomes modeling can be used to eliminate construction and analysis costs (Schreiber et al., 2002). The spacewalker's specifications are used in design construction to produce a comprehensive design.

The synthesized machine must be abstractly described in the spacewalker specification. To build the design, this function builds lower-level component library components automatically. These components with structural properties, like bit width, are parameterized, the generator instantiates them automatically. Component design, optimization, and latency, power, and area characterization are done by hand (Satyanarayanan et al., 2009). An evaluator is also required for this structure in order to compute metrics for assessing the design specification, the detailed design, and both. Utilizing static estimation and simulation, PICO calculates the design's cost in gates and silicon area as well as its performance in cycles to carry out the application.

2.3 Hierarchical Design

To simplify design, designers break gadgets into subsystems do not communicate much. System design space is saved by combining the best subsystem designs to form complete systems. The system is divided into subsystems by hierarchical system design. For every framework, a spacewalker, constructor, evaluator, and subsystems specification space template are needed (Vaughan, 2019). Developers must determine the decomposition approach of a system framework before automating its design. However, system composition is a component of design space ex+ploration. The system model, constructor, along with evaluator structural decomposition are all included in a system decomposition. Subsystem evaluators must report system performance, but they may use various measurements than system evaluators. Without the eliminating Pareto design system-level subsystems, evaluators should reject weak subsystems. Only systems constructed from Pareto-optimal subsystems are allowed to occupy the planned space at each level in a hierarchical design. The design of the system's space is significantly reduced by this limitation (Ai et al., 2018).

2.4 Pico-Design Flow

PICO locates and removes each kernel after receiving a C program with several computer-intensive loops or kernels. The optimal Non-Programmable Accelerator (NPA) produces outputs that the PICO-NPA spacewalker then identifies on a continuous basis (Mao et al., 2017). In addition to generating Registration Transmit Level (RTL) design or VLIW coding to repeatedly initialize as well execute the NPA, the NPA constructor setups the kernel for parallelism and main memory bandwidth. The PICO-VLIW spacewalker maintains the best designs and continuously determines a VLIW processor. This system-level PICO spacewalker uses suitable VLIW, cache, and NPA designs to generate Pareto-optimal system designs (Mach & Becvar, 2017). VLIW kernels of information must be used as NPAs instead of software, according to the results of hardware-software dividing and codesign using PICO.

2.5 *NPA Design*

For a C loop nest with a range of performance restrictions and external memory bandwidth, PICO-NPA produces a Pareto set of NPAs. For every Pareto design, PICO creates a Verilog/VHDL structure, determines an NPA at its register transfer level, and initializes the external host processor's invocation code (Suarez-Albela et al., 2018). Interior memories (IMs), a memory external interface, and the memory-mapped the host processor interface with loop-specific instruction sequencers in addition to the pipeline function units (FUs) with adjustable widths can all be included in the architecture;

- Personalized interconnects for loop nests.
- Generalized set up walls in per-register read and write access, avoiding able to be addressed files.

The data path disperses loop nest operations among FUs in place of centralizing instruction storage. Data can be stored in register memory, internal memory, or external memory through loop-nest transformations and optimizations (Park et al., 2018). To determine each operation's issue and then assign it to a FU so that it's operates every II turns without resource conflicts, the constructor employs modulo scheduling. The constructor's heuristics minimize hardware costs both before and after scheduling and binding by clustering similar width operations or optimizing storage and connectivity. By simulating the initial source code, this NPA creation process enables engineers to identify PICO software and input specification problems. To each macro cells library component, parameterized formulas are used estimate design chip area and gates in the NPA cost evaluation. The NPA is developed into a parameter steady-state efficiency by the constructor. To determine NPA performance, the evaluator takes into consideration memory-induced stalls as well as time needed to fill and drain PE pipeline (Wachter, 2019).

3 Next-Generation Computing Systems: s2

3.1 *Edge Computing*

The development of electronics and communication can be linked to Edge computing (EC). After recognizing the difficulties with cloud-based computing, EC was suggested using cloudlets. Despite being similar the cloud, fog, cloudlet, and edge are different. Using clouds (a virtual computer for the Dell Optiplex 9010) rather than the internet can cut down on power consumption and latency in mobile devices by 42% and 51%, respectively. Additionally, EC also means by the "Working on the Edge" could cloudify the wireless network and make it a crucial component of it. The EC is a "marriage" between IT and telecom (Mujica et al., 2018). EC is a novel paradigm for data processing that tackles the challenges of cloud computing. Thus, EC attempts to use the cloud. Data is handled locally by EC devices, that only gather information and connect to other EC devices and a data center when necessary. Thus, in the organized cloud-based paradigm, EC is anticipated to address memory and latency constraints. The EC device terms are flexible because of the IoT's diversity and the EC field's youth. Still, to satisfy an EC device's growing performance and memory requirements, improvements in both hardware and software are required. EC devices process data before sending it to data centers in an unprocessed state. EC nodes are expected to decrease upload bandwidth or compute complexity for the cloud by implementing "real-signal or processing of images, combined optimization, agent-based modeling, analysis of big data, as well as more." Energy efficiency is influenced by smooth decision-making, efficient control, and secure services. Continuous innovation is required due to the power of cloud storage and computing solutions (such as Google Cloud Platform and Amazon's Elastic Compute Cloud (Cha et al., 2018), a growing number, type, or space dispersion of devices, along with the enormous volumes and even types of data generated. Data processing may demand different information, efficiency, or security owing to various IoT applications.

Cloud Computing, fog, or edge computing technology, their hierarchies, and the ways integration might be required for the best IoT services were recently covered in "multi-tier computing for intelligent IoT." Sensors with sophisticated hardware and software for gathering data are used all over the world. One simple sensing use case is the detection of one or as arrays of amounts from a terrestrial or just interplanetary environment by a single sensor or a network of intricately integrated sensors (Avgeris et al., 2019). Think of a sensor for Bacteria, Fungi, Methane, or Mercury. The sensor generates an initial signal that corresponds to the desired biological and chemical species when it is calibrated. Distributed sensing, such as in oil and gas, should benefit from EC. The vast array of EC applications that are quickly developing into different service levels demonstrates their combined influence. Solar cell efficiency can be maximized with spectral variation or environmental sensors to track temperature, humidity, pressure, wind speed, cloud movement, etc., potentially producing significant financial gains. Traffic lights can be made smart locally using hyperspectral images, chem-bio sensors, and weather sensors, along with signatures of the local population and infrastructure. Easy access to personal medical information, smart monitoring of health and healthcare framework, flexible framework for rapid detection of fires, disaster recovery strategies, accelerometers for structure monitoring of health, micro-seismic monitoring platform for the hydraulic fractures, enhanced multimedia communications (Nguyen et al., 2018), and a cyber-security FPGA system, Customs and Border Protection (CBP) officers who handle illegal substances and contraband require edge devices so they can make selections without interacting with other data center. There will be more sensors as a result of the opioid crisis, and EC machines with sufficient processing power and a quick connection can make critical decisions.

4 Edge Sensing to Edge Computing

4.1 Carbon Nanotube CPU and Edge Device

An essential nanosystem for EC and nanoelectronics is the carbon nanotube (CNTs). Many transistors made from silicon wafers were shrunk in size during the chip-making process, led to information loss, transistor failure, and unreal energy dissipation. Conductivity losses cause silicon transistors to lose energy and produce heat as they shrink. Compared to silicon, CNTs can use less energy because of their superior electron transport. Compactly arranged and integrated into new computer systems, carbon nanotubes (CNTs) are broad aspect-ratio quantum cables with ensuring electrical, plasmonic, thermal, and mechanical (Zhu et al., 2018). CNTs are advantageous in terms of heat, optics, and electricity. Single-walled carbon nanotubes (CNTs) with cylindrically rolled-up graphene layers have a useful electricity structure of bands with transport characteristics. CNTs use photoluminescence, electroluminescence, and illumination to produce light. Light intensity in excitonic states can be enhanced by current manipulation and cavity connection. In plasmonic and photonic design of environments, including a nanophotonic crystal cavity, the photonic density of wave states can be engineered to achieve optimal modal volume, optical quality factors, as well as Purcell enhancement. Although n-type semiconductors can be produced by electron doping, CNTs are likewise p-type semiconductors. By taking advantage of CNTs transport characteristics (electrons and the phonons in quantized or ballistic the conductance), a transistor based on nanotubes or device may perform better than silicon-based the transistor on the same scale. According to the International Technology Roadmap for Semiconductor materials (ITRS) (Zhang et al., 2018), "Miniaturization of logic transistors by technology "nodes, " with fewer suggesting newer technologies for faster and smaller devices." These requirements were recently satisfied by a 40-nanometer CNTs transistor. CNTs will therefore be essential to next-generation sensors and CPUs, that will make powerful EC devices possible.

4.1.1 Topological Edge States to Aid Edge Computing

To new optical signals electrical, and quantum hardware, topological materials hold promise. These materials have intrigued the surface and edge states that may allow for quick, low-power communication and computation. Certain states (such as edge states) allow electron conductance as well as photon the emission without being impacted by defects or disturbances, exhibiting topological behavior in electronic, photonic, or atomic systems (Sun et al., 2018). Time-reversal symmetry and topology affect the electrical and photonic reactivity of materials like topological insulation, which transfers charge on their surface and insulates within. Due to their Geometry, Topological materials' electrical band structures allow unexpected surface states. High-symmetry Brillouin zone sites are revealed by computing the constant features of nonmagnetic topological electrical materials. Like mathematical topology, the topological characteristics that materials have in their electrical and photonic band forms connect preserved values with ongoing deformations to produce new materials. Wavevector-space structure can support photonic states that allows light propagates on edges or boundaries without disturbance, while specific designs in electrical band forms allow for significant transport features (Athavale & Krishnan, 2018). The topological characteristics of low-dimensional resources are particularly interesting since graphene and carbon nanotubes are being researched for next-generation processors. To increase processor size and performance, topological materials, like CNTs, must be produced and integrated on a large scale.

4.1.2 Nanophotonic and Plasmonic to Aid Edge Devices

In many communications applications, such as long-distance and high-speed information network use, photons outperform electrons. For the development of processors, logical or compressed quantum states of light are also being studied. Intriguing near- and far-field phenomena can be produced by photons connecting to other quasiparticles in materials to gather electronic density oscillations, such as plasmonic excitation. Reading, Amplification, Modulation, and other processes can be facilitated by plasmonic devices, can result in on-chip nearfield devices and features like plasmon-induced transparency (Oueida et al., 2018). Single-photon quantum light sources and detectors incorporating nanophotonic elements enable all-optical processors, secure rapid communications, and metrology. The development of EC devices will be accelerated by these capabilities, which will also benefit AI hardware platforms and quantum network edge nodes. For ultra-small high-bandwidth components like frequency synthesizers, chip signal processors, and electro-optic modulators, we developed integrated microwave photonics. On-chip differential equation integrators show photonics' potential for high-speed computing and data processing. Adding cellular automata, computing functionality, and chip-to-chip interconnects (Rosario et al., 2018). Since photon propagation is controlled by the natural periodic structure, photonics provides greater versatility in crystals. In materials with synthetic periodic patterns, propagation over specific spectral bands can be controlled or forbidden. Photonic bandgap materials help optical processors (switches, modulators) just like silicon-based semiconducting materials. The challenges of localizing and confining photons outweigh these photonic advantages, making device integration and miniaturization challenging. In the majority of metal-based nanotechnology that facilitates collectively electronic excitation, the nonradiative decay of plasmons produces heat and might not be appropriate for tight integration (Rodriguez et al., 2018).

4.1.3 Quantum Processor and Computing to Aid Edge Devices

Quantum sensors may be used by quantum edge devices to produce data that quantum computers or algorithms can process and exchange via the quantum internet. Dispersed quantum sensing, which employs quantum correlations across various sensors to determine unknown characteristics more accurately than unentangled sensors, may be advantageous for EC (Chen & Chuang, 2017). It is necessary to investigate their special synergy in order to fully utilize the energy of EC with the quantum detecting, whether distributed or not. The growth of EC devices raises serious security concerns, and quantum effects could provide effective answers. Applying quantum superposition states and photons, quantum key distribution can provide unrestricted communication security beyond distant parties. For EC to process data more quickly, an additional hardware platform is needed. Cellular communication makes use of EC. The EC enhances a ground-based 5G networks mobile communication network's delay, computational capacity, and resource utilization. A 5G satellites EC architecture has been suggested as an improvement to terrestrial 5G EC. One example of how quantum effects can improve satellite communication and assist EC in overcoming cloud computing challenges is single-photon qubit teleportation to ground-to-satellite interface. Another great example is the use of the distribution of quantum keys for secure satellite-to-ground communication (Idrees et al., 2018). Global quantum networks could therefore appear soon. Therefore, at the nascent stages of quantum technologies, quantum networks with edge nodes are being researched and developed. The photon sensors, the transmitters, readers, buffers, memory, and other quantum subsystems can be housed on these network nodes. These nodes are necessary for connecting to optical quantum circuits in quantum sensing and other quantum information systems. In quantum networks, edge nodes use a network of app architecture to deliver quantum application layer peer-to-peer services. This design distributes workloads and responsibilities for quantum applications among peers with equal privilege. A quantum edge device might encourage research, quantum analyzing for EC requires the development of Quantum Information Theory (Ferrandez Pastor et al., 2018).

4.1.4 Neuromorphic Computing and Edge Devices

The limitations of silicon technology might be overcome by neuromorphic computing. The intricate dynamics and responses of biological systems are being exploited in various fields, including sensing and computing. Neural circuits and early brain-like technologies are briefly discussed in "neuromorphic engineering" (Lau et al., 2018). In contrast to serial digital computers, neuromorphic or brain-inspired computing seeks to produce massive parallel analog computations at high speed and low power. This current "evolvable natural electrochemical transistor" that imitates the biological synapse is one of the fundamental components needed to build neuromorphic devices. In its operating environment, the first synaptic device a transistor channel composed of electropolymerized conductive polymer created new synapses that resemble biological synapses. Learning and memory are made possible by the brain synapse, an active memory unit. A sensible method for developing neuromorphic processors is synapse emulation, but this approach requires a scalable and effective physical foundation that can be included in a combined architecture (Park et al., 2017). Memristive devices are used in sparse coding and dimension reduction for neuromorphic computing. Neurones have been simulated using Nonlinear, memory-preserving nanoscale spintronic oscillators. Other recent developments include the electrochemically based neuromorphic factors natural inexpensive plastic device, neuromorphic circuits that simulate connection, memristive reaction employing ionic interactions in MoS2, quantum phenomena in superconductors in in photonic integrated-circuit synapses, the quantum dots-based, oxide-based, along with single flux (Pyatkov et al., 2016). The development of next-generation computing systems leverages diverse picosystems, including carbon nanotubes, quantum dots, and topological materials, each tailored for specific applications such as qubits, interconnects, and transistor channels (Table 1). For instance, GaAs and silicon photonics components exploit excitons and spin states for quantum computing, while superconductors and trapped ions serve as foundational qubit platforms.

Table 1 Development of the next-generation computing systems

Picosystem	Typical excitation	Application	References
Carbon nanotubes	Electron and hole transport	Transistor channel, vias, cooling, and connectors	He et al. (2018)
Both nanowires and nanoantenna	Plasmons	Qubit, connection, along with interconnect	Yan et al. (2011)
GaAs, doped and undoped Si, and additional quantum dots	Excitons	Qubit	Petersson et al. (2012)
Components of silicon photonics	Electron and hole donor states, charge, and spin	Qubits, transistor material, and quantum computing	Hafezi et al. (2013)
Organic materials	Charge	Transistor material	Crone et al. (2000)
The nitrogen vacancy	Spin states	Qubits and quantum computing	Siampour et al. (2017)
Trapped ions	-	Qubits and quantum computing	Monroe and Kim (2013)
Superconductors	Electron	Qubits and quantum computing	Zhang et al. (2018)
Multilayers	Bloch surface waves	Interconnect	Silva et al. (2014)
Topological materials	Photons, plasmons, phonons, and electrons	Components for a transistor, qubit, connection, and interconnect	Zhang et al. (2019)

5 Internet of Medical Things (IoMT): s11

To evaluate and suggest medications, sensors, an internet of medical things (IoMT), and other medical equipment gather patient data. In an intensive care unit, the right sensors were taken into account (Xu et al., 2017).

5.1 IoMT Sensor Monitoring Circuits

The Medical Internet of Things (IoMT) has a lot to offer the Healthcare Industry. The products and services provided by this industry enhance healthcare, lessen the workload for physicians' patients can receive care from nurses and at home. Cloud computing and IoT integration in medicine are made easier by the IoMT application framework. It also outlines the way to supply a medical service network with patient data from numerous sensors and medical devices (Chen et al., 2006). Sphygmomanometers (blood pressure tracks), endoscope monitoring, pulse oxygen meters, EEGs, ECGs, EMGs, and other devices are used to keep an eye on patient conditions.

5.1.1 Glucometer Monitoring

The gadget records a person's blood Glucose levels. The Glucometer measures the amount of glucose in the blood by placing a drop of human blood on a single test strip. Consequently, Diabetes develops after blood sugar levels remain high for a long time (Zhang et al., 2019). It is regarded to be one of our most common diseases in the human population. The three primary types of diabetes that are commonly found are Type 1 and 2 Diabetes, along with Gestational diabetes. The disease and its various forms can be identified using three tests,

(a) The oral tolerance to glucose test
(b) A fasting blood glucose
(c) Random plasma glucose tests

The most common diagnostic method for detecting diabetes is "Fingerpicking," that is followed by blood glucose testing. Recent developments in IoT technology have led to the development of a several non-invasive, comfortable, helpful, and safe wearables blood glucose screening devices (Xue et al., 2019). An m-IoT-based non-invasive the Glucometer for monitoring blood glucose levels in real time.

5.1.2 Temperature Sensor Monitoring

This gadget lets patients measure their body temperature because it can be used to detect many diseases that change over time (Vergniory et al., 2019). Based on the conditions, the doctor or as physician can then administer treatment. Human body temperature is used in many diagnostic procedures to show homeostasis. Body temperature can change due to trauma, infection, and other factors. By monitoring a patient's temperature changes, doctors can evaluate their health in a number of conditions. A mouth, ear, and rectal thermometer are used to take traditional temperature readings. To measure the temperature of the Tympanic membrane core, a wearable, 3D-printed gadget with an infrared sensor was suggested (Siampour et al., 2017). The gadget has an information processing unit and a wireless sensor module. Physical exertion and environmental conditions have no effect on the recorded temperature.

5.1.3 Blood Pressure Sensor Monitoring

The sensor measures the systolic and diastolic blood pressure. It assists physicians in keeping an eye on patients' cardiac issues (Savage et al., 2012). Blood pressure is measured by this circuit's pressure sensor. Good linearity, low noise, and little external stress are provided by the ceramic chip with nylon plastic. To increase accuracy, stability, and repeatability, it also makes use of temperature correction and the internal standard approach. Blood pressure testing must be a part of any diagnostic procedure. The data must be maintained by someone using the conventional blood pressure method. IoT and various sensing technologies have changed blood pressure monitoring. In a wearable cuffless device that measures the diastolic and systolic pressure were suggested. Cloud storage can hold collected data. A multi-biosensor, designed to be wearable and non-invasive, provides continuous blood pressure monitoring for clinical use. Arterial impedance and Plethysmography waveform performed the best among the various models tested for pulse transit time (PTT)-BP assessment. This cutting-edge blood pressure monitoring system organizes and stores health data in the cloud using the Internet of Things (Zhao et al., 2019).

5.1.4 Airflow Sensor Monitoring

By tracking the patient's respiration rate, this sensor assists the physician in diagnosing the condition and prescribing treatment. A noninvasive method of measuring oxygen saturation, pulse oximetry, is essential for healthcare analysis. Pulse oximetry has been improved by IoT that has multiple medical uses. Heart rate, pulse, and blood oxygen saturation are all measured by the non-invasive tissue oximeter (Welser et al., 2018). Wi-Fi and Zigbee can also be used to send data to the server. To continuously track physiological indicators like blood pressure and body temperature, Researchers designed a wireless and flexible biosensor patch using electrocardiography. This patch is reliable and durable for arched skin because it is optimized for both mechanical stretchability and adaptability. The IoMT platform, cloud server, mobile app, and web-based remote monitoring were demonstrated for health management with diagnostics (Lereu et al., 2013).

5.1.5 ECG Sensor Monitoring

The Electrocardiogram (ECG/EKG) sensor is used to diagnose cardiac electrical and muscle activity. One of the most significant medical checkups, ECG/EKG helps assess cardiac tests such Myocardial Ischemia, Infarction, Heart Failure, Arrhythmia. An electrocardiogram (ECG) shows the heart's electrical activity from atria and ventricle depolarization and repolarization. ECGs diagnose cardiac problems and reveal heart muscle rhythms. ECG monitoring by IoT technologies may help identify heart disorders (Lereu et al., 2008). This t-shirt-mounted ECG monitor was small and low-power. A biopotential chip provided high-quality ECG data. Bluetooth transmitted recorded data to users. A smartphone app could display ECG data.

5.1.6 EMG Sensor Monitoring

Muscle electrical activity is measured by Electromyography (EMG) sensors both during contraction and during rest. EMG signals are used in various biological and clinical fields. Mostly used to diagnose kinesiology, motor control abnormalities, lower back pain, and neuromuscular disorders. EMG signals are used to operate prosthetic hand, arms, lower limbs, and other devices (Liao et al., 2017). Along with the elbow-top reference electrode the muscle electrodes that are measuring E1 and E2 entered the preamps that are at three inches from bicep top and bottom. The oscilloscope connects to the output for confirmation. DC voltage is impeded by a capacitor, and the fourth operational amplifier serves as a high-pass filter between the oscilloscope probe and the differential amplifier. It is possible to assess the output (Zhuang et al., 2018).

5.1.7 Breathing Rate Sensor Monitoring

Breathing issues are brought on by chronic illnesses like asthma that restrict airflow. In asthma, airway edema can cause airways to narrow. The wheezing coughing, chest pain, and shortness of breath are symptoms of numerous medical conditions. The only an inhaler or nebulizer can save an asthmatic from an attack, that can occur at any time. Numerous IoT-based asthma tracking systems have been developed recently (Yan et al., 2019). A sensor in a discrete IoT asthma solution tracked respiratory rate. An LM35 temperature sensor is used by his respiratory monitoring or alarm system to determine breathing rate (Cheng et al., 2017). Air intake and exhalation were measured.

5.1.8 Mood Monitoring

Mood tracking provides vital emotional information and supports mental health. Additionally, it helps doctors treat bipolar disorder, stress, depression, and other conditions. Interpretation of mental states is improved by personal psychological monitoring. According to a CNN network, moods can be categorized as happy, excited, sad, calm, upset, and furious. Real-time emotion was measured by the interactive “Meezaj” (Van de Burgt et al., 2018). Additionally, by demonstrating the significance of happiness in decision-making, the software assisted policymakers identify important components of happiness. The heart rate can predict anxiety using a state-of-the-art machine learning system. Additionally, the technology can alert the patient to stress (Indiveri & Douglas, 2000). The nano-biosensors measure blood pressure, serotonin, and pulse sensor detect abnormal cardiac rate, along with Galvanic Skin Response (GSR) sensors detect skin temperature variations.

5.1.9 COVID-19 Monitoring

Through fever tests, IoT technology assists in identifying the infectious agent, taking into account its symptoms, and preventing its spread through social distancing. It also keeps an eye on smart parking, occupancy control, pollution, including air quality management, and remote health surveillance. Numerous targets are identified by fever screening without the need for human interaction. Cameras and sensors give images and calibrated color temperatures. The first line defense against viruses is pre-screening patients, employees, and disaster evacuees (Bartolozzi, 2018). The respiratory virus SARS-CoV-2 causes COVID-19. Researchers in suggest a device that measures respiratory system volume and breathing rate using a strain sensor. These sensors track patients’ breathing and notify doctors via an Internet of Medical Things (IoMT) connection. An early coronavirus infection is detected by an IoT-based system. To automate attention-based deep 3D multi-instance learner (AD3D-MIL) was developed for COVID-19 chest CT testing (Zhu et al., 2019). A distinct medical administration network is produced by the Internet of Things (IoT) from a number of networked devices. To keep patients safe, it checks and warns of any illnesses. Doctors can treat coronavirus patients more quickly thanks to IoT-enabled devices. For IoT devices, an automated categorization system based on data is required.

6 Ultra-Dense Network: s19

6.1 Small Cell Type

6.1.1 Macrocells

Compared to microcells, macrocells are more extensive. At a height that is visible to nearby buildings and assets, macrocells antennas are supported by stanchions, roofs, and other ground structures. Furthermore, the average emission from macrocells is tens of watts. High BSs are used in rural areas to cover greater distances, and macrocells are used to define the largest cell range size (Kumar et al., 2018). These BSs can be found in highway or rural areas. High transmission power is required from the macro-cell BS for optimal coverage. Operators consistently provide minimal coverage for voice calls outside of these areas due to the small number of customers and the high cost of powerful BSs.

6.1.2 Microcell

In densely populated areas, such as highways and urban skyscrapers, microcells boost system capacity. They are frequently used to create high-density mobile phone networks that are less expensive than Picocells. Microcells use power management to restrict the coverage radius as well. Although microcells expand channel capacity, they make managing radio resources more difficult. Microcells are more negatively impacted by traffic and obstacles from short-term to long-term changes (new buildings) than macrocells (Page et al., 2021).

6.1.3 Picocell

A 100 meter insurance area can be covered by operators using picocells, which are tiny BSs. A Picocell transfers internet users from macrocells to help tens of active customers in heated environments, whether indoors or out (Vimbi et al., 2023). Picocells are appropriate for a variety of applications due to their 33 dBm transmission strength. Picocell backhauling offers an ideal connection with minimal latency and substantial bandwidth, much like macrocells (microwave or fiber lines).

6.1.4 Femtocell

Femtocells are user-installed, indoor base stations for homes, offices, and meeting spaces that support a limited number of users (Nguyen Gia et al., 2019). To consumers with a strong indoors signal, the majority of data traffic is generated by a femtocell. Femtocells are connected to the network via DSL, cable, or fiber internet connections.

6.1.5 Relay

Relays installed by the operator enhance macrocells edge execution to cover dead areas. Macrocells use wireless backhaul to transmit and receive user data. Relays and picocells are similar in terms of coverage and transmit power, but they are different in three ways (Bhat & Bhat, 2017). Picocells are complete BSs, relays are extensions of macrocells. Picocells relay coverage and supply capacity. Microcells in mobile phone the networks are serviced by low-power cell sites.

6.2 *Vertical and Horizontal Densification*

As tiny cells proliferate, wireless networks get denser. There are two types of network densification: Horizontal and Vertical. Hotspots and streets have a horizontal density (Gia et al., 2017). Houses, offices, conference rooms, and buildings with BSs are all examples of vertical densification. From this classification, three key components specifically, densification scheme modeling can be distinguished. To depict the network's reality, a 3D vertically densified network is required. A backhaul can be replaced by the vertically dense network. Additionally, they contributed to the development of 3D fading channels (Ota et al., 2017). The vertical and horizontal performance of network densification differs. Spectral efficiency is increased in tall buildings with multiple vertical compression levels.

6.2.1 G Ultra-Dense Cellular Network

Gigabit wireless traffic is transmitted by hundreds of large-scale MIMO antennas in the fundamental architecture of 6G cellular networks (Pradhan et al., 2021). Each antenna's transmit power needs to be reduced if the 6G BS connection power is restricted to the 5G BS share power. The 6G BS radius ought to drop by an order of magnitude as the antenna transmit power drops. 6G cellular networks depend on millimeter waves. These technologies can help small cellular networks transition to 6G networks. As a result, 6G networks will be extremely dense (Phan et al., 2022).

6.2.2 Architecture of 6G Ultra-Dense Cellular Network

Many small cells, large MIMO antennas, or millimeter wave technology for communication will all be used in ultra-dense 6G wireless networks (Lingaiah et al., 2013). Therefore, the first challenge is designing the 6G ultra-dense cell network. These extremely dense wireless networks are spread by both single and multiple gateways.

6.2.3 Conventional Cellular Network Architecture

A typical Cell Network Architecture consists of a tree network with a gateway for all backhaul traffic and B administrators in the fundamental network managing each macrocells base station (Agustine et al., 2018). Femtocell, Picocell, or hotspot microcell deployments are supported by a hybrid architecture. In this hybrid network, the microcell network is designed as a tree network. Both macrocellular and microcellular coverage overlap. Microcell BS backhaul traffic is sent to the core network via fiber cables or broadband Internet. Unlike macrocells, microcell base stations can wirelessly communicate at high speeds both indoors and in hotspots. This design facilitates macrocells networks that transmit subregions wirelessly at high speeds (indoor or hotspot environments) (Phan et al., 2022).

6.2.4 Distribution Architecture

The high-density use of small cells in 6G mobile networks is characterized by large MIMO antennas and millimeter-wave communication technology (Tekeste et al., 2019). Since millimeter-wave technology limits wireless transmission range, small-cell base stations frequently aren't able to send wireless backhaul traffic straight to a specific gateway. As a wireless backhaul communication must be transmitted to the gateway via multi-hop. A viable approach for ultra-dense 6G wireless networks is distributed network architecture. As a result, networks of macrocells are not enhanced by networks of small cells. Rather, small cells and macrocells are also a part of ultra-dense 6G mobile networks (Bathilde et al., 2018).

6.2.5 Energy Efficiency

Energy efficiency (EE) will be given more priority in future wireless communication technologies than it was in 5G and 6G. Global concerns about indirect energy use, ecological harm, and reliance on natural resources have made sustainable/green communication technologies more attractive. There is a need for new energy-efficient wireless network frameworks and techniques (Adamovich et al., 2005). Although the concept of sustainable 6G communication is arbitrary, the technology can lessen environmental harm and energy consumption. Energy-efficient sleeping or switching modes, resource allocation, and energy storage may all contribute to the development of environmental-friendly communication networks. Creating wireless networks for communication that optimize energy use in small-scale access points, integrate energy harvesting devices, and distribute limited resources efficiently can all help to increase energy efficiency (Qi et al., 2018). Methods for optimizing energy efficiency were selected without taking into account their cost or economic feasibility.

7 Pico-hazards and Safety: s4

Human health is at risk because remote locations frequently have safety regulations and are disregarded (Singh et al., 2021). Since electrical connections carry electricity and have the potential to injure users, they are the most crucial Pico-safety precautions. Because of their strength, durability, safety, and low current leakage, users should therefore use high-quality cables. In Pico-hydro systems, cables can be as long as 1.5 km, that makes them costly. Most PICO users have experienced energy loss, breakage, and difficulty identifying non-insulated wires with small diameters that hang low on bamboo poles or rest on the ground (Holma et al., 2021). Pedestrians run the risk of dying or receiving an electric shock they come into contact with uninsulated cables. To make Pico-hydro safer and more effective, the following safety factors are crucial:

- Set up the load controller: Although an ELC is somewhat expensive, it offers constant voltage and safeguards appliances and electrical equipment (Alam et al., 2021). As a result, the generator will last longer and cost less to replace electrical appliances.
- The Residual Current Device (RCD) should be installed: An RCD is necessary for PICO users to prevent electric shock. Not all users will be turned off by an RCD; only those with broken equipment and devices will be (Solyman & Yahya, 2022).
- Take into account utilizing premium cables: Select the appropriate cable and avoid compromising quality for cost in order to prevent electric shock both indoors and outdoors (Holma et al., 2021).
- Due to a lack of experience, end users should be trained and involved in the installation of PICO hydropower projects to educate them about electrical hazards (Ravikumar et al., 2024).

8 Applications of Picotechnology

Picotechnology could make it possible to produce stronger, lighter, and more fuel-efficient materials that use less energy, generate less waste, and operate better in ships, airplanes, and spacecraft (Xu et al., 2021). Picocoatings can shield translucent or opaque surfaces from radiation, abrasion, and corrosion. New photoelectric materials for affordable solar energy panels, Molecular-semiconductor hybrid gadgets for the upcoming information age revolution and picoscale electrical, magnetic, or mechanical systems and devices enabling unprecedented information processing chemical, photochemical, or biological sensors to safety, health, production, or the environment are all feasible. Significant improvements can be made in terms of health, safety, quality of life and environmental protection (Joud et al., 2018). We may live longer to PICO medical technology in two ways. It can reverse aging, repair our cells, lengthen telomeres, and assist medical professionals in curing deadly illnesses like Cancer, Heart Attacks, Strokes, and HIV. Picotechnology can extend life expectancy past the previous century and treat serious illnesses. Picotechnology in healthcare could lower the number of deaths from cardiovascular disease and cancer over the next 10 years. Numerous research initiatives are focused on these techniques (Ge et al., 2016). By affixing magnetic particles to cancer cells in the bloodstream and eliminating them before they develop into new tumors, medical Pico-technology may be able to cure cancer. Similar research initiatives are looking into ways to combat heart disease, is another major cause of death. There are several projects in progress. "Santa Barbara University researchers created a nanoparticle that targets plaque in artery walls to deliver drugs." Another objective of medical Pico-technology is to repair cells to increase longevity. The most intriguing application, perhaps. Pico-technology can repair our cells (Alhammadi et al., 2019).

9 Conclusion

Picotechnology is a huge leap forward in computing and communication that could change many in industries. By working at an incredibly small scale, it creates new technologies that were once only fiction. PICO show how we can design custom, high-performance computer for specific needs. And also works well with next gen networks like 6G, helping to build faster, and more reliable connections using their tiny cells and advanced antennas. In Healthcare, this Picotechnology makes tools like wearable sensors and non-invasive medical devices even better, allowing for real-time health monitoring. Their ability to monitor vital signs continuously and transmit data securely via 6G networks could revolutionize patient care, making more proactive and personalized. As research and development continue to push the boundaries across disciplines like environmental science, biotechnologies, and so on will unleash their complete potential. There upgrade is a complete game changer. Picotechology opens doors to breakthrough that are just mere starting.

References

Adamovich, S. V., Merians, A. S., Boian, R., Lewis, J. A., Tremaine, M., Burdea, G. S., Recce, M., & Poizner, H. (2005). A virtual reality-based exercise system for hand rehabilitation post-stroke: Presence teleoperators. *Virtual Environment, 14*, 161–174.

Agustine, L., Muljono, I., Angka, P. R., Gunadhi, A., Lestariningsih, D., & Weliamto, W. A. (2018). Heart rate monitoring device for arrhythmia using pulse oximeter sensor based on Android. In *2018 International Conference on Computer Engineering, Network and Intelligent Multimedia (CENIM), Surabaya, Indonesia* (pp. 106–111).

Ai, Y., Peng, M. G., & Zhang, K. C. (2018). Edge computing technologies for internet of things: A primer. *Digital Communications and Networks, 4*, 77–86.

Alam, M. S., Kurt, G. K., Yanikomeroglu, H., Zhu, P., & Đào, N. D. (2021). High altitude platform station based super macro base station constellations. *IEEE Communications Magazine, 59*, 103–109.

Alhammadi, A., Roslee, M., Alias, M. Y., Shayea, I., Alriah, S., & Abas, A. B. (2019). Advanced handover self-optimization approach for 4G/5G HetNets using weighted fuzzy logic control. In *15th international conference on telecommunications (ConTEL), Graz, Austria* (pp. 1–6).

Athavale, Y., & Krishnan, S. (2018). A device-independent efficient actigraphy signal-encoding system for applications in monitoring daily human activities and health. *Sensors, 18*, 2966–2970.

Avgeris, M., Spatharakis, D., Dechouniotis, D., Kalatzis, N., Roussaki, I., & Papavassiliou, S. (2019). Where there is fire there is SMOKE: A scalable edge computing framework for early fire detection. *Sensors, 19*, 639–655.

Balasubramanian, P. (2023). Automation in data science, software, and information services. In S. Y. Nof (Ed.), *Springer handbook of automation* (pp. 989–1014). Springer.

Bartolozzi, C. (2018). Neuromorphic circuits impart a sense of touch. *Science, 360*, 966–967.

Bathilde, J. B., Then, Y. L., Chameera, R., Tay, F. S., & Zaidel, D. N. A. (2018). Continuous heart rate monitoring system as an IoT edge device. In *2018 IEEE sensors applications symposium (SAS), Seoul, Republic of Korea* (pp. 1–6).

Bhat, G. M., & Bhat, N. G. (2017). A novel IoT based framework for blood glucose examination. In *2017 International conference on electrical, electronics, communication, computer, and optimization techniques (ICEECCOT), Mysuru, India* (pp. 205–207).

Bolonkin, A. A. (2009). Converting of matter to nuclear energy by AB generator. *American Journal of Engineering and Applied Sciences, 2*, 683–693.

Bolonkin, A. A. (2010). Femtotechnology: Design of the strongest AB matter for aerospace. *Journal of Aerospace Engineering, 23*, 281–292.

Calvanese Strinati, E., Barbarossa, S., Choi, T., Pietrabissa, A., Giuseppi, A., De Santis, E., & Kim, I. (2020). 6G in the sky: On-demand intelligence at the edge of 3D networks. *Electronics and Telecommunications Research Institute Journal, 42*, 643–657.

Cha, H. J., Yang, H. K., & Song, Y. J. (2018). A study on the design of fog computing architecture using sensor networks. *Sensors, 18*, 3633–3640.

Chen, C. L., & Chuang, C. T. (2017). A QRS detection and R point recognition method for wearable single-lead ECG devices. *Sensors, 17*, 1969–1976.

Chen, Z. H., Appenzeller, J., Lin, Y. M., Sippel-Oakley, J., Rinzler, A. G., Tang, J. Y., Wind, S. J., Solomon, P. M., & Avouris, P. (2006). An integrated logic circuit assembled on a single carbon nanotube. *Science, 311*, 1735–1742.

Cheng, Z., Rios, C., Pernice, W. H. P., Wright, C. D., & Bhaskaran, H. (2017). On-chip photonic synapse. *Science Advance, 3*, e1700160.

Crone, B., Dodabalapur, A., Lin, Y. Y., Filas, R. W., Bao, Z., LaDuca, A., Sarpeshkar, R., Katz, H. E., & Li, W. (2000). Large-scale complementary integrated circuits based on organic transistors. *Nature, 403*, 521–523.

Ferrandez Pastor, F. J., Garcia Chamizo, J. M., Nieto Hidalgo, M., & Mora Martinez, J. (2018). Precision agriculture design method using a distributed computing architecture on internet of things context. *Sensors, 18*, 1731–1740.

Ge, X., Pan, L., Tu, S., Chen, H. H., & Wang, C. X. (2016). Wireless backhaul capacity of 5G ultra-dense cellular networks. In *IEEE 84th vehicular technology conference (VTC-fall), Montreal, Canada* (pp. 1–6).

Gia, T. N., Ali, M., Dhaou, I. B., Rahmani, A. M., Westerlund, T., Liljeberg, P., & Tenhunen, H. (2017). IoT-based continuous glucose monitoring system: A feasibility study. *Procedia Computer Science, 109*, 327–334.

Hafezi, M., Mittal, S., Fan, J., Migdall, A., & Taylor, J. M. (2013). Imaging topological edge states in silicon photonics. *Nature Photonics, 7*, 1001–1005.

He, X., Htoon, H., Doorn, S. K., Pernice, W. H. P., Pyatkov, F., Krupke, R., Jeantet, A., Chassagneux, Y., & Voisin, C. (2018). Carbon nanotubes as emerging quantum-light sources. *Nature Materials, 17*, 663–670.

Holma, H., Viswanathan, H., & Mogenses, P. (2021). *Extreme massive MIMO for macro cell capacity boost in 5G-advanced and 6G*. White paper, Nokia Bell Labs.

Idrees, Z., Zou, Z., & Zheng, L. (2018). Edge computing based IoT architecture for low-cost air pollution monitoring systems: A comprehensive system analysis, design considerations and development. *Sensors, 18*, 3021–3030.

Indiveri, G., & Douglas, F. (2000). Robotic vision—Neuromorphic vision sensors. *Science, 288*, 1189–1190.

Joud, M., García-Lozano, M., & Ruiz, S. (2018). User specific cell clustering to improve mobility robustness in 5G ultra-dense cellular networks. In *14th Annual Conference on Wireless On-demand Network Systems and Services (WONS), Isola, France* (pp. 45–50).

Kumar, M., Abbas, S., & Kim, J. (2018). All-oxide-based highly transparent photonic synapse for neuromorphic computing. *ACS Applied Materials & Interfaces, 10*, 34370–34376.

Lau, C., Srimani, T., Bishop, M. D., Hills, G., & Shulaker, M. M. (2018). Tunable n-type doping of carbon nanotubes through engineered atomic layer deposition HfOX films. *ACS Nano, 12*, 10924–10931.

Lereu, A. L., Passian, A., Farahi, R. H., van Hulst, N. F., Ferrell, T. L., & Thundat, T. (2008). Thermoplasmonic shift and dispersion in thin metal films. *Journal of Vacuum Science and Technology, 26*, 836–841.

Lereu, A. L., Farahi, R. H., Tetard, L., Enoch, S., Thundat, T., & Passian, A. (2013). Plasmon assisted thermal modulation in nanoparticles. *Optics Express, 21*, 12145–12158.

Liao, S.-K., Cai, W.-Q., Liu, W.-Y., Zhang, L., Li, Y., Ren, J.-G., & Yin, J. (2017). Satellite-to-ground quantum key distribution. *Nature, 549*, 43–47.

Lingaiah, T. B., Kumar, D. H., & Nagaraja, C. (2013). Measurement of pulse rate and SPo2 using pulse oximeter developed using LabVIEW. *IOSR Journal of Electrical and Electronics Engineering, 8*, 22–26.

Mach, P., & Becvar, Z. (2017). Mobile edge computing: A survey on architecture and computation offloading. *IEEE Communications Surveys & Tutorials, 19*, 1628–1656.

Mao, Y. Y., You, C. S., Zhang, J., Huang, K. B., & Letaief, K. B. (2017). A survey on mobile edge computing: The communication perspective. *IEEE Communications Surveys & Tutorials, 19*, 2322–2358.

Meier, T., & Fischer, G. (2011). *Assessment of the Pico and Micro-hydropower market in Rwanda*. LIRE, Vientiane.

Miorandi, D., Sicari, S., De Pellegrini, F., & Chlamtac, I. (2012). Internet of things: Vision, applications and research challenges. *Ad Hoc Networks, 10*, 1497–1516.

Monroe, C., & Kim, J. (2013). Scaling the ion trap quantum processor. *Science, 339*, 1164–1169.

Mujica, G., Rodriguez-Zurrunero, R., Wilby, M., Portilla, J., Gonzalez, A. B. R., Araujo, A., Riesgo, T., & Diaz, J. J. V. (2018). Edge and fog computing platform for data fusion of complex heterogeneous sensors. *Sensors, 18*, 3630–3640.

Nguyen Gia, T., Dhaou, I. B., Ali, M., Rahmani, A. M., Westerlund, T., Liljeberg, P., & Tenhunen, H. (2019). Energy efficient fog-assisted IoT system for monitoring diabetic patients with cardiovascular disease. *Future Generation Computer Systems, 93*, 198–211.

Nguyen, V. C., Dinh, N. T., & Kim, Y. (2018). A distributed NFV-enabled edge cloud architecture for ICN-based disaster management services. *Sensors, 18*, 4136–4143.

Nitkin, P. I., Gorshkov, B. G., Nitkin, E. P., & Ksnevich, T. I. (2005). Sensors and actuators B: Chemical. *Sensors and Actuators B: Chemical, 111*, 500–504.

Ota, H., Chao, M., Gao, Y., Wu, E., Tai, L.-C., Chen, K., & Matsuoka, Y. (2017). 3D printed "earable" smart devices for real-time detection of core body temperature. *ACS Sensors, 2*, 990–997.

Oueida, S., Kotb, Y., Aloqaily, M., Jararweh, Y., & Baker, T. (2018). An edge computing based smart healthcare framework for resource management. *Sensors, 18*, 4307–4315.

Page, M. J., McKenzie, J. E., Bossuyt, P. M., Boutron, I., Hoffmann, T. C., Mulrow, C. D., & Shamseer, L. (2021). The PRISMA 2020 statement: An updated guideline for reporting systematic reviews. *British Medical Journal, 372*, 71–83.
Park, H., Kim, H., Joo, H., & Song, J. (2016). Recent advancements in the internet-of-things related standards: A oneM2M perspective. *ICT Express, 2*, 126–129.
Park, R. S., Hills, G., Sohn, J., Mitra, S., Shulaker, M. M., & Wong, H. S. P. (2017). Hysteresis-free carbon nanotube field-effect transistors. *ACS Nano, 11*, 4785–4791.
Park, D., Kim, S., An, Y., & Jung, J. Y. (2018). LiReD: A light-weight real-time fault detection system for edge computing using LSTM recurrent neural networks. *Sensors, 18*, 2110–2116.
Petersson, K. D., McFaul, L. W., Schroer, M. D., Jung, M., Taylor, J. M., Houck, A. A., & Petta, J. R. (2012). Circuit quantum electrodynamics with a spin qubit. *Nature, 490*, 380–383.
Phan, D. T., Nguyen, C. H., Nguyen, T. D. P., Tran, L. H., Park, S., Choi, J., Lee, B., & Oh, J. (2022). A flexible, wearable, and wireless biosensor patch with internet of medical things applications. *Biosensors, 12*, 139–145.
Pradhan, B., Bhattacharyya, S., & Pal, K. (2021). IoT-based applications in healthcare devices. *Journal of Health Engineering, 21*, 6632599.
Pyatkov, F., Futterling, V., Khasminskaya, S., Flavel, B. S., Hennrich, F., Kappes, M. M., Krupke, R., & Pernice, W. H. P. (2016). Cavity-enhanced light emission from electrically driven carbon nanotubes. *Nature Photonics, 10*, 420–427.
Qi, J., Yang, P., Waraich, A., Deng, Z., Zhao, Y., & Yang, Y. (2018). Examining sensor-based physical activity recognition and monitoring for healthcare using internet of things: A systematic review. *Journal of Biomedical Informatics, 87*, 138–153.
Ravikumar, S., Sekar, S., Sirenjeevi, P., & Deepa, R. (2024). Optimizing resource allocation in ultra-dense networks with UAV assistance: A levy flight-based approach. *Expert Systems with Applications, 235*, 120954.
Rodriguez, A., Valverde, J., Portilla, J., Otero, A., Riesgo, T., & de la Torre, E. (2018). FPGA-based high-performance embedded systems for adaptive edge computing in cyber-physical systems: The ARTICo3 framework. *Sensors, 18*, 1877–1885.
Rosario, D., Schimuneck, M., Camargo, J., Nobre, J., Both, C., Rochol, J., & Gerla, M. (2018). Service migration from cloud to multi-tier fog nodes for multimedia dissemination with QoE support. *Sensors, 18*, 329–335.
Satyanarayanan, M., Bahl, P., Caceres, R., & Davies, N. (2009). The case for VM-based cloudlets in mobile computing. *IEEE Pervasive Computing, 8*, 14–23.
Savage, K. J., Hawkeye, M. M., Esteban, R., Borisov, A. G., Aizpurua, J., & Baumberg, J. J. (2012). Revealing the quantum regime in tunnelling plasmonics. *Nature, 491*, 574–577.
Schlansker, M. S., & Rau, B. R. (2000). EPIC: Explicitly parallel instruction computing. *Computer, 33*, 37–45.
Schreiber, R., Rau, B. R., & Schlansker, M. S. (2002). PICO-NPA: High-level synthesis of nonprogrammable hardware accelerators. *Journal of Signal Processing Systems, 31*, 127–142.
Siampour, H., Kumar, S., & Bozhevolnyi, S. I. (2017). Nanofabrication of plasmonic circuits containing single photon sources. *ACS Photonics, 4*, 1879–1884.
Silva, A., Monticone, F., Castaldi, G., Galdi, V., Alu, A., & Engheta, N. (2014). Performing mathematical operations with metamaterials. *Science, 343*, 160–163.
Singh, S., Hamidon, M. N., Zuber, M., & Ahmad, K. A. (2021). Wireless sensing technology with IoMT approach for continuous monitoring of breathing rate and volume during COVID-19. *Frontiers in Sustainable Cities, 3*, 6–13.
Solyman, A. A. A., & Yahya, K. (2022). Evolution of wireless communication networks: From 1G to 6G and future perspective. *International Journal of Electrical and Computer Engineering, 12*, 3943–3950.
Spanner, K., & Gloss, R. (2010). New challenges in nanopositioning technologies. In *Proceedings of the International Conference and Exhibition on New Actuators and Drive Systems, Bremen, Germany* (pp. 258–263).

Suarez-Albela, M., Fraga-Lamas, P., & Fernandez-Carames, T. M. (2018). A practical evaluation on RSA and ECC-based cipher suites for IoT high-security energy-efficient fog and mist computing devices. *Sensors, 18*, 3868–3875.

Sun, J., Wang, X., Wang, S., & Ren, L. (2018). A searchable personal health records framework with fine-grained access control in cloud-fog computing. *PLoS One, 13*, e0207543.

Tekeste, T., Saleh, H., Mohammad, B., & Ismail, M. (2019). Ultra-low power QRS detection and ECG compression architecture for IoT healthcare devices. *IEEE Transactions on Circuits and Systems I: Regular Papers, 66*, 669–679.

Van de Burgt, Y., Melianas, A., Keene, S. T., Malliaras, G., & Salleo, A. (2018). Organic electronics for neuromorphic computing. *Nature Electronics, 1*, 386–397.

Vaughan, O. (2019). Working on the edge. *Nature Electronics, 2*, 2–6.

Vergniory, M. G., Elcoro, L., Felser, C., Regnault, N., Bernevig, B. A., & Wang, Z. J. (2019). A complete catalogue of high-quality topological materials. *Nature, 566*, 480–485.

Vimbi, V., Shaffi, N., Mahmud, M., Subramanian, K., & Hajamohideen, F. (2023). Application of explainable artificial intelligence in Alzheimer's disease classification: A systematic review. *Research Square, 14*, 126–137.

Wachter, S. (2019). Data protection in the age of big data. *Nature Electronics, 2*, 6–10.

Welser, J., Pitera, J. W., & Goldberg, C. (2018). Future computing hardware for AI. In *2018 IEEE international Electron devices meeting (IEDM), San Francisco, CA, USA* (pp. 1–5).

Xu, J. L., Dai, R. X., Xin, Y., Sun, Y. L., Li, X., Yu, Y. X., Xiang, L., Xie, D., Wang, S. D., & Ren, T. L. (2017). Efficient and reversible electron doping of semiconductor-enriched single-walled carbon nanotubes by using decamethyl cobaltocene. *Scientific Reports, 7*, 1735–1743.

Xu, T., Pan, Z., Zhang, H., Zou, Q., & Bao, C. (2021). Modeling and analysis of millimeter wave 5G cellular networks based on 3-D spatial model. *Journal of Physics Conference Series, 1944*, 012025.

Xue, H. R., Yang, Y. H., Gao, F., Chong, Y. D., & Zhang, B. L. (2019). Acoustic higher-order topological insulator on a kagome lattice. *Nature Materials, 18*, 108–112.

Yan, H., Choe, H. S., Nam, S. W., Hu, Y. J., Das, S., Klemic, J. F., Ellenbogen, J. C., & Lieber, C. M. (2011). Programmable nanowire circuits for nanoprocessors. *Nature, 470*, 240–244.

Yan, L., Cao, S., Gong, Y., Han, H., Wei, J., Zhao, Y., & Yang, S. (2019). SatEC: A 5G satellite edge computing framework based on microservice architecture. *Sensors, 19*, 831–840.

Zhang, X., Lin, J., Chen, Z., Sun, F., Zhu, X., & Fang, G. (2018). An efficient neural-network-based microseismic monitoring platform for hydraulic fracture on an edge computing architecture. *Sensors, 18*, 1828–1835.

Zhang, T. T., Jiang, Y., Song, Z. D., Huang, H., He, Y. Q., Fang, Z., Weng, H. M., & Fang, C. (2019). Catalogue of topological electronic materials. *Nature, 566*, 475–479.

Zhao, W. J., Qi, J. W., Lu, Y., Wang, R. D., Zhang, Q., Xiong, H., Zhang, Y. Q., Wu, Q., & Xu, J. J. (2019). On-chip plasmon-induced transparency in THz metamaterial on a LiNbO3 subwavelength planar waveguide. *Optics Express, 27*, 7373–7383.

Zhu, L., Fu, Y., Chow, R., Spencer, B. F., Park, J. W., & Mechitov, K. (2018). Development of a high-sensitivity wireless accelerometer for structural health monitoring. *Sensors, 18*, 262–270.

Zhu, X. J., Li, D., Liang, X. G., & Lu, W. D. (2019). Ionic modulation and ionic coupling effects in MoS2 devices for neuromorphic computing. *Nature Materials, 18*, 141–148.

Zhuang, Q. T., Zhang, Z. S., & Shapiro, J. H. (2018). Distributed quantum sensing using continuous-variable multipartite entanglement. *Physical Review A, 97*, 032329.

Picotechnology for Biological Science

K. Gayathri, P. Saranraj, Riyaz Sayyed, L. Charlie Jelura, and S. Monisha

1 Introduction

Picotechnology is the technical manipulation of matter at a Picoscale (10^{12}) or fraction of a meter. This is less than the 10^9 scale of Nanotechnology. Matter is influenced on a smaller scale than atoms. Picotechnology is 1000 times smaller than nanotechnology, even if the nanometers represent the billionth of a meter and help determine the direction of the field. Actually, it is working with atoms to build complex atomic structures (Prathyusha et al., 2018). Picotechnology is essentially driven by the fundamental phenomena of electron manipulation. It means changing the energy levels of atoms to control their structural and chemical characteristics so as to produce fresh and better features. This technology is not new; rather, its roots may be traced in the building of natural systems since main biological activities in nature usually occur at the atomic or even subatomic level. Together with microscopy, recent developments in Picotechnology have allowed scientific communities to produce and investigate materials and events including enhanced surface area and quantum effects resulting from their handling and organization at the nano- and picoscales.

Many metal solutions are known to have considerable activity when their concentration in solution lowers near picomoles. Homeopathy is predicated on the physical premise that, in physiological settings (more than 108 times) comparable

K. Gayathri · P. Saranraj (✉) · L. C. Jelura · S. Monisha
PG and Research Department of Microbiology, Sacred Heart College (Autonomous), Tirupattur, Tamil Nadu, India

R. Sayyed
Department of Biological Sciences and Chemistry, College of Arts and Science, University of Nizwa, Nizwa, Oman

S. Kulkarni (ed.), *A Journey from Nanotechnology to Picotechnology*, Lecture Notes in Nanoscale Science and Technology 47,
https://doi.org/10.1007/978-3-032-07228-3_12

metal ions "amplify" the bioreaction of the same metals in opposite order upon decreasing metal given concentration. This perspective is generating interest in reactions that require cofactor concentrations of calcium, iron, magnesium, zinc, etc. at picomole concentration in an attempt to understand how "very low metal administration" could activate the intracellular battery of proteins and protein regulating factors either poorly understood or unknown. Arsenic oxide, calcium phosphate, ferrum ox (iron oxide), and natrum mur are the best examples. The interaction between proteins and metal cofactor is essential in biomolecules, so new technical developments are needed to increase their picomolar range. Actin-myosin and microtubule proteins involve calcium; sodium is involved in neurofilament proteins. Conversely, by increasing enthalpy and entropy, ultrasmall ion concentrations are more likely to induce thermodynamics and energy activation than they are to generate a molecular conformational trigger (Mostafavi et al., 2019).

Our technological capacity has advanced significantly from the nanoscale to the picometer level, and this gives whole new directions for biological research. Picotechnology explores the domain of individual atoms, bond lengths, and even subatomic interactions while nanotechnology, addressing structures and devices usually ranging from 1 to 100 nanometers, has already made significant contributions to areas like drug delivery and diagnostics. Unlocking some of the most basic and elusive secrets of biological systems requires this capacity to explore and control matter at such a beautiful degree of detail (Sharma, 2016). Imagine being able to exactly arrange individual atoms within a protein to create new functions or directly see the dynamic interaction of electrons during an enzyme reaction. Picotechnology brings once-fancious concepts right under our reach. Nowadays, advanced microscopy methods allowing scientists to see single molecules and their conformational changes in real time are pushing the boundaries of resolution beyond what was formerly thought possible. Picometer-scale sensitivity is being sought by means of techniques including Atomic Force Microscopy (AFM) with specialized tips and Near-field Scanning Optical Microscopy (NSOM), so enabling researchers to map the topography and even the chemical identity of biological surfaces with hitherto unheard-of accuracy (Sharma, 2016).

Picotechnology goes beyond vision to include the creation of picometer-scale precisely engineered tools and materials. This covers the creation of ultra-sensitive biosensors able to identify single molecules, so providing fresh opportunities for early illness detection and tailored medication. Advanced tissue engineering and implantable devices with improved biocompatibility are made possible by engineers designing materials at the atomic level, therefore enabling extremely specialized interactions with biological systems. Moreover, the ability to precisely control molecules with picometer accuracy could result in revolutionary developments in drug discovery and enable the design of therapeutics aiming at particular biomolecules with unmatched accuracy and lowest off-target effects. Picotechnology has consequences going beyond simple observation and manipulation. Probing biological systems at this basic level helps us to better grasp the fundamental physical ideas controlling life events. This covers investigating the function of quantum events in biological events like electron transport in photosynthesis and avian navigation sensitivity to the magnetic field of Earth. Solving these challenging biological riddles will depend critically on theoretical modeling and computational methods as well as experimental validation at the picometer scale (Alpaslan & Webster, 2014).

2 Picotechnology Versus Nanotechnology

In many different fields, nanotechnology is flourishing; for example, engineered nanoparticles find extensive application in agriculture biomedicine, cosmetics, electrical, pharmaceutical, and textile industries. Furthermore a promising approach for wastewater treatment systems including household wastewater treatment that might eliminate pollutants including phosphorous and others is this one. By means of advances in illness prevention, diagnosis, and treatment as well as nano-based pharmaceutical and medical preparations approved by regulatory agencies, nanotechnology has lately transformed the pharmaceutical and medical sectors. Particularly used as drug delivery carriers are several types of nanomaterials including polymeric drug conjugates, liposomes, and micelles. With nanotechnologically based treatments, cancer can be cured; specifically, smaller than 100 nm nanoparticles can adhere to tumor locations and destroy them. Actually, nanodrugs enter tumor tissues more easily via particular kinds of receptors and antigens overexpressed on the surface of malignant cells. These developments have made Picotechnology much more able to transform the medical field. Regarding regenerative medicine, several interactions at atomic and subatomic levels control biological events including intracellular activities (Dvir et al., 2011). Linked with this quite low level, picotechnology is a wise choice in material synthesis, imaging, medicine, and biological applications. It may also create opportunities for more sophisticated and individualized therapies. When given inside a body, picomolecules (PM) have great degree of magnitude for physiological reaction activation than do nanomolecules. Furthermore considered is the idea that a margin of activities of many drugs increases with decreasing size dimensions down to picoscale. For instance, following their physical distribution, the bioreaction capacity of metallic elements such as iron oxide, calcium phosphate, sodium phosphate, and arsenic declined (typically by ten times). Linked to the quantum events controlling matter at the picoscale is the adjustment of physical and chemical properties, such color, electrical conductivity, magnetism, and melting point changes in line with nanotechnological interventions. Applying currents, chemical reactions, or magnetic fields comparable to what occurs in biological cells helps PMs acquire new capabilities by a more sophisticated reorganization of atomic structures.

The tenability idea of characteristics helps to clearly show the quantum influence of picoscale. Just changing the particle size helps to clarify additional tuning via means of this modification, a scientist can correct several characteristics of materials, including the fluorescence hue, which subsequently finds application not only in specifically identifying that particle but also in labeling various materials (via fluorescent markers) for varied uses (Ramaseshan et al., 2006). Apart from that, quantum impact of picoscale also provides the tunneling feature that benefits flash memory and scanning tunneling of microscope in computer applications. Picomaterials are valuable in drug delivery systems since their higher surface area and improved reactivity help them to be functionalized. Picotechnology presents the possibility to modify surfaces without any toxicity problems as those of nanotechnology since it integrates science and engineering of a new generation. Red or purple gold picoparticles seem more viable than gold nanoparticles (NPs) because of the limited mobility trends of electrons in gold-NPs. While Gold-PPs have better selectivity and specificity, Gold-NPs react differently to light depending on limited electron mobility; hence, they can gather in tumor cells without disturbing surrounding normal cells, enabling one to precisely bio-image and aim laser elimination of cancerous tissues. This method has potential to find the atomic locations of materials in high-resolution, picoscale images and is more efficient than earlier imaging methods. Thanks to one-dimensional nanomaterials with a great surface area and efficient pore interconnectivity nanofibers, air filtering is feasible during pollution control activities.

Particularly useful and reasonably priced for removing troublesome materials like chemical and biological pollutants, Volatile Organic Compounds (VOCs), and other contaminants present in water and air systems are electrospun nanofibers. Manganese Oxide (MnO) nanofiber catalyst also allows one to remove Volatile Organic Compounds (VOCs) from industrial stacks. Furthermore, compared to traditional materials, small-diameter Carbon Nano-Tubes (CNTs) with a greater surface-volume ratio have shown efficiency as adsorptive materials. For a spectrum of analytes, 48–360 ppb, DNA-CNT sensors show adjoining potential toward many, including isovaleric acid and dimethyl sulfone (Webster & Favi, 2015).

3 Promises of Picotechnology

Using atomic level materials gives an opportunity to profit from unique events occurring at the smallest possible sizes. Picotechnology is one of the main technologies that will greatly enhance our future since it enables individuals to investigate structures at the pico-molecular level. Picotechnology is three orders of magnitude smaller than nanotechnology while emphasizing the picometer-scale structure of materials. Part of this developing field is the study of molecules as tiny as a trillionth of a meter, including biomolecules (Ismail Beigi, 2017). More Picotechnology reorders and structures complex atomic formations by means of atom and molecule manipulation. The structural features of an atom can be altered by varying the energy levels of its electrons. Picoscale technology is fundamental in research on the building of atomic structures. For instance, while studying the interactions between neighboring atoms, changes at the picoscale significantly affect the force between them. In the framework of regenerative medicine, atomic and subatomic interactions influence many biological events including intracellular activities. Among the fields researchers could improve using Picotechnology include biosystems, health, imaging, and material synthesis. Picotechnology thus helps to produce more exact and customized treatments and therapies. Like nanotechnology, the picoscale quantum processes controlling matter produce unexpected changes in the physical and chemical characteristics of systems. At the picoscale, for example, particle size influences properties including color, magnetic, electrical conductivity, melting point. Consequently, the atomic level manipulation of matter and the application of its unique properties can be rather advantageous in many disciplines. Moreover, atomic formations can be rearranged to provide Picomaterials new characteristics by including currents, magnetic fields, or chemical processes to replicate those occurring in cells naturally (Schakenraad, 2006).

Picotechnology is increasingly advocated as a particular instrument for the diagnosis and treatment of many kinds of diseases nowadays. By means of interactions between biomaterials and their physiological environment, these special characteristics can induce tissue development and control inflammation and infection. Picomaterials with the necessary surface properties have lately been produced using several picoscale engineering techniques. These techniques have let researchers modify atom-to-atom bond lengths and orientations, therefore influencing atom arrangement and, hence, material properties. Surface free energy's effect on cell spreading has been exhaustively studied (Lutolf & Hubbell, 2005). Higher surface energy has been found to encourage cell spreading both with and without serum proteins. In vivo studies have also investigated the effect of surface free energy of implanted materials on tissue development and repair surrounding implants. Higher free energy surfaces revealed noticeably higher rates of cellular growth and activity in several investigations. Furthermore shown to be related with surface charge density is cell adhesion (Kudryasheva et al., 2002).

More specifically, there was noted a parabolic link between surface charge and cell adhesion. According to this study, surface charge has an ideal range that one can further adjust to enhance biological reactions. This surface modification can boost tissue adherence and help to lower infections at the implant site. One can modify electron distributions inside an atom to reach a desired charge and sur face energy for improved cell development. Improved biological reactions including cell migration, proliferation, adhesion, extracellular matrix development, and infection inhibition have also been shown by manipulation of atom charge concentrations. Picotechnology, as a new generation of material engineering, is especially fascinating since it allows surface alteration without maybe generating toxicity concerns with nanoparticles. Through control of cellular microtubules (MTs), Picotechnology can also be applied to control cell mobility. Microtubules, structures made from tubulin proteins, are vital for cell movement. Microtubule tips coupled with end-tracking proteins (+TIPs) function as dynamic regulatory sites that are essential for key cellular processes, including cell migration, division, and polarity (Baier et al., 2006). Fluorescent agents exhibit a comet-like pattern in the presence of which one interprets accelerated protein turnover on MT plus ends and dissociation of 1TIPs from MT ends. Thus, using external stimulation to MTs having end-binding proteins results in regulated cell motions using Picotechnology (Steiner & Lana, 2013).

4 Biosystem

4.1 Neurofilaments

The neurofilament unit network consists of one hundred picometers. It clarifies our understanding of proteins in nerve cells, especially those including neurons expressing neurofilament. Designed for light, medium, and heavy accordingly, the three neurofilament (NF) proteins that form the intermediate filaments are NF-L, NFM, and NF-H. Important are also alpha-internexin, IF, and nestin; neuronal members of the plakin family seem to anchor neurofilaments to actin filaments and MTs (Bolonkin, 2010). Two types of intermediate filaments found in nervous system cells are electron micrographs. Freeze-etched images displaying significant cross-linking via protein cross-bridges show neurofilaments in a nerve cell axon. This configuration is supposed to give notable tensile strength in long cell process. Long, nonhelical extensions generated by the carboxyl terminus of the largest neurofilament protein birth the cross-links. (B) A freeze-etch picture displaying glial cells' limited number of cross-bridges and smoothness of their filaments. Transgenic mice generating too high levels of NF-L or NF-H (Dragestein, 2008) showed the etiological relevance of neurofilaments for motor neuron disease.

4.2 Actin

It is a 7 nm diameter cytoskeletal protein with a polar shape that forms three-dimensional actin cytoskeleton networks. The hydrolysis of tubulin and the actin monomers' binding with ATP produce dynamic instability, which is crucial for their roles. Tropomyosin forms a cooperative relationship with actin filaments to generate cell surface movements, such as phagocytosis, cytokinesis, and cell motility (Liu et al., 2003).

4.3 Microtubules (MTs)

Involved in chromosomal separation during mitosis, intracellular organelle transport, and cell motions are the polymers known as MTs. Comprising tubulins, MTs have a cylindrical form with a 25 nm radius (Shi et al., 2010). Dynamic instability results when MT plus end-binding proteins, sometimes known as "plus end-tracking proteins can surf" the dynamic ends of MTs. APC, CLASP2, LIS1, CLIP 170, EB1, EB3, CLIP 115, Dynactin complex. When +TIPs are generated as GFP-fusions, the fluorescence decreases in intensity toward the minus end of the MT and is strongest at the developing tip of the MT, hence producing a "comet tail." Originally, this particular link of +TIPs with MT distal ends was explained using tread milling.

Another element is "Delivery (or deposition) of Kinesin." GFP-TIP speculates on MT dynamic behavior during depolymerization. Using various GFP-tagged MT plus end-binding proteins as markers has helped to somewhat explain MT behavior. Fluorescence in time-lapse Most MTs, at least in CHO cells, develop consistently from the cell center toward the cell periphery, microscopy visualization of GFP-TIPs in cells revealed; all fusion proteins travel in a comet-like manner (Ferrari, 2005).

5 Picotechnology in Tissue Application

One partially less dangerous method employed to encourage tissue development and generate the following generation of tissue engineering materials is Picotechnology. Picotechnology is the control of electron distribution around atoms to generate certain properties. By means of electron dispersion, surface energy can be greatly changed, thereby affecting the binding behavior of proteins to a surface. By activating or reorganizing electrons around atoms, one can influence several biological processes, including adhesion, growth, intracellular transport to organelles, cell motility, and the synthesis of extracellular matrix (ECM). Though Picotechnology has great promise, practically much research in this field has been conducted (Colon et al., 2006). The idea of manipulating cellular M = microtubules (MTs) with Picotechnology is really fascinating. Having a 25 nm diameter, MTs are tubulin-based cylindrical cellular formations. Sometimes referred to as "plus end-tracking proteins," dynamic instability occurs from MT plus end-binding proteins' ability to "surf" the dynamic endpoints of MTs. Recent studies show that the fluorescence is strongest at the MT and decreases in intensity toward the minus end of the MT when tips are produced as green fluorescent proteins, creating a comet tail. Picotechnology seems to be able to stimulate from the outside the MT and end-binding proteins, therefore encouraging cell migration (Pattison et al., 2005). This may be a less damaging approach to change surface energy to encourage tissue development since electron distributions can be changed for numerous macro-, micro-, or nanomaterials (Shakhssalim et al., 2012).

6 Applications of Picotechnology in Biomedical Sciences

Despite various developments in therapeutic treatments, some clinical diseases especially those related with cancer are growing more common; so, thorough treatment plans are required. The ideal therapy of these diseases depends on promising drug delivery systems that can target tumor cells selectively; so, appropriate distribution of anti-cancer medications is essential. Since, nanotechnology promises a great variety of possible applications in the treatment of several medical disorders, so ensuring the efficient target drug delivery of medications. Apart from providing many benefits for the treatment of diseases, constant nanotechnology research and development has opened the path for Picotechnology (Anitua et al., 2008). Doctors can now effectively create new gadgets and build inventive treatment plans safer, more successful, and more efficient than those utilized in the past by leveraging the natural framework of Picotechnology. For example, the bio-barcode test is a vibrant Picotechnology tool that greatly helps in the identification of complicated disease agents even in minute levels in living entities. Northwestern University (USA) did verify fundamental approach that syndicates the Gold picoparticles with DNA amplifiers and recognition particles while working on the Prostectomy of prostate cancer biomarker. Combining biological systems with engineering intervention, the biotechnological approach and multidisciplinary field of tissue engineering and regenerative medicine only aims to generate functional replacements for treating diseased and damaged tissues by improving, repairing, or preserving their potential to function. The tissue engineering and regenerative medicine technique generates tissue engineering structures using three main components: growth factors, scaffolds, and stem cells. Nanotechnology has become a trademark technology thanks to the application of nanomaterials with the features of outstanding drug delivery systems: their traverse ability to access several nanostructures inside bones, their capacity to attach to the targeted disease sites, and their continuous drug release (Bieling et al., 2007). These nanoparticles also offer very useful fine structure scaffolds for tissue regeneration, hence transforming tissue engineering treatments. In terms of Picotechnology, its worth is shown by its consistency in consideration while controlling cellular microtubules (MTs). Usually with a diameter of 25 nm, the complex cellular structures known as MTs are basically composed of tubulins. Fluorescent proteins have green tips, hence fluorescence is brightest at microtubules and progressively fades as one approaches the negative ends of these structures. By letting MTs and end-binding proteins to be triggered by an outside stimulation source, the flexible picotechnological discipline is meant to enhance cellular motility. Usually utilized in biochemical-based research, a PerkinElmer Zeemen 3030 spectrophotometer evaluates electro-thermal atomic absorption spectrophotometry (Jariwala et al., 2013). Additionally, an efficient method to evaluate thyroid-stimulating hormones is chemiluminescent; another excellent approach to keep an eye on thyroid hormones is enzyme-linked immunosorbent assay. One instance of how this technology has tremendously helped research is the recently developed Picoscope biosensor for real-time investigation of numerous biological components. This approach clearly detects multiple variants (at the picolevel) in the thickness of several bio-recognition spots observed on the surface of a biochip. Thus, the effective and comprehensive development of Picotechnology will increase the usefulness of research tools on this modern scale (Kumar, 2018).

Usually, the picoscope stabilized an antibody on the surface of a biochip by means of a BS bridge mechanism. Picoscope can detect bacterial organisms using a biotinylated glass slide. Streptavidin is regarded as the first biotinylated antibody used against microbiological organisms. One can detect variations in the antigen layer thickness on sensor surfaces (Kumar & Tripathi, 2014). Surface plasmon resonance, or BIAcore 2000, is a very powerful biosensing alternative depending just on fluorescent labels, same like a picoscope. With its multi-spot biochip and charged pair device camera, the picoscope which detects several fluid agents including food pathogens has many advantages including enhanced output screening (Goldberg et al., 2007). The main approach applied to evaluate the kinetics of binding trends is a two-channel picoscope intervention. Furthermore very important are several picochips in the development stage meant to build and immobilize various antibodies and biotin-streptavidin bridges.

Electroacupuncture (EA) is getting increasing favor in scientific clinical settings. This is especially true for cardiac uses, such pacemaker installation for patients. This is the reason a great product from Pico-Technology Limited, the Picoscope TM with model number 3425, which incorporates a four-channel differential amplifying input, has demonstrated to be effective in simulating currents (Kumar, 2018). EA was distributed using Cefar acus4TM in two experimental setups meant to evaluate the success of the above described approach. First, the results revealed that EA given by two acupuncture surgery needles may precisely predict the paths and areas of influence of the simulation currents. Moreover, when these needles were placed near acupuncture points in a limb, neither the leg nor the chest showed any clear spreading patterns. For cardiothoracic surgery operations, these tools are absolutely useful. Picotechnology may also be totally or partially transferred to the push to change medical sciences, including preventive and diagnostic processes as well as the treatment of varied illness effects (Mohammadpour et al., 2015). Promising possibilities for medical systems that arose from nanotechnology and may come from Picotechnology in the years to come include many coating materials with newly adjustable properties based on many criteria, including electrical, magnetic, mechanical, and optical (Lee et al., 2002).

Apart from effectively delivering drugs to the targeted diseased tissues, initiatives are under progress to produce picoscale particles capable of reducing the adverse effects of chemotherapy. By extending telomeres, which can undo aging and many life-threatening diseases including cancer, heart attacks, and strokes, picotechnology may help extend life (Tang & Cao, 2012). Another way picosecond technology as used in laser tattoo removal is enhancing dermatological treatments. In contrast to nanosecond lasers, photo-acoustic impact-based picos usually generate ultra-short pulses. Especially considering fractionated picos, these techniques could also be applied to pigment, wrinkles, scars, and other problems. PicoWay's clinical studies for acne scars and wrinkles produced interesting and aesthetically pleasing results, claims (Basu et al., 2009). Furthermore very important globally are South Korean technologies derived from picosecond lasers. Live cells can be distributed thanks to polypico dispersion method. Micro-well plates have been arranged Chinese hamster ovary cells using this technique. Furthermore, the resources of the technique have been applied to assess the stem cell distribution on well plates. By applying environmentally friendly technology in the microgravity environment, aiming at picotechnological-based medication delivery systems has helped efforts to eradicate several chronic diseases including HIV, cancer, and Alzheimer's disease. With modern, reduced systemic toxicity and lower dose costs, picoparticles offer hope for the efficient delivery of drugs to the brain and hippocampal given their great surface area (Sato, 2006). Particularly experts at the Aphios Corporation are hopeful that these pico-based drug delivery techniques could show to be a sensible approach for halting the development of Alzheimer is disease (Lam et al., 2004).

7 Picotechnology in Future

In the fight against climate change, nanotechnology has demonstrated to be a helpful weapon in some measure (Andersson & Nilsson, 2010). It influences the fight against air pollution quite significantly. Nanosensors detect contaminants, humidity other dangerous bacteria at very low concentrations as well as other elements. Picoscale sensor development will become more efficient and advanced if we are to track concentrations even at the ultrasmall level (Thapa et al., 2003). The present use of nano-structured membranes effective in removing carbon dioxide and methane from exhaust will inspire research communities to develop pico-structured membranes more effective than nanotechnology in eliminating greenhouse gasses (Zhang & Webster, 2009). Picotechnology will help to realize the broad impossible by controlling molecules and structures usually out of control. Eventually, the functionalization and operational management of this level of technology will totally alter human interaction with even the most basic equipment. Future recognition for picochips and other microdevices will be especially in environmental and biomedical uses (Webster & Ying, 2001). The possibility for Picotechnology to enable cellular and subcellular tissue restoration represents maybe the most fascinating use of this technical development. Furthermore, the synthesis of picolevel materials will enhance the contribution of nanotechnology in improving fuel consumption efficiency by means of nanocatalysts (Cao et al., 2016). Consequently, good Picotechnology research and application will help Femtotechnology to grow (Peran et al., 2013).

8 Conclusion

Tissue engineering has quickly embraced nanotechnology over the past 10 years. Many in vitro and in vivo studies find the use for nano-structured materials. Synthesis and application of new structures into biological molecules have produced a biomimetic environment for tissue development. Of course, there still have to be addressed problems including those related to environmental and human toxicity. Enough study should be done before applying inorganic nanomaterials (like carbon nanotubes) in therapeutic environments. Thus, functionalizing such nanoparticles is essential to minimize their toxicity instead of giving up on them should too great toxicity be found. Any macro-, micro-, or nanomaterial can cause electrons; so, future treatments might use Picotechnology instead of nanotechnology to lower the toxicity. By changing the electron distribution and hence the subsequent redistribution of charges, surface energetics can be adjusted to modify the adsorption of particular proteins and biological activities.

Along with advancing some fascinating disciplines including chemical analysis, computing interventions, imaging, medicine, and material synthesis, effective research and development in the field of Picotechnology is expected to lead even more great knowledge of biological systems. If this assumption is realized, the scientific community will be able to efficiently employ a range of materials either naturally existing or generated at this size. One can control the energy states of electrons to vary the structural arrangement of different atoms, hence producing complex materials (picomaterials) with remarkable features. In biomolecules, protein–metal cofactor interactions can be somewhat important; hence, new technical developments are required to grasp their picomolar range. Usually, the main objectives of these developments in the medical sciences are lower inflammation, lower incidence of infectious disorders, and encouragement of tissue development. Even at very low levels, materials based on Phototechnology are rather effective in detecting some hormones, metabolites, and nucleotides. Through advances made possible by present research and development, Picotechnology may thus become a new modern technology with great computational capacity in data simulation and wide-ranging applications in biological and environmental engineering. Furthermore important to keep in mind is that the advancements discussed could only mark the start of Femtotechnology and more. Eventually, developments in picotechnology will change human life by allowing studies on picoscale systems and structures.

References

Alpaslan, E., & Webster, T. J. (2014). Nanotechnology and Picotechnology to increase tissue growth: A summary of in vivo studies. *International Journal of Nanomedicine, 9*, 7–12.

Andersson, S., & Nilsson, S. I. (2010). Influence of PH and temperature on microbial activity, substrate availability of soil-solution bacteria and leaching of dissolved organic carbon in a mor humus. *Soil Biology and Biochemistry, 33*, 1181–1191.

Anitua, E., Sanchez, M., Orive, G., & Andia, I. (2008). Delivering growth factors for therapeutics. *Trends Pharmacology Science, 29*, 37–41.

Baier, R., et al. (2006). Surface properties determine bioadhesive outcomes: Methods and results. *Journal of Biomedical Materials Research, 18*(4), 337 55.

Basu, B., Katti, D. S., & Kumar, A. (2009). *Advanced biomaterials: Fundamentals, processing, and applications*. Wiley.

Bieling, P., Laan, L., Schek, H., Munteanu, E. L., Sandblad, L., Dogterom, M., & Surrey, T. (2007). Reconstitution of a microtubule plus-end tracking system in vitro. *Nature, 7172*, 1100–1110.

Bolonkin, A. (2010). Femtotechnology: Design of the strongest AB matter for aerospace. *Journal of Aerospace Engineering, 23*(4), 281–292.

Cao, H., Chen, R., Wang, L., Jiang, L., Yang, F., Zheng, S., Wang, G., & Lin, X. (2016). Soil pH, Total phosphorus, climate and distance are the major factors influencing microbial activity at a regional spatial scale. *Scientific Reports, 6*, 25815.

Colon, G., Ward, B. C., & Webster, T. J. (2006). Increased osteoblast and decreased *Staphylococcus epidermidis* functions on nanophase ZnO and TiO_2. *Journal of Biomedical Materials Research. Part A, 78*(3), 595–604.

Dragestein, K. A. (2008). Dynamic behavior of GFP CLIP-170 reveals fast protein turn over on microtubule plus ends. *The Journal of Cell Biology, 180*(4), 729–737.
Dvir, T., Timko, B. P., Kohane, D. S., & Langer, R. (2011). Nanotechnological strategies for engineering complex tissues. *Nature Nanotechnology, 6*, 13–20.
Ferrari, M. (2005). Cancer nanotechnology: Opportunities and challenges. *Nature Reviews Cancer, 5*(3), 161–171.
Goldberg, M., Langer, R., & Jia, X. (2007). Nanostructured materials for applications in drug delivery and tissue engineering. *Journal of Biomaterials Science. Polymer Edition, 18*(3), 241–268.
Ismail Beigi, S. (2017). Picoscale materials engineering. *Nature Reviews Materials, 2*(11), 17060–17065.
Jariwala, D., Sangwan, V. K., Lauhon, L. J., Marks, T. J., & Hersam, M. C. (2013). Carbon nanomaterials for electronics, optoelectronics, photovoltaics, and sensing. *Chemical Society Reviews, 42*, 2824–2860.
Kudryasheva, N., Vetrova, E., Kuznetsov, A., Kratasyuk, V., & Stom, D. (2002). Bioluminiscence assays: Effects of Quinones and phenols. *Ecotoxicology and Environmental Safety, 53*, 198–203.
Kumar, D. (2018). Carbon nanomaterials-based nanocomposite as emerging field for pollution control. *Nanotechnology in Environmental Science, 15*, 125–152.
Kumar, P., & Tripathi, L. (2014). Challenges in pain assessment: Pain intensity scales. *Indian Journal of Pain, 28*, 61–70.
Lam, C. W., James, J. T., McCluskey, R., & Hunter, R. L. (2004). Pulmonary toxicity of single-wall carbon nanotubes in mice 7 and 90 days after intratracheal instillation. *Toxicological Sciences, 77*(1), 126–134.
Lee, S., Owen, K. E., Choi, H. K., Lee, H., Lu, G., & Wengler, G. (2002). Identification of a protein binding site on the surface of the Alphavirus nucleocapsid and its implication in virus assembly. *The Structurist, 4*(5), 531–541.
Liu, Y. H., Wu, X. Y., Fang, J. G., & Xie, W. (2003). *Herbal Drugs Central South Pharmacy, 1*, 302–305.
Lutolf, M. P., & Hubbell, J. A. (2005). Synthetic biomaterials as instructive extracellular microenvironments for morphogenesis in tissue engineering. *Nature Biotechnology, 23*(1), 47–55.
Mohammadpour, M., Jabbarvand, M., Hashemi, H., & Delrish, E. (2015). Prophylactic effect of topical silica nanoparticles as a novel antineovascularization agent for inhibiting corneal neovascularization following chemical burn. *Advanced Biomedical Research, 4*, 124–135.
Mostafavi, E., Soltantabar, P., & Webster, T. J. (2019). Nanotechnology and Picotechnology: A new arena for translational medicine. *Biomaterials in Translational Medicine, 15*, 191–212.
Pattison, M. A., Wurster, S., Webster, T. J., & Haberstroh, K. M. (2005). Three-dimensional, nanostructured PLGA scaffolds for bladder tissue replacement applications. *Biomaterials, 26*(15), 2491–2500.
Peran, M., Garcia, M. A., Lopez-Ruiz, E., Jiménez, G., & Marchal, J. A. (2013). How can nanotechnology help to repair the body? Advances in cardiac, skin, bone, cartilage and nerve tissue regeneration. *Materials, 6*(4), 1333–1359.
Prathyusha, K., Reddy, J. G. M., Venkataswamy, M., & Ramesh, A. (2018). Picotechnology: Instruments used and applications in pharmaceutical field. *Research Journal of Pharmacy and Technology, 10*, 34–40.
Ramaseshan, R., Sundarrajan, S., Liu, Y., Barhate, R., Lala, N. L., & Ramakrishna, S. (2006). Functionalized polymer nanofibre membranes for protection from chemical warfare stimulants. *Nanotechnology, 17*, 2947–2953.
Sato, M. (2006). *Nanophase hydroxyapatite coatings for dental and orthopedic applications*. Purdue University.
Schakenraad, J. (2006). The influence of substratum surface free energy on growth and spreading of human fibroblasts in the presence and absence of serum proteins. *Journal of Biomedical Materials Research, 20*(6), 773–784.

Shakhssalim, N., Dehghan, M. M., Moghadasali, R., Soltani, M. H., Shabani, I., & Soleimani, M. (2012). Bladder tissue engineering using biocompatible nanofibrous electrospun constructs: Feasibility and safety investigation. *Urology Journal, 9*(1), 410–419.
Sharma, V. (2016). Nanotechnology and its applications in energy sector. *International Journal of Technology, 6*(2), 1–8.
Shi, J., Votruba, A. R., Farokhzad, O. C., & Langer, R. (2010). Nanotechnology in drug delivery and tissue engineering: From discovery to applications. *Nano Letters, 10*(9), 3223–3230.
Steiner, F., & Lana, M. C. (2013). Effect of pH on boron adsorption in some soils of Paraná, Brazil. *Chilean Journal of Agricultural Research, 73*, 28–29.
Tang, S., & Cao, Z. (2012). Adsorption and dissociation of ammonia on graphene oxides: A first-principles study. *The Journal of Physical Chemistry, 116*, 8778–8791.
Thapa, A., Miller, D. C., Webster, T. J., & Haberstroh, K. M. (2003). Nano-structured polymers enhance bladder smooth muscle cell function. *Biomaterials, 24*(17), 2915–2926.
Webster, T., & Favi, P. (2015). Using nanotechnology and Picotechnology to increase tissue growth and reduce bacterial infections. *The FASEB Journal, 29*(1), 208–215.
Webster, T. J., & Ying, J. Y. (2001). *Advances in chemical engineering* (pp. 125–166). Academic Press.
Zhang, L., & Webster, T. J. (2009). Nanotechnology and nanomaterials: Promises for improved tissue regeneration. *Nano Today, 4*(1), 66–80.

Picotechnology: Advancing Precision in Bioimaging and Disease Diagnosis

Kevin Philip Sabu and Rony Rajan Paul

1 Introduction

The persistent push for accuracy in bioimaging and disease diagnosis has driven amazing improvements in science and technology. High-resolution observation of biological processes is important for considerably advancing our comprehension of cellular structures, complex molecular interactions, and detailed disease mechanisms. Bioimaging technology has progressed from large-scale to smaller-scale imaging and this improvement enables the observation of many molecular and an important number of atomic phenomena. Nanotechnology, a remarkably promising field, is among the most rapidly developing areas driving this deeply transformative change; it is exceptionally well-positioned to considerably improve our capacity for highly sensitive disease observation and diagnosis (Szczyglewska et al., 2023).

Picotechnology, working at the incredibly small scale of a trillionth of a meter, offers exciting new opportunities in bioimaging and diagnostics. This considerably improves the precision of current nanoscale technologies by exceeding their limits in spatial resolution and detection sensitivity. Nanotechnology greatly affects many areas by working at the nanometer scale (10^{-9} meters). Picotechnology is advancing this field by examining structures and processes at scales smaller than a nanometer. This unmatched precision enables unmatched observation and comprehension of biological processes. This leads to breakthroughs in the field (Pashazadeh-Panahi & Hasanzadeh, 2019).

K. P. Sabu
Centre for Immunology & Cancer Research, Faculty of Medicine, Hebrew University of Jerusalem, Jerusalem, Israel

R. R. Paul (✉)
Department of Chemistry, CMS College, Kottayam, Kerala, India
e-mail: rony@cmscollege.ac.in

S. Kulkarni (ed.), *A Journey from Nanotechnology to Picotechnology*, Lecture Notes in Nanoscale Science and Technology 47,
https://doi.org/10.1007/978-3-032-07228-3_13

The emergence of picotechnology can be linked to developments in imaging methods and material science. The potential for accurate diagnostics and bioimaging applications has increased as scientists improve techniques for working with and measuring matter at such small sizes (Prathyusha et al., 2018). Picotechnology promises advances in fields like targeted drug administration, early illness diagnosis, and personalized treatment by utilizing the characteristics of materials like graphene, quantum dots, and metal nanoparticles (Shafi et al., 1998).

The remarkable sensitivity with which picotechnology can identify and quantify biomarkers is among its most alluring features. Once-difficult-to-diagnose conditions like early-stage cancer, neurological conditions, and cardiovascular problems can now be detected at the molecular stage, long before outward symptoms appear (Bhagawati et al., 2023). In the end, better patient outcomes result from early intervention and more effective therapy made possible by this accuracy. Additionally, the use of picotechnology in bioimaging can offer dynamic, real-time insight into cellular and molecular processes, leading to a more precise comprehension of the course of disease and the effectiveness of treatment.

This chapter explores the fundamental ideas of picotechnology and how it has developed over time, highlighting scientific discoveries that have made it a useful tool for illness diagnostics and bioimaging. The benefits of picotechnology over traditional nanoscale technologies will also be discussed, emphasizing how its capacity to resolve structures and interactions at the picometre scale results in diagnostic techniques that are earlier, more precise, and non-invasive. Picotechnology has the potential to significantly alter the medical field, and as it advances, a new era of precision healthcare may be ushered in (Pan et al., 2019).

1.1 Overview of Picotechnology

At the picometre (μm) scale, where the dimensions of interest reach one trillionth of a meter, picotechnology functions. Compared to nanotechnology, which usually works with structures that are between 1 and 100 nm in size, this scale is smaller (Rush, 2022). Unmatched precision and control are made possible by the development of picotechnology, which permits manipulation and analysis at the atomic and subatomic levels (Moriarty, 2022). In domains like bioimaging, where molecular interactions and structural complexities greatly impact diagnostic results, such precision is especially beneficial (Umadevi et al., 2024).

The fundamental idea behind picotechnology is its direct interaction with quantum-level processes. Advanced instruments like as scanning probe techniques, high-resolution electron microscopy, and state-of-the-art spectroscopic approaches are used to accomplish this interaction. Picotechnology makes it possible to visualize and measure biological processes that were previously unattainable through nanotechnology by utilizing the quantum characteristics of materials (Khorasaninejad et al., 2016).

1.2 Comparison with Nanotechnology in Biomedicine

Even though nanotechnology has greatly improved biomedicine by making it possible to create drug delivery systems, biosensors, and high-resolution imaging modalities, its resolution and specificity are limited. By functioning at a scale that allows for atomic-level precision, picotechnology fills these gaps. The following is a summary of how the two technologies differ from one another:

- **Resolution:** Nanotechnology achieves resolutions down to the nanometer scale, whereas picotechnology enables visualization at the picometer level, enhancing the ability to study atomic bonds and quantum interactions (Crous & Abrahamse, 2021).
- **Applications:** While nanotechnology focuses on creating materials and devices for targeted drug delivery, bioimaging, and diagnostics, picotechnology excels in elucidating the structural dynamics of biomolecules, such as protein folding and DNA interactions, with unparalleled accuracy (Woźniak et al., 2022).
- **Tools and Techniques:** Picotechnology employs sophisticated methodologies such as attosecond spectroscopy and advanced quantum computing algorithms, surpassing the capabilities of conventional nanoscale tools (Jain, 2010).

By complementing nanotechnology, picotechnology enriches the toolkit available to biomedical researchers, paving the way for breakthroughs in precision medicine and disease diagnostics.

1.3 Historical Milestones

Important turning points in physics, materials science, and biology have paved the way for picotechnology's rise as a scientific field (Hanany, n.d.). Among the noteworthy advancements are:

- **Early Quantum Theories (1920s–1930s):** Foundational work by physicists such as Werner Heisenberg and Erwin Schrödinger laid the groundwork for understanding quantum mechanics, which is central to picotechnology (Gargini, 2004).
- **Advances in Microscopy (1980s–1990s):** The invention of scanning tunneling microscopy (STM) and atomic force microscopy (AFM) enabled visualization of surfaces at the atomic level, bridging the gap between nanotechnology and picotechnology (Moriarty, 2022).
- **Emergence of Picotechnology (2010s):** The term "picotechnology" gained prominence with breakthroughs in high-resolution imaging and the manipulation of quantum states for biomedical applications.
- **Bioimaging Innovations (2020s):** Recent advances include the use of picometer-scale resolution in cryo-electron microscopy to study protein structures as well

as the application of quantum dots in molecular diagnostics (Fitzpatrick et al., 2013).

These achievements highlight the ongoing technological advancements that have allowed researchers to access previously unthinkable dimensions. The historical track of picotechnology encourages hope for its revolutionary potential in biomedicine as it develops further.

2 Fundamentals of Picotechnology

With its emphasis on sub-nanometer matter manipulation, picotechnology is a step up from nanotechnology. The concepts, instruments, and material characteristics that characterize this innovative field are explored in this section (Rogers et al., 2013).

2.1 Principles of Sub-Nanometer Precision

Achieving accuracy at the picometre scale is the cornerstone of picotechnology, which necessitates overcoming formidable obstacles related to quantum processes, thermal fluctuations, and measurement constraints. Among the fundamental ideas are:

- **Quantum-Level Manipulation:** At the picometer scale, interactions are governed by quantum mechanics. Understanding and utilizing phenomena such as electron tunneling, wave-particle duality, and quantum entanglement are crucial for picometer-level control (Acín et al., 2018).
- **Atomic-Level Accuracy:** Techniques like scanning tunneling microscopy (STM) rely on atomic forces to achieve sub-nanometer positioning accuracy. This capability is vital for tasks such as imaging molecular structures and manipulating single atoms (Prathyusha et al., 2018).
- **Minimization of Noise:** Thermal and mechanical noise can significantly impact precision at this scale. Advanced cooling systems, vibration isolation, and noise-canceling algorithms are employed to mitigate these challenges (Al Mortuza et al., 2017).

These principles enable researchers to visualize and manipulate matter with precision previously deemed unattainable, opening avenues for breakthroughs in biomedicine and material science.

2.2 Tools and Techniques for Manipulation at the Pico-Scale

In order to fully realize the potential of picotechnology, instrumentation advancements have been essential. Important instruments and methods consist of:

- **High-Resolution Electron Microscopy:** Transmission electron microscopy (TEM) and scanning transmission electron microscopy (STEM) provide atomic-level imaging and structural analysis capabilities (Jin et al., 2017).
- **Scanning Probe Techniques:** STM and atomic force microscopy (AFM) allow researchers to visualize and manipulate surfaces at the atomic level with picometer precision (Wright et al., 2011).
- **Spectroscopy Methods:** Techniques like attosecond spectroscopy enable the study of ultrafast processes at the atomic scale, revealing dynamic interactions within biological systems (Blasco & Pico, 2011).
- **Quantum Dots and Nanostructures:** These tools are used for targeted imaging and diagnostics, exploiting their unique optical and electronic properties at the pico-scale (Kumar et al., 2018).

The integration of these tools has transformed our ability to study and manipulate biological and material systems at unprecedented levels of detail.

2.3 Material Properties and Characterization

Understanding material properties at the picometer scale is essential for advancing picotechnology. Key aspects include:

- **Atomic and Molecular Interactions:** At this scale, van der Waals forces, hydrogen bonding, and covalent interactions dominate, influencing material behavior and reactivity (Mostafavi et al., 2019).
- **Surface Properties:** Surface energy, roughness, and reactivity play critical roles in determining material interactions and their suitability for biomedical applications (Sugita et al., 2011).
- **Advanced Characterization Techniques:** Methods such as X-ray diffraction (XRD), nuclear magnetic resonance (NMR), and synchrotron-based techniques provide detailed insights into atomic arrangements and electronic properties (Pico et al., 2019).

The ability to characterize materials at this scale enables the development of novel biomaterials, targeted therapies, and precision diagnostic tools, showcasing the transformative potential of picotechnology in addressing complex biomedical challenges.

3 Picotechnology in Bioimaging

Bioimaging has seen tremendous breakthroughs to picotechnology, which has made it possible to investigate biological systems with previously unheard-of resolution and accuracy (Fig. 1). Researchers can see molecular and subcellular structures with remarkable clarity by working at the picometre scale. The revolutionary effects of picotechnology on bioimaging methods, real-time imaging, and the multiplexed study of intricate biological systems are examined in this section (Qasim et al., 2020).

3.1 High-Resolution Imaging Techniques Enabled by Picotechnology

3.1.1 Cryo-EM and AFM Advancements

The development of picotechnology has led to notable improvements in atomic force microscopy (AFM) and cryo-electron microscopy (Cryo-EM). Picometer-level precision has been added to cryo-EM, which is well-known for its capacity to produce high-resolution pictures of biological macromolecules in their natural state (Fig. 2). This enables:

- Visualization of atomic details within protein complexes.
- Structural elucidation of dynamic biological processes.

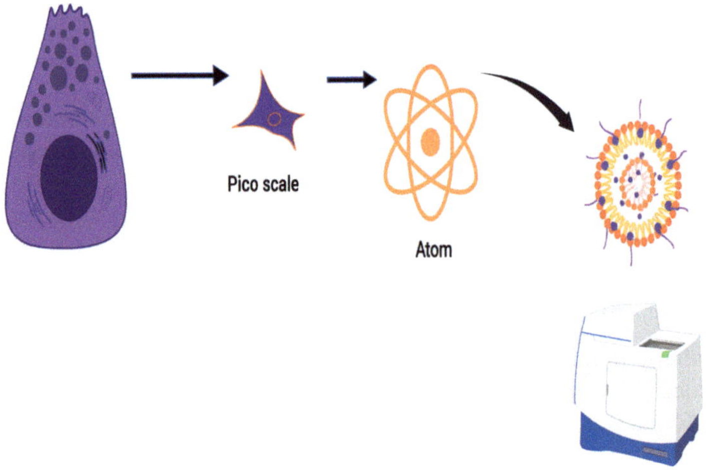

Fig. 1 Application of picotechnology in bioimaging. This figure illustrates the integration of pico-scale particles and contrast agents (e.g., quantum dots, nanoparticles) into biological cells, enhancing the resolution and sensitivity of imaging

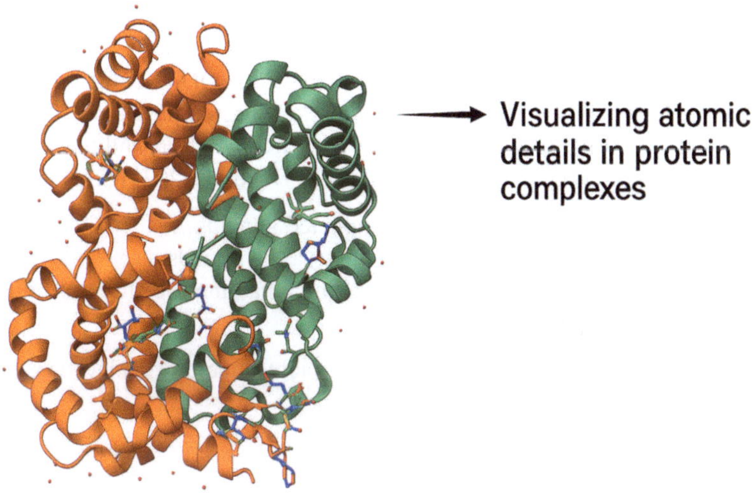

Fig. 2 Visualization of atomic details within protein complexes using picotechnology. This figure demonstrates the application of picometer-scale imaging techniques to reveal atomic-level structural information of protein complexes. Enhanced resolution achieved through pico-scale methods allows for the precise identification of atomic interactions, conformational changes, and subatomic details, providing valuable insights into protein function and molecular mechanisms at an unprecedented level of detail

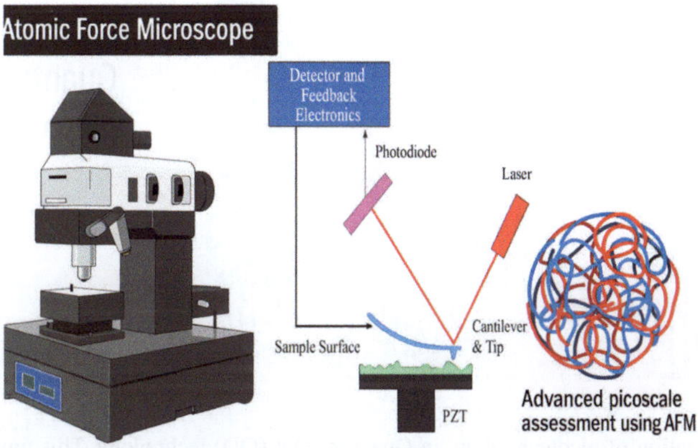

Fig. 3 Atomic Force Microscopy (AFM) leveraging picotechnology to achieve enhanced surface resolution. This figure illustrates how AFM, when coupled with picotechnology, enables the observation of surface structures at the atomic scale. The improved resolution allows for the detailed mapping of surface morphology, interactions, and mechanical properties of materials, providing crucial insights into nanoscale topography and the behavior of molecular assemblies

AFM, however, uses picotechnology to increase the resolution of the surface (Fig. 3). This development enables topographical mapping of protein assemblies, cell membranes, and biomolecular interactions at hitherto unheard-of scales (Marty et al., 2024).

3.1.2 Quantum Dot Technology

The remarkable optical characteristics of quantum dots (QDs), including their great brightness, photostability, and adjustable emission spectra, have made them effective instruments in bioimaging. QDs now have improved capabilities due to the incorporation of picotechnology (MacDougall & Donthula, 2024) (Fig. 4), such as:

- Precision labeling of biomolecules for targeted imaging.
- Reduced photobleaching for extended observation periods.
- Improved resolution for subcellular imaging, enabling the study of organelles and molecular complexes.

These advancements underscore the role of picotechnology in elevating imaging techniques to new levels of accuracy and detail.

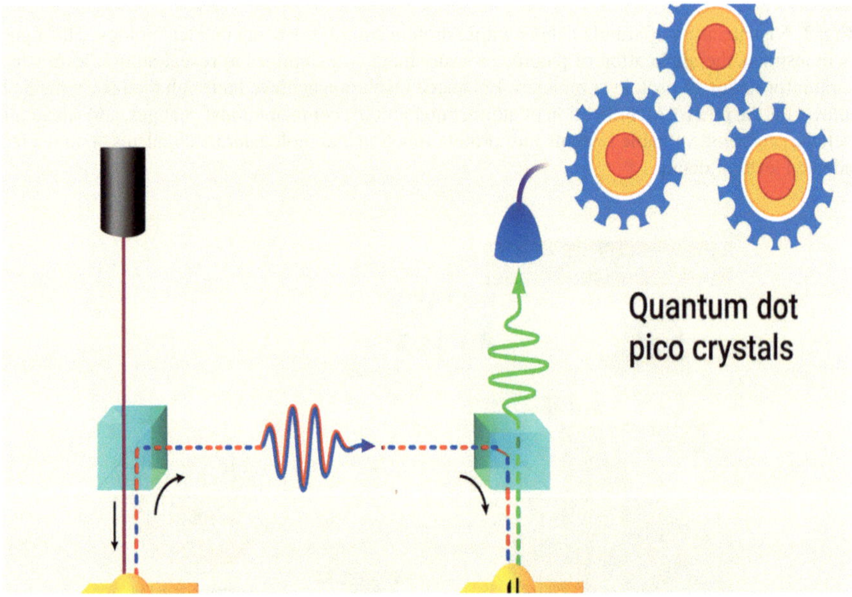

Fig. 4 Application of picotechnology in Quantum Dot (QD) technology. This figure showcases how picotechnology enhances the size, stability, and functionalization of quantum dots for bioimaging and diagnostics. By achieving precise control at the picometer scale, quantum dots exhibit improved fluorescence properties, allowing for high-resolution imaging of cellular structures and molecular interactions. The enhanced optical properties of QDs at the pico-scale enable more accurate detection and tracking in biological systems

3.2 *Real-Time Imaging at Sub-Cellular Levels*

Understanding dynamic biological processes including cellular signaling, chemical transport, and enzyme activity requires real-time imaging. Real-time imaging is made possible by picotechnology (Li et al., 2012):

- Enhancing the speed and sensitivity of imaging systems.
- Enabling the observation of molecular interactions and structural changes at picometer resolutions.
- Supporting the development of ultrafast imaging modalities, such as single-molecule fluorescence microscopy and real-time Cryo-EM.

These developments give scientists the ability to record fleeting moments and acquire a better understanding of the processes that underlie cellular activity and the development of disease.

3.3 *Multiplexed Imaging for Complex Biological Systems*

Complex biological systems often involve intricate networks of molecules and interactions. Picotechnology enables multiplexed imaging (Fig. 5), allowing simultaneous visualization of multiple targets within a single sample (Giménez-Gómez et al., 2020). Key benefits include:

- **High Throughput:** The ability to analyze numerous biomolecules in parallel accelerates data acquisition and interpretation.
- **Enhanced Specificity:** Improved resolution ensures accurate identification of closely spaced or overlapping signals.
- **Comprehensive Analysis:** Multiplexed imaging provides a holistic view of biological systems, aiding in the study of cellular heterogeneity and disease mechanisms.

Examples of multiplexed imaging enabled by picotechnology include advanced fluorescence imaging techniques, such as (Chen et al., 2025):

- Super-resolution microscopy for detailed spatial mapping.
- Spectrally distinct quantum dots for multicolor labeling.
- Combined modalities, such as Cryo-EM with fluorescence imaging, for integrated structural and functional analysis.

By combining precision, speed, and specificity, picotechnology transforms bioimaging into a powerful tool for exploring the complexities of life at an atomic scale.

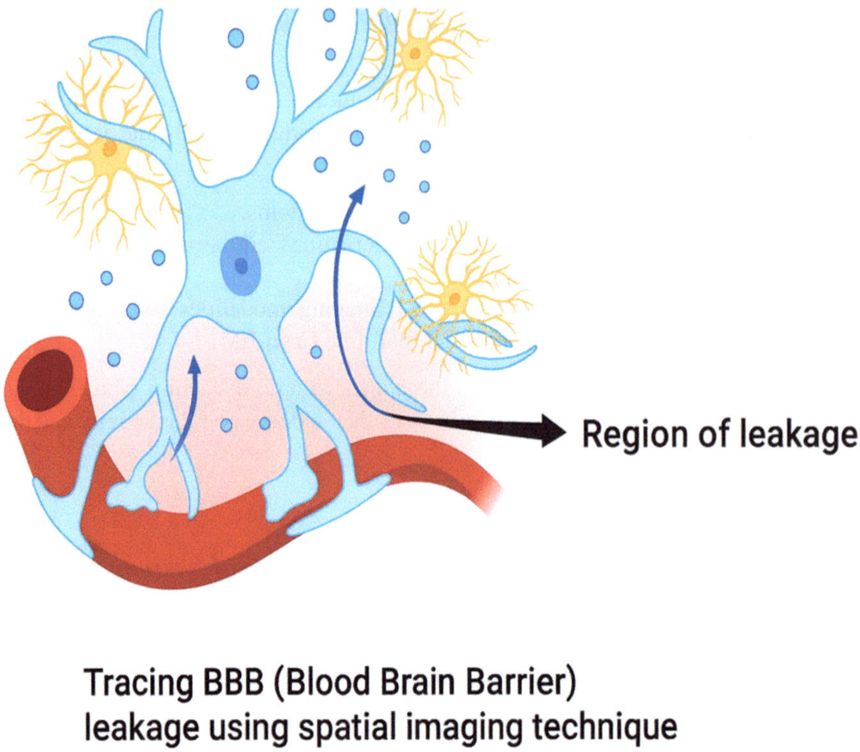

Fig. 5 Spatial mapping enhanced by picotechnology. This figure demonstrates the application of picotechnology in high-resolution spatial mapping, where picometer-scale precision enables the detailed localization of molecules, cellular structures, or nanoparticles within complex biological systems. The use of picotechnology allows for unprecedented accuracy in mapping the distribution and interactions of biomolecules, facilitating advanced studies in cellular architecture, tissue organization, and molecular dynamics

4 Picotechnology in Disease Diagnosis

This technique has enormous applications when it comes to diagnosis, which involves.

4.1 Disease Biomarkers at the Pico-Scale

The detection and analysis of disease biomarkers, which are molecular indicators of physiological or pathological processes, have achieved previously unheard-of levels of sensitivity and specificity due to picotechnology (Pico et al., 2019; Hosnijeh et al., 2024; Hartmann et al., 2021). Notable developments include:

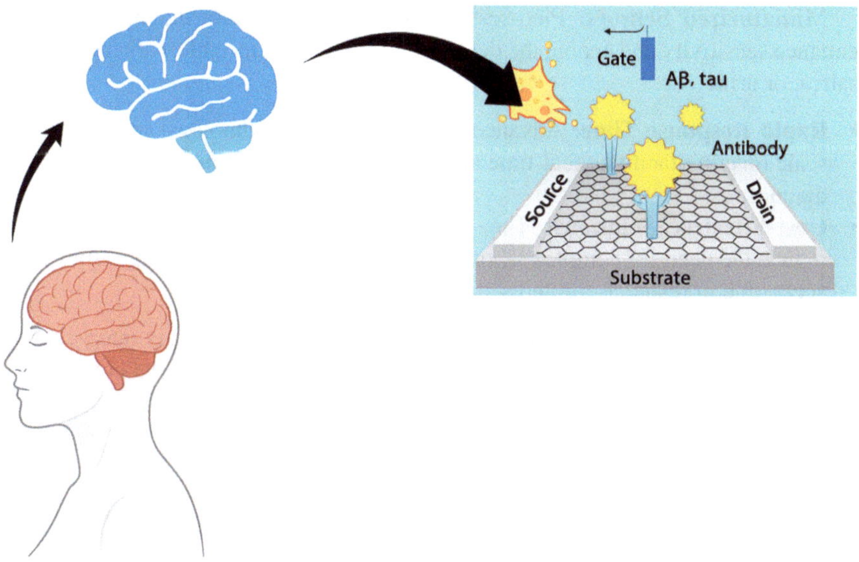

Fig. 6 Disease biomarkers as molecular indicators in picotechnology. This figure illustrates how picotechnology enhances the detection and analysis of disease biomarkers at the picometer scale. By utilizing highly sensitive picotechnology tools, molecular indicators of physiological or pathological processes can be identified with unparalleled accuracy. The figure highlights how picotech-enabled techniques improve the sensitivity, specificity, and resolution of biomarker detection, enabling early diagnosis and monitoring of disease progression at the molecular level

- **Single-Molecule Detection:** Tools such as atomic force microscopy (AFM) and quantum-enhanced sensors enable the detection of biomarkers at the single-molecule level, significantly improving diagnostic accuracy.
- **Picometer-Scale Quantification:** The ability to measure minute structural changes in proteins, nucleic acids, and other biomolecules helps identify disease states at their earliest stages.
- **High-Fidelity Analysis:** Picotechnology allows the differentiation of molecular isoforms and post-translational modifications, offering insights into complex biological processes and disease mechanisms (Fig. 6).

By leveraging these capabilities, researchers and clinicians can detect diseases such as cancer, neurodegenerative disorders, and infectious diseases with greater confidence and precision.

4.2 *Picotechnology-Enhanced Point-of-Care Diagnostics*

The goal of point-of-care (POC) diagnostics is to deliver precise and timely test findings at or close to the patient care location. These systems' performance has been greatly improved by picotechnology through (Vandenryt et al., 2017; Rabiee et al., 2021):

Miniaturized Sensors: Pico-scale sensors integrated into diagnostic devices enhance sensitivity and enable the detection of low-abundance biomarkers in blood, saliva, or urine.

- **Rapid Response Time:** Advances in quantum dot technology and picometer-scale imaging facilitate real-time analysis, reducing turnaround times for critical diagnostics.
- **Cost-Effective Solutions:** The precision and efficiency of picotechnology reduce the need for complex laboratory infrastructure, making diagnostics more accessible in resource-limited settings.

These enhancements contribute to early disease detection, timely interventions, and improved patient outcomes, particularly in remote or underserved areas.

4.3 Applications in Oncology, Neurology, and Infectious Diseases

The application of picotechnology spans diverse medical fields, addressing some of the most challenging diagnostic scenarios (Mostafavi et al., 2019; Yang et al., 2018). Key examples include:

- **Oncology:** Picotechnology enables the detection of circulating tumor cells (CTCs), DNA mutations, and protein markers associated with cancer. Techniques such as cryo-electron microscopy (cryo-EM) provide detailed insights into tumor microenvironments and drug interactions at the atomic level (Fig. 7).

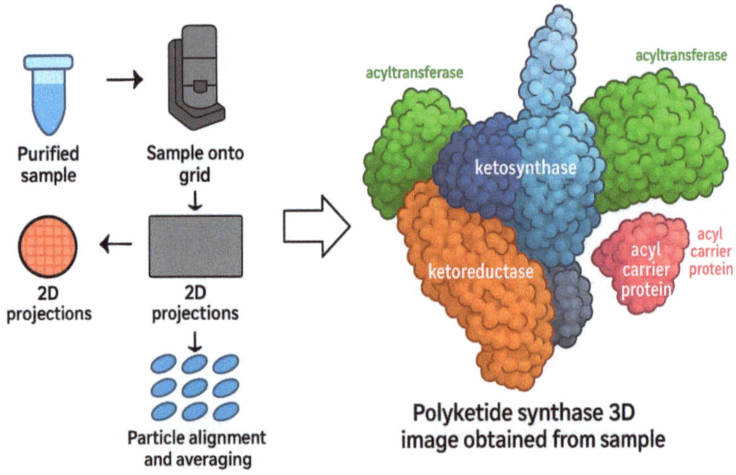

Fig. 7 Techniques such as cryo-electron microscopy (cryo-EM) providing detailed insights into tumor microenvironments and drug interactions at the atomic level. This figure demonstrates how cryo-EM, when paired with picotechnology, enables the visualization of tumor microenvironments and drug interactions with atomic-level precision. The technique reveals the structural dynamics of proteins, cell membranes, and drug molecules within complex tumor systems, offering valuable information for targeted therapies and understanding molecular mechanisms of cancer progression

- **Neurology:** Diagnosing neurodegenerative diseases, such as Alzheimer's and Parkinson's, is greatly enhanced by picotechnology, which facilitates the detection of misfolded proteins and abnormal aggregates with sub-nanometer precision.
- **Infectious Diseases:** The technology aids in identifying viral particles, bacterial strains, and specific antigen-antibody interactions, enabling the development of rapid diagnostic assays and personalized treatment strategies.

By addressing these diverse applications, picotechnology demonstrates its potential to revolutionize disease diagnostics, improving accuracy, reducing costs, and fostering personalized medicine.

5 Integration with Other Emerging Technologies

Unprecedented breakthroughs in bioimaging and illness diagnostics are being fuelled by the combination of picotechnology and other cutting-edge technologies. Picotechnology is expanding the scope of biomedical research and clinical applications by fusing its accuracy with the computing capacity of artificial intelligence, the adaptability of theranostics, and the comprehensive understanding of multi-omics (Prathyusha et al., 2018).

5.1 Synergy with Artificial Intelligence and Machine Learning

Artificial intelligence (AI) and machine learning (ML) have transformed data analysis in many scientific domains. Their combination improves the accuracy and speed of bioimaging and diagnostics when combined with picotechnology (Levkin, 2024):

- **Data Analysis and Interpretation:** AI algorithms process vast datasets generated by picometer-scale imaging systems, identifying patterns and anomalies that might be imperceptible to human analysis.
- **Predictive Modeling:** ML models predict disease progression and treatment outcomes based on high-resolution data obtained through picotechnology.
- **Automation:** AI-driven platforms automate image acquisition and analysis, increasing throughput and reducing errors.

This integration enables real-time decision-making and accelerates the discovery of novel biomarkers and therapeutic targets.

5.2 Role of Picotechnology in Theranostics

Theranostics, the combination of therapy and diagnostics, is a rapidly growing field that benefits immensely from picotechnology. The flowchart of the working is given below (Fig. 8) Key contributions include (Wan et al., 2023; Saravanan et al., 2021):

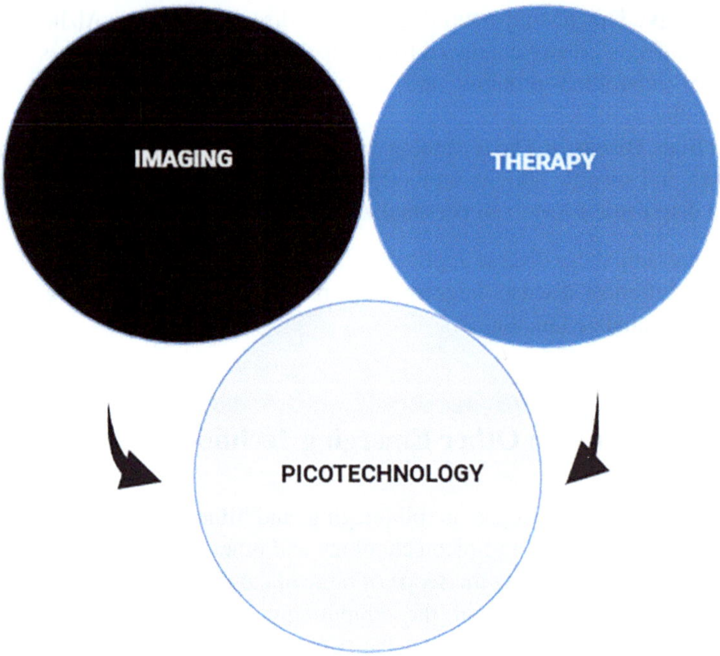

Fig. 8 Flowchart illustrating the concept of theranostics, which combines diagnostic and therapeutic approaches for personalized medicine. The diagram outlines the integration of imaging technologies, and targeted therapy, enabled by picotechnology, to achieve precise disease diagnosis and treatment. This approach enhances real-time monitoring, treatment efficacy, and patient-specific therapeutic strategies

- **Targeted Drug Delivery:** Picotechnology enables the design of nanoscale drug carriers with picometer-level precision, ensuring accurate delivery to specific cellular or molecular targets.
- **Personalized Medicine:** By identifying patient-specific biomarkers, picotechnology facilitates the development of tailored therapeutic strategies.
- **Simultaneous Diagnosis and Treatment:** Advanced imaging agents that operate at the pico-scale allow simultaneous monitoring of treatment efficacy and disease progression.

The fusion of picotechnology with theranostics enhances treatment precision, minimizes side effects, and improves patient outcomes.

5.3 *Multi-Omics Approaches and Precision Medicine*

The integration of picotechnology with multi-omics—genomics, proteomics, metabolomics, and others—is paving the way for precision medicine. This collaboration offers (Ide et al., 2022; Saharkhiz, 2022):

- **Comprehensive Insights:** Picotechnology's ability to analyze molecular interactions at the pico-scale complements multi-omics data, providing a complete view of biological systems.
- **Enhanced Biomarker Discovery:** Combining pico-scale analysis with omics technologies aids in identifying novel biomarkers with high specificity and sensitivity.
- **Tailored Therapeutic Interventions:** Precision medicine benefits from picotechnology's detailed profiling of molecular and cellular changes, enabling the development of highly targeted therapies.

By bridging the gap between molecular insights and clinical applications, the convergence of picotechnology and multi-omics approaches is driving innovations in disease diagnosis, treatment, and prevention.

6 Challenges and Limitations

While picotechnology offers groundbreaking advancements in bioimaging and disease diagnosis, several challenges and limitations must be addressed to fully realize its potential. This section discusses the primary obstacles, including issues of scalability, material biocompatibility, safety concerns, and ethical and regulatory considerations.

6.1 Scalability and Cost-Effectiveness

One of the most significant challenges in adopting picotechnology is its scalability for widespread application:

- **High Production Costs:** The development of pico-scale devices and instruments involves advanced techniques and materials, leading to high manufacturing costs.
- **Complex Fabrication Processes:** Scaling up production while maintaining precision and quality is a complex and resource-intensive task.
- **Limited Accessibility:** The high cost of picotechnology-based tools and infrastructure restricts their availability to well-funded institutions, limiting broader adoption.

Efforts to optimize production methods and reduce costs are essential to make picotechnology more accessible and practical for diverse applications.

6.2 Material Biocompatibility and Safety Concerns

The use of picotechnology in biological systems requires careful consideration of material biocompatibility and safety:

- **Toxicity Risks:** Certain materials used in pico-scale devices or probes may have toxic effects on living tissues, posing risks to patient health.
- **Long-Term Effects:** The long-term biocompatibility of pico-engineered materials remains under-researched, raising concerns about their safety in clinical applications.
- **Environmental Impact:** The disposal and degradation of picotechnology-based materials could have adverse environmental consequences.

Addressing these concerns through rigorous testing, sustainable material design, and thorough impact assessments is crucial for the safe integration of picotechnology in healthcare.

6.3 Ethical and Regulatory Considerations

The development and application of picotechnology also raise important ethical and regulatory issues:

- **Data Privacy:** Advanced bioimaging technologies enabled by picotechnology can generate detailed patient data, necessitating robust privacy protections.
- **Equity and Access:** Ensuring equitable access to picotechnology-based diagnostics and treatments is critical to avoid exacerbating healthcare disparities.
- **Regulatory Challenges:** Establishing clear guidelines and standards for the use of picotechnology in medical and research settings is essential to ensure safety and efficacy.
- **Ethical Implications:** The potential for misuse of advanced diagnostic tools, such as in non-consensual surveillance or enhancement, requires careful ethical consideration.

By addressing these challenges, stakeholders can ensure that the benefits of picotechnology are realized while minimizing potential risks and ethical dilemmas.

7 Future Prospects

The potential of picotechnology to revolutionize bioimaging and disease diagnosis extends far into the future. This section highlights emerging innovations, implications for healthcare and personalized medicine, and predictions for the next decade of biomedical advancements.

7.1 *Emerging Innovations in Picotechnology*

Continuous advancements in picotechnology are driving novel applications across various disciplines:

- **Quantum Computing Integration:** Utilizing pico-scale precision to enhance quantum computing capabilities for medical data analysis.
- **Advanced Imaging Modalities:** Development of hybrid imaging techniques that combine pico-scale resolution with other imaging technologies, such as fluorescence and Raman spectroscopy.
- **Next-Generation Probes:** Creation of ultra-sensitive probes and sensors for real-time molecular diagnostics (Passian & Imam, 2019).

These innovations promise to enhance diagnostic accuracy, therapeutic efficacy, and overall patient outcomes.

7.2 *Implications for Healthcare and Personalized Medicine*

Picotechnology holds transformative potential for healthcare systems and personalized medicine (Chen & Marco, 2022,):

- **Targeted Therapies:** Pico-scale tools enable precise delivery of therapeutic agents to diseased cells while minimizing off-target effects.
- **Early Detection:** Enhanced biomarker detection at the pico-scale facilitates earlier and more accurate diagnosis of diseases.
- **Personalized Diagnostics:** Integration of pico-scale devices into multi-omics platforms supports individualized treatment strategies based on a patient's unique molecular profile.

Such advancements will empower clinicians with tools to make informed decisions tailored to each patient's needs.

7.3 *Envisioning the Next Decade of Biomedical Advancements*

The next decade is poised to witness remarkable progress in biomedical research and clinical applications driven by picotechnology (Madera Anaya et al., 2018):

- **Comprehensive Disease Mapping:** Integration of pico-scale imaging with AI-driven analytics to create comprehensive disease maps at the molecular level.
- **Global Health Impact:** Deployment of cost-effective pico-scale diagnostic tools in resource-limited settings to address global health disparities.

- **Interdisciplinary Collaborations:** Increased collaboration among physicists, biologists, and data scientists to expand the horizons of picotechnology applications.

As these advancements unfold, picotechnology is set to redefine the landscape of modern medicine, offering unprecedented precision, efficiency, and accessibility in healthcare.

8 Conclusion

Picotechnology offers a revolutionary change in how we view and comprehend biological systems, and it is at the forefront of bioimaging and illness diagnostics. Researchers can now examine the finer intricacies of cellular structures, molecular interactions, and disease mechanisms with a level of accuracy never before possible only because of picotechnology, which operates at the picometre scale. It has consequently become a potent instrument for furthering both fundamental biomedical research and therapeutic applications (Alpaslan & Webster, 2014).

One of the most impressive contributions of picotechnology is its ability to detect and measure biomarkers with exceptional sensitivity, enabling the early diagnosis of diseases at their molecular stages. This is a significant departure from traditional diagnostic methods, which often detect diseases only after they are visible or symptomatic. The review of recent advancements in picotechnology highlights its immense potential across various aspects of biomedicine. By using picotechnology, clinicians can identify diseases like cancer, neurodegenerative disorders, and cardiovascular conditions at much earlier stages, greatly increasing the likelihood of successful intervention and improving patient prognosis (Tamanna & Chopra, 2023).

In the domain of bioimaging, picotechnology has opened up new opportunities for observing dynamic biological processes in real-time. Its ability to record high-resolution images at the picometer scale enables researchers to examine the activity of individual molecules, witness cellular connections, and track disease progression in ways that were previously not feasible. This degree of expertise not only broadens our comprehension of biological systems but also offers a stronger foundation for creating individualized and targeted remedies.

Additionally, picotechnology has great potential for drug development and delivery, allowing for more accurate tracking of drug interactions at the molecular level, which provides a valuable tool for improving drug design and treatment efficacy. Picotechnology can also help develop new diagnostic platforms, like point-of-care devices that can provide fast and accurate results, which could revolutionize healthcare delivery by offering more affordable and accessible solutions (Sharma et al., 2011).

Picotechnology has a lot of potential, but its application in therapeutic settings is still in its infancy. The creation of appropriate materials, the requirement for additional imaging technique improvement, and the creation of standardized protocols for clinical application are some of the obstacles that still need to be overcome. These issues should be resolved as research advances, though, and picotechnology will become an essential component of therapeutic approaches and medical diagnostics.

In conclusion, picotechnology represents a paradigm shift in the way we approach the study and treatment of diseases. By providing a level of precision and sensitivity that surpasses current technologies, it offers a powerful means of advancing our understanding of complex biological processes and diagnosing diseases at their earliest and most treatable stages. As the field continues to grow, picotechnology has the potential to revolutionize biomedical research and clinical applications, ultimately leading to more effective, personalized, and accessible healthcare solutions.

8.1 Summary of Current Trends and Breakthroughs

Ultra-high-resolution imaging, real-time sub-cellular analysis, and accurate disease biomarker detection are just a few of the notable advances brought about by the evolution of picotechnology. Other innovations like quantum dot technology, cryo-EM, and pico-level manipulations have created new opportunities for the study of biological processes and the diagnosis of diseases with previously unheard-of precision. Additionally, the combination of picotechnology with AI, machine learning, and multi-omics techniques has accelerated its use in precision medicine and theranostics.

8.2 Paving the Way for Picotechnology in Clinical Applications

Picotechnology has enormous clinical promise in the future. Picotechnology has the potential to revolutionize healthcare by enhancing early disease identification and facilitating tailored therapeutic approaches. To ensure its widespread adoption and success, it will be imperative to overcome issues with scalability, cost-effectiveness, biocompatibility, and ethical considerations.

As we look to the future, the ongoing fusion of picotechnology and new fields holds the potential to open up revolutionary possibilities that will spur innovation and enhance patient outcomes in ways that were previously unthinkable.

References

Acín, A., Bloch, I., Buhrman, H., Calarco, T., Eichler, C., Eisert, J., & Wilhelm, F. K. (2018). The quantum technologies roadmap: A European community view. *New Journal of Physics, 20*(8), 080201.

Al Mortuza, A., Pervez, M. F., Hossain, M. K., Sen, S. K., Mia, M. N. H., Basher, M. K., & Alam, M. S. (2017). Pico-current measurement challenges and remedies: A review. *Universal Journal of Engineering Science, 5*(4), 57–63.

Alpaslan, E., & Webster, T. J. (2014). Nanotechnology and picotechnology to increase tissue growth: A summary of in vivo studies. *International Journal of Nanomedicine, 9*(sup1), 7–12.

Bhagawati, H., Kashyap Sarma, M., Talukder, A., Das, S., & Nath, D. (2023). Picoscience: Picotechnology's advent in health and medicine. *Current Drug Therapy, 18*(4), 312–322.

Blasco, C., & Pico, Y. (2011). Determining nanomaterials in food. *TrAC Trends in Analytical Chemistry, 30*(1), 84–99.

Chen, B., & Marco, G. (2022). Recent advances in the medical field with the revolution of nanotechnology. *Reviews of Adhesion and Adhesives, 10*(2), 1.

Chen, D., Bryden, W. A., McLoughlin, M., Ecelberger, S. A., Cornish, T. J., Moore, L. P., & Regan, K. M. (2025). digitalMALDI: A single-particle–based mass spectrometric detection system for biomolecules. *Journal of Mass Spectrometry, 60*(2), e5110.

Crous, A., & Abrahamse, H. (2021). Innovations in nanotechnology for biomedical sensing, imaging, drug delivery, and therapy. *Nanotoxicology*, 21–42.

Fitzpatrick, A. W., Lorenz, U. J., Vanacore, G. M., & Zewail, A. H. (2013). 4D cryo-electron microscopy of proteins. *Journal of the American Chemical Society, 135*(51), 19123–19126.

Gargini, P. A. (2004). Silicon nanoelectronics and beyond. *Journal of Nanoparticle Research, 6*(1), 11–26.

Giménez-Gómez, P., Rodríguez-Rodríguez, R., Ríos, J. M., Pérez-Montero, M., González, E., Gutiérrez-Capitán, M., & Jiménez-Jorquera, C. (2020). A self-calibrating and multiplexed electrochemical lab-on-a-chip for cell culture analysis and high-resolution imaging. *Lab on a Chip, 20*(4), 823–833.

Hanany, S. (n.d.). *PICO: Probe of Inflation and Cosmic Origins.*

Hartmann, A., Hartmann, C., Secci, R., Hermann, A., Fuellen, G., & Walter, M. (2021). Ranking biomarkers of aging by citation profiling and effort scoring. *Frontiers in Genetics, 12*, 686320.

Hosnijeh, F. S., Liu, H. H., Denholm, N., Wilkes, E., & Levick, B. (2024). RWD86 heterogeneity in biomarker data collection, recording, and availability across European sources of Oncology RWD & Potential Impacts on Pan-EU HTA assessment. *Value in Health, 27*(12), S589.

Ide, K., Nishikawa, Y., Maruyama, T., Tsukada, Y., Kogawa, M., Takeda, H., & Takeyama, H. (2022). Targeted single-cell genomics reveals novel host adaptation strategies of the symbiotic bacteria Endozoicomonas in Acropora tenuis coral. *Microbiome, 10*(1), 220.

Jain, K. K. (2010). Advances in the field of nanooncology. *BMC Medicine, 8*, 1–11.

Jin, L., Barthel, J., Jia, C. L., & Urban, K. W. (2017). Atomic resolution imaging of YAlO3: Ce in the chromatic and spherical aberration corrected PICO electron microscope. *Ultramicroscopy, 176*, 99–104.

Khorasaninejad, M., Zhu, A. Y., Roques-Carmes, C., Chen, W. T., Oh, J., Mishra, I., & Capasso, F. (2016). Polarization-insensitive metalenses at visible wavelengths. *Nano Letters, 16*(11), 7229–7234.

Kumar, D. S., Kumar, B. J., & Mahesh, H. M. (2018). Quantum nanostructures (QDs): An overview. In *Synthesis of inorganic nanomaterials* (pp. 59–88). Elsevier.

Levkin, D. (2024). Towards sustainable IoT: Current and voltage data analysis in NB-IoT systems.

Li, P. H., Ting, H., Chen, Y. C., & Urban, P. L. (2012). Recording temporal characteristics of convection currents by continuous and segmented-flow sampling. *RSC Advances, 2*(32), 12431–12437.

MacDougall, P. J., & Donthula, K. K. (2024). Using quantum atomics and machine learning to advance picotechnology. *Theoretical Chemistry Accounts, 143*(10), 1–9.

Madera Anaya, M., Franco, J. V. A., Ballesteros, M., Sola, I., Urrutia Cuchi, G., & Bonfill Cosp, X. (2018). Evidence mapping and quality assessment of systematic reviews on therapeutic interventions for oral cancer. *Cancer Management and Research*, 117–130.

Marty, A. P. M., Ward, E. N., Lamb, J. R., van Tartwijk, F. W., Peck, L. S., Clark, M. S., & Kaminski, C. F. (2024). A high-resolution microscopy system for biological studies of cold-adapted species under physiological conditions. *Small Methods, 9*, 2401682.
Moriarty, P. (2022). *Nanotechnology: A very short introduction*. Oxford University Press.
Mostafavi, E., Soltantabar, P., & Webster, T. J. (2019). Nanotechnology and picotechnology: A new arena for translational medicine. In *Biomaterials in translational medicine* (pp. 191 212). Academic Press.
Pan, Y., Liu, M., Lian, C., Xia, Y., & Shen, D. (2019). Disease-image specific generative adversarial network for brain disease diagnosis with incomplete multi-modal neuroimages. In *Medical image computing and computer assisted intervention–MICCAI 2019: 22nd international conference* (pp. 137–145).
Pashazadeh-Panahi, P., & Hasanzadeh, M. (2019). Revolution in biomedicine using emerging of picomaterials: A breakthrough on the future of medical diagnosis and therapy. *Biomedicine & Pharmacotherapy, 120*, 109484.
Passian, A., & Imam, N. (2019). Nanosystems, edge computing, and the next generation computing systems. *Sensors, 19*(18), 4048.
Pico, Y., Alfarhan, A., & Barcelo, D. (2019). Nano-and microplastic analysis: Focus on their occurrence in freshwater ecosystems and remediation technologies. *TrAC Trends in Analytical Chemistry, 113*, 409–425.
Prathyusha, K., Reddy, J. G. M., Venkataswamy, M., & Ramesh, A. (2018). Pico technology: Instruments used and applications in pharmaceutical field. *Research Journal of Pharmaceutical Dosage Forms and Technology, 10*(1), 34–41.
Qasim, M. Z., Hammad, H. M., Abbas, F., Saeed, S., Bakhat, H. F., Nasim, W., & Fahad, S. (2020). The potential applications of picotechnology in biomedical and environmental sciences. *Environmental Science and Pollution Research, 27*, 133–142.
Rabiee, N., Ahmadi, S., Fatahi, Y., Rabiee, M., Bagherzadeh, M., Dinarvand, R., & Webster, T. J. (2021). Nanotechnology-assisted microfluidic systems: From bench to bedside. *Nanomedicine, 16*(3), 237–258.
Rogers, B., Adams, J., & Pennathur, S. (2013). *Nanotechnology: The whole story*. CRC Press.
Rush, B. E. (2022). *Using bioimaging techniques as muscle quality biomarkers for sarcopenia and cachexia diagnosis and treatment*. The University of Wisconsin-Madison.
Saharkhiz, N. (2022). *Development and optimization of a clinical harmonic motion imaging system for breast tumor characterization and neoadjuvant chemotherapy response assessment*. Columbia University.
Saravanan, M., Barabadi, H., Vahidi, H., Webster, T. J., Medina-Cruz, D., Mostafavi, E., & Periakaruppan, P. (2021). Emerging theranostic silver and gold nanobiomaterials for breast cancer: Present status and future prospects. In *Handbook on nanobiomaterials for therapeutics and diagnostic applications* (pp. 439–456). Elsevier.
Shafi, K. V., Gedanken, A., Prozorov, R., & Balogh, J. (1998). Sonochemical preparation and size-dependent properties of nanostructured CoFe2O4 particles. *Chemistry of Materials, 10*(11), 3445–3450.
Sharma, R., Sharma, A., & Chen, C. (2011). Emerging trends of nanotechnology towards picotechnology: Energy and biomolecules. *Nature Precedings, 1*, 1.
Sugita, Y., Ishizaki, K., Iwasa, F., Ueno, T., Minamikawa, H., Yamada, M., & Ogawa, T. (2011). Effects of pico-to-nanometer-thin TiO_2 coating on the biological properties of microroughened titanium. *Biomaterials, 32*(33), 8374–8384.
Szczyglewska, P., Feliczak-Guzik, A., & Nowak, I. (2023). Nanotechnology–general aspects: A chemical reduction approach to the synthesis of nanoparticles. *Molecules, 28*(13), 4932.
Tamanna, T., & Chopra, L. (2023, May). Pico technology–an advancement in dye degradation. *AIP Conference Proceedings, 2535*(1), 030009.

Umadevi, K., Sundeep, D., Vighnesh, A. R., Misra, A., & Krishna, A. G. (2024). Current trends and advances in nanoplatforms-based imaging for cancer diagnosis. *Indian Journal of Microbiology, 65*, 1–40.

Vandenryt, T., Van Grinsven, B., Eersels, K., Cornelis, P., Kholwadia, S., Cleij, T. J., & Wagner, P. (2017). Single-shot detection of neurotransmitters in whole-blood samples by means of the heat-transfer method in combination with synthetic receptors. *Sensors, 17*(12), 2701.

Wan, F., Thakur, A., & Thakur, P. (2023). Carbon nanomaterials in the field of theranostics. In *Integrated nanomaterials and their applications* (pp. 81–105). Springer.

Woźniak, M., Płoska, A., Siekierzycka, A., Dobrucki, L. W., Kalinowski, L., & Dobrucki, I. T. (2022). Molecular imaging and nanotechnology—Emerging tools in diagnostics and therapy. *International Journal of Molecular Sciences, 23*(5), 2658.

Wright, C. D., Aziz, M. M., Shah, P., & Wang, L. (2011). Scanning probe memories–technology and applications. *Current Applied Physics, 11*(2), e104–e109.

Yang, L., Bhaduri, S., & Webster, T. J. (Eds.). (2018). *Biomaterials in translational medicine*. Academic Press.

Picotechnology in Material Science

M. Manigandan, Shrikaant Kulkarni, F. Fareedhul Fahmitha, R. Yuvaraj, and P. SivaSakthivelan

1 Introduction

Picotechnology is a new branch of science and technology that is devoted to the control and manipulation of matter at the picometer range, or a trillionth of a meter (10^{-12} meters). While Nanotechnology is concerned with structures in the nanometer range (10^{-9} meters), Picotechnology proceeds three orders higher into the scale of atomic and subatomic scale. At this scale, the very building blocks of matter—atoms and even their electron orbits—can be controlled with great accuracy, making possible a vast array of physics, materials science, biotechnology, electronics, and quantum computing applications. The novelty of Picotechnology is in its capacity to deal with and manipulate interactions at the atomic level. It allows scientists and engineers to surpass manipulating particles or molecules and start directly changing atomic placement, bonding configurations, and even quantum states (Prathyusha et al., 2018). This kind of precision is essential in furthering areas like quantum materials, high-accuracy sensors, and future semiconductors. Single-atom transistors, for instance, which operate at the boundary of quantum limits, depend on the same precepts that Picotechnology strives to codify. In contrast to Nanotechnology, in which devices and materials tend to be fabricated by bottom-up assembly or

M. Manigandan (✉) · F. F. Fahmitha · R. Yuvaraj
PG and Research Department of Microbiology, Sacred Heart College (Autonomous), Tirupattur, Tamil Nadu, India

S. Kulkarni
Sanjivani University, Kopargaon, India

Faculty of Business, Victorian Institute of Technology, Melbourne, VIC, Australia

P. SivaSakthivelan
Agricultural Research Station, Tamil Nadu Agricultural University, Erode, Tamil Nadu, India

S. Kulkarni (ed.), *A Journey from Nanotechnology to Picotechnology*, Lecture Notes in Nanoscale Science and Technology 47,
https://doi.org/10.1007/978-3-032-07228-3_14

top-down fabrication methods, Picotechnology tends to involve direct atomic-level alterations facilitated by Scanning Tunneling Microscopes (STM), Atomic Force Microscopes (AFM), and quantum probes. These tools not only image atomic configurations but can also pull and push single atoms with precision, thus facilitating atomic-scale building and engineering. This results in the creation of exotic materials with highly customized properties like superconductivity, ultra-hardness, or novel magnetic behaviors—none of which are achievable with traditional materials design techniques. In the biomedical field, Picotechnology holds new promise for precision diagnostics and therapeutic interventions. For example, picometer-scale sensors might be employed to track biological processes in real-time at the molecular level, enabling earlier detection of disease and more targeted treatments. DNA-based nanodevices, which can operate with pico-level precision, are also being investigated for their potential to interact with cells and tissues in highly specific manners, serving as both diagnostics and drug delivery systems. Environmental science also has a lot to gain. Picotechnology can enable detection and cleanup of pollutants at record levels of precision (Sharma et al., 2011). Materials synthesized based on picotech concepts could be used as extremely sensitive environmental sensors or as catalysts for the degradation of toxic substances into harmless byproducts, leading to clean and sustainable technology solutions. One of the most thrilling frontiers where Picotechnology might prove transformative is quantum computing. Quantum bits (qubits), the heart of quantum information, are liable to atomic-scale disruptions. Precision control of atomic position and of quantum dots or crystal defects (such as nitrogen-vacancy centers in diamond) on the picometer level is required in order to create scalable, error-protected quantum computers. But the road to practical Picotechnology is fraught with difficulties (Bolonkin, 2009). Operating at such a submicroscopic dimension demands exceedingly high environmental control, sophisticated theoretical models, and highly advanced instrumentation that can provide and sustain pico-level accuracy. Further, quantum phenomena like tunneling, entanglement, and decoherence become prevalent and must be fully comprehended and controlled in order to develop dependable pico-scale systems. Finally, Picotechnology is a fast-developing and very promising field that develops from the groundwork established by Nanotechnology but takes it a step further to the atomic and quantum domains. Its prospective applications span from groundbreaking developments in electronics, materials, and energy to new breakthroughs in medicine, environmental conservation, and computing. As technique and comprehension advance, Picotechnology could form the basis of a new technological paradigm one in which ultimate control of matter makes possible the solution to some of the world's most challenging science and engineering problems (Nitkin et al., 2005).

2 Picotechnology Contrasting with Nanotechnology

Nanotechnology is flourishing in many different fields; for example, engineered nanoparticles are extensively applied in agriculture biomedicine, cosmetics and electrical, pharmaceutical, and textile sectors. In wastewater treatment methods, it is also a potential technology for the removal of pollutants including phosphorous, etc., for example, in home wastewater treatments. Through advances in prevention, diagnosis, and treatment of illness problems, nano-technology has revolutionized medicine and pharmaceutical sectors in recent years; moreover, nano-based pharmaceutical and medical preparations have been licensed by regulating authorities. Particularly used as drug delivery carriers are several types of nanomaterials including polymeric drug conjugates, liposomes, and micelles. In cancer treatment, nanotechnological-based preparations are feasible; specifically, smaller nanoparticles lesser than 100-nm size can follow tumor locations and cause death. Particular kinds of antigens and receptors expressed on the surface of malignant cells are exploited to facilitate the simple entrance of nanodrugs into tumor tissues. These developments have provided even more opportunities for Picotechnology to transform the medical field. Regarding regenerative medicine, several interactions at atomic and subatomic levels control biological events including intracellular activities. Linked with this quite low level, Picotechnology is a wise choice for material synthesis, imaging, medicine, and biological applications (Singh & Prasad, 2017). It could also provide avenues for more advanced and individualized treatments. When administered by body, Picomolecules (PM) have a much higher degree of physiological reaction activation than nanomolecules. Furthermore, the margin of activities of many drugs increases when their size dimension lowers to Picoscale. Actually, the tenability idea of characteristics helps to clearly show the quantum effect of picoscales. One can also clearly define tunability by changing the particle size. Via means of this modification, a scientist can correct several aspects of materials, including the fluorescence color, which subsequently finds application not only in specifically identifying that particle but also in tagging various materials (via fluorescent markers) for varied uses (Xu et al., 2019). Apart from that, quantum effect of picoscale also provides the tunneling feature that makes adventures such as flash memory and scanning tunneling of microscope in computing relevant. Picomaterials' greater reactivity and bigger surface area help to promote their functionalization, therefore rendering them valuable for use in drug delivery systems. Apart from that, Picotechnology offers to change surfaces without possible toxicity issues as those of Nanotechnology by combining science and engineering of new generation. Compared to gold Nanoparticles (NPs), gold picoparticles with red or purple color show more promise since the mobility trends of electrons in gold-NPs are limited. Gold-NPs so behave differently in light exposure depending on limited electron mobility. Conversely, Gold-PPs have better selectivity and specificity so they can build up tumor cells without disturbing surrounding normal cells. Living cells allows one to precisely apply targeted laser annihilation of tumor tissues and bio-imaging. This image processing technique resolves the atomic structure of materials with high precision, directly mapping the location of individual atoms in picoscale resolution images. Its superiority over earlier techniques is demonstrated by its ability to clearly visualize a gold nanoparticle's atomic lattice, including specific atomic sites and contraction patterns. During pollution control activities, one-dimensional nanomaterials with high surface area and effective pore interconnectivity are suitable for air filtration (Sankar et al., 2018). Volatile organic compounds (VOCs), biological and chemical pollutants, and other contaminants in air and water systems can all be effectively and affordably cleaned by electrospun nanofibers. Analogous viability for VOC removal from industrial stacks is provided by Manganese Oxide (MnO) nanofiber catalyst. Carbon Nano Tubes (CNTs) with a small diameter and a higher surface-volume ratio have demonstrated effective adsorptive behavior when compared to conventional materials. The discovery of gaseous analysis, related to the evolution of several operational sensors. Graphene is one of the few materials with good mechanical strength, flexibility, thermal conduction, electrical conductivity, and stability. Excellent adsorbent for the detection of hazardous gaseous chemicals are graphene oxide and other related compounds. These potential properties of graphene are attributed to the presence of massive active sites, including carbon atoms on graphitic layers and epoxy and hydroxyl functional groups, which allow for a significant amount of NH3 adsorption. The development of new materials through Picotechnology that will, through an even longer surface-volume ratio, perform better against the monitoring and detection of significant air pollution will eventually benefit greatly from the advancements in Nanotechnology and its applications, such as nanofibers and nanosensors (Kumar, 2018).

3 Special Feature of Picotechnology

Most biological phenomena take place at atomic and subatomic levels. It is by working at the Picoscale that researchers can innovate new processes capable of augmenting biological systems, medicine, imaging and many more areas as it is at atomic and subatomic levels. Picotechnology is not just operating at progressively smaller sizes but operating on the picoscale allows researchers to harness the unique physical, chemical and optical properties of materials that occur. The properties of the particles are not greatly different when solid matter visible particle sizes are compared to what is visible to the naked eye under a typical optical microscope. Picoscale gold exhibits the unique characteristics at the picoscale. Picoscale gold particles may look red or purple; they are not the yellow color to which we are familiar. The gold's electrons are confined motion at the nanoscale. This motion is confined, therefore gold nanoparticles respond differently when exposed to light compared to larger-scale gold particles.

One intriguing and accurate outcome of quantum phenomena at the picoscale is the concept of tunability of properties. In other words, a scientist can effectively optimize the material's characteristic (such as shifting fluorescence color) by altering the particle size. On the other hand, a particle's fluorescence color can be used to identify it, and different materials may be labeled with fluorescent markers for different purposes. Tunneling, a phenomenon that makes computer flash memory and scanning tunneling microscopy possible is another extremely significant quantum effect of the picoscale. At the picoscale, a large portion of cellular activity naturally takes place. For example, the protein hemoglobin, which transports oxygen around the body, has only a 5.5.5 nm diameter. One of the building blocks of human life, a DNA strand is merely about 2 nm across, a great deal wider than a picometer. Numerous clinical researchers are working on creating tools, treatments, and technologies more precise and tailored than traditional ones and that are potentially usable earlier in the progression of a disease and have less dangerous side effects on employing the natural picoscale of biology (Tang & Cao, 2012). The bio barcode test is just one of the medical applications of Picotechnology, a comparatively low cost method of detecting disease-specific biomarkers in blood, even in a sample containing extremely few of them. First demonstrated at Northwestern University for a prostate cancer biomarker after prostatectomy, the underlying process which binds recognition|| particles and DNA amplifiers onto gold picoparticles was first demonstrated. Modified to almost any molecular target, the bio-barcode assay has proven much more sensitive than conventional assays for the same target biomarkers (Nitkin et al., 2005). Picoscale biomolecular structures are affecting fields beyond medicine as growing knowledge of them expands. By applying picoscale biological concepts of molecular self-assembly, self-organization, and quantum physics, some scientists envision constructing novel computing platforms. But if that 1 cubic centimeter volume were packed with 1 mm cubes, there would be 1000 mm-sized cubes (10 × 10 × 10), each having a surface area of 6 square millimeters, for a total surface area of 60 square centimeters roughly the same as one side of two-thirds of a 3-series ×5-series note card (Durand et al., 2012). The catalytic converter on a car, which reduces the engine's fumes' toxicity, is a familiar example of catalysis. Increased reactivity at the picoscale can be employed by Picodesigned atteries, fuel cells, and catalysts to produce cleaner, safer, and more economically viable methods of generating and storing energy. To cover applications from the administration of medicines to the insulating of garments, it further facilitates functionalization—that is, the addition of particles for specific purposes of Picoscale material surfaces (Li et al., 2025a, b).

4 Picodevices

Picodevices are hypothetical or ultra-miniature devices that function at the Picometer scale (10^{-12} meters) a step below Nanotechnology and near the size of individual atoms and subatomic structures. They are a central concept in Picotechnology, which is the manipulation of matter at this extremely small scale, with electron shell, atomic orbital, and quantum control.

Functional Picodevices are not yet available in commercial or common scientific use a number of experimental and theoretical developments are providing the groundwork for this technology.

Table 1 Types of picodevices

S.No	Technology	Function	Status	Reference
1	Single-atom transistors	Implantable devices	Theoretical to early stage	Chen et al. (2024)
2	Quantum dots as picosensors	Molecular and neuronal detection	In development	Ghanbarlou et al. (2025)
3	Molecular machines	Drug delivery and cell penetration	Experimental	Ariga (2023)
4	Pico-scale logic gates	Biomolecule analysis	Used in research	Wang et al. (2025)
5	DNA-based nanodevices (with Pico-precision)	Smart drug delivery and diagnostics	Pre-clinical stage	Kansara et al. (2025)

4.1 Types of Picodevices in Picotechnology

Fully functioning Picodevices like sensors, processors, or machines composed of moving atomic parts at Picometer scale do not exist outside of simulations or early lab based prototypes. There are different types of Picodevices and some of them are under process to develop and they are presented in table 1.

4.1.1 Single-Atom Transistors

Single-atom transistors are among the most advanced and promising instances of Picotechnology, operating on the edge of atomic precision. Unlike conventional transistors, consisting of hundreds of atoms, single-atom transistors use a single atom most commonly phosphorus, silicon, or gold to control electron flow. Such miniaturization takes device design to the picometer scale, where atomic orbitals, quantum tunneling, and electron conductance must be precisely known and controlled. Single-atom transistors represent a significant stepping stone on the path to ultra-efficient electronics and quantum computing, which have the potential to reduce size and power consumption while increasing processing speed. In Picotechnology, such devices require atomic-scale positioning tools such as Scanning Tunneling Microscopes (STM) and Atomic Force Microscopes (AFM) and manipulation of quantum states with high precision. Although promising, single-atom transistors are hindered by their monumental issues such as quantum instability, thermal noise, and manufacturability scalability. However, since they can operate at an incredibly tiny scale, they are the key foundation of the atomic-scale logics circuitry of the future, quantum processors, and second-generation sensors. Single-atom transistors could potentially be building blocks in a next generation of atomically engineered devices once Picotechnology is developed (Chen et al., 2024).

4.1.2 Quantum Dots as Picosensors

Quantum dots (QDs) are nanocrystalline semiconductors having sizes ranging most often from 2–10 nm but showing quantum features and electron confinement in the sub-nanometer (picometer) scale range and are therefore very useful in the realm of Picotechnology. Quantum dots have enormous sensitivity toward external environment variations of pH, temperature, magnetic field, and also the presence of certain biomolecules when they function as picosensors. This sensitivity is due to their discrete energy levels and size-dependent optical and electronic properties, which are controlled by quantum confinement effects at the atomic level. In Picotechnology, quantum dots are being investigated for ultra-sensitive detection and diagnostics, especially in biomedical applications. For example, QDs can be designed to bind to specific cellular targets or molecular markers, where they emit light, allowing real-time imaging and single-molecule detection. Their tunable emission spectra, stability, and picometer-scale interaction sensitivity make them suitable for applications in biosensors, medical imaging, and targeted therapy (Ghanbarlou et al., 2025).

4.1.3 Molecular Machines

Engineered Molecular machines are nanoscale or sub-nanoscale structures made up of single molecules that are engineered to carry out mechanical-like operations through controlled motion or conformational change. In Picotechnology, these machines work at atomic or picometer scale precision, where atom-atom interactions and molecular bonds govern motion and function. They are able to do things like turning, changing, carrying freight, or reacting to certain chemical or physical stimuli. The 2016 Nobel Prize in Chemistry honored trailblazing research in the field, highlighting the possibility of molecular machines as the foundation of the intelligent, responsive materials and nanoscale systems of the future. In Picotechnology, molecular machines are central building blocks for the construction of atomically precise devices in medicine, computing, and materials science. For example, in drug targeting, molecular machines may be designed to traverse biological surroundings and release therapeutic agents at precise locations, prompted by chemical triggers or environmental signals. Their capability to operate with picometer-scale resolution enables unprecedented selectivity and efficiency on the cellular and molecular level (Ariga, 2023).

4.1.4 Pico-Scale Logic Gates

Theoretical Pico-scale logic gates are a visionary future of Picotechnology, where logic operations are performed with components made up of single atoms or molecules. Ultra-miniature gates function at or near the picometer scale (10^{-12} meters), considerably above the state of the art of current Nanotechnology. On this scale, normal electronic behaviors yield to quantum phenomena such as tunneling, electron orbital interactions, and discrete energy level transitions. The vision of pico-scale logic gates is to attain extremely high computational density, extremely low power consumption, and record processing rates through the utilization of atomic and subatomic phenomena for logical operations. Some of the research in this field involves employing single-atom transistors, molecular switches, and quantum dots to execute logic operations. For instance, logic gates made up of a handful of atoms by scanning tunneling microscopy (STM) have shown simple operations such as AND, OR, and NOT. Furthermore, molecular logic gates, chemical inputs that result in a binary output due to molecular reorganization have shown promise for bio-computing and in vivo diagnostics (Wang et al., 2025).

4.1.5 DNA-Based Nanodevices (with Pico-Precision)

DNA nanodevices are a quickly evolving category of molecular devices designed employing the foreseeable base-pairing characteristics of DNA to form well-defined nanoscale objects and functional devices. Although normally a part of the field of Nanotechnology, the devices are also being viewed with growing interest because of their pico-precision—measuring spatial resolution on the sub-nanometer (picometer) scale, thus making them an integral component of Picotechnology. By employing methods like DNA origami, scientists are able to fold DNA strands into complex two- and three-dimensional structures at an atomic level of precision, and thus prepare smart materials, sensors, and delivery systems that can work with incredible accuracy. These nanodevices may be engineered to execute specific functions like molecular recognition, drug delivery, biosensing, and logic operations based on environmental stimulation (e.g., pH fluctuations, enzyme activity, or molecular targets). Their capacity to make controlled conformational changes allows dynamic behavior similar to molecular machines, with motion and function controlled at the picometer level. In biomedical applications, DNA nanodevices have the potential for targeted therapy, early disease diagnosis, and single-cell diagnostics, where accuracy and biocompatibility are essential (Kansara et al., 2025).

5 Comparing Displacements at the Picometer Level

In precision engineering and nanometrology, the comparison of displacements at the picometer level was given in Fig. 1.

At this scale, previously thought to be negligible measurement errors become substantial, and the material and system physical behavior needs to be known with very high precision. A comparable error of locating the center frequency of the Fabry-Perot resonance replaces fringe interpolation errors in a Michelson interferometer, another possible error source familiar when undertaking high precision interferometry.

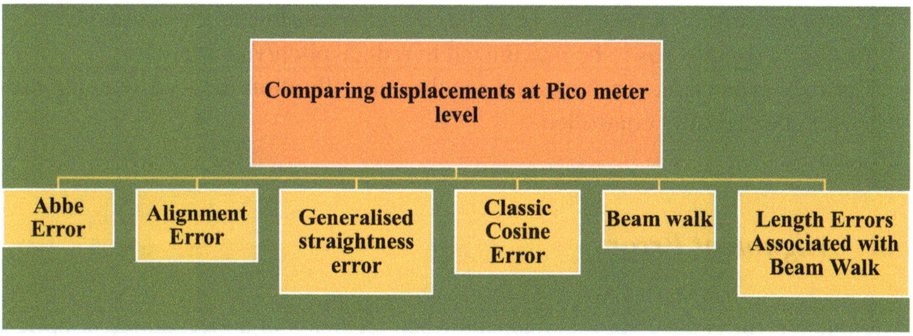

Fig. 1 Comparing displacements at picometer level

5.1 Abbe Error: Abbe Offset, Pitch, and Yaw

The requirements for picometre measurements are rather strict because abbe error is one of the main reasons why any dimensional measurement can be inaccurate. If the central interferometer mode is exactly at the geometric center of the outer three interferometer modes, then the Abbe offset would be zero. This paper will make use of small-angle approximations; the Abbe error is d, where d is the Abbe offset and represents a slight angular variation brought on by the pitch and yaw of the stage that moves the baseplate. Abbe Error results from Abbe's Principle, developed by Ernst Abbe during the nineteenth century. It says that in order to provide linear measurement accurately, the measuring scale (or sensor) should be aligned collinear with the measurement axis. When the scale is offset from this axis, angular displacements like pitch and yaw create linear displacement errors, which become more and more critical as measurement resolutions reach the sub-nanometer range. The Abbe Offset is the distance between the point of measurement and the measurement sensor or scale axis at a right angle. As a component tends to rotate slightly around an axis caused by pitch (rotation about the horizontal axis) or yaw (rotation about the vertical axis), the offset converts the angular motion to an apparent linear displacement error. The equation for the Abbe Error is:

$$\text{Abbe Error}\left(\Delta L\right) = d \times \theta$$

where:

ΔL = linear error incurred,

D = Abbe offset (distance between scale and measurement axis),

Θ = angular error (in radians, caused by pitch or yaw).

This equation demonstrates that even tiny angular deviations result in substantial linear errors if d is considerable or if the angular motion is magnified as a result of environmental or mechanical instability. Within the picometer regime, a few nanoradians of angular error with millimeter-sized Abbe offsets generate several picometers of error (Xu et al., 2025). The sensitivity thus places severe demands on machine design:

- Metrology sensors have to be located as close as possible to the action point.
- Structural rigidity has to be maximized to reduce pitch/yaw motions.
- Thermal expansion, air flow, and mechanical vibration-induced angular drift have to be carefully controlled.

5.2 Alignment Error

Alignment error is the error that happens when elements of a mechanical or optical system are not in their desired geometric relationship. Such an error can occur in any system where accurate positioning is important, including coordinate measuring machines (CMMs), optical systems, CNC machines, and robot systems. Alignment errors usually stem from improper assembly, thermal expansion, vibration, or long-term wear and tear. These errors can appear as linear displacements or angular misalignments (pitch, yaw, or roll) and result in decreased accuracy and repeatability of measurements or operations. In high-precision applications—like semiconductor fabrication or scientific instruments even micron or picometer-level alignment errors can result in considerable performance degradation. To prevent alignment errors, engineers use techniques such as laser alignment tools, interferometry, and software error compensation. Periodic calibration and environmental controls also have important roles in ensuring accurate alignment over time (Altaqui et al., 2025).

5.3 Classic Cosine Error

Cosine Error is another important consideration in high-accuracy displacement measurement. It occurs when the actual motion direction is not exactly along the sensor's measuring axis. The measured length is then the projection of the true displacement vector onto the sensor axis. If a component travels a distance L at an angle θ to the measurement axis, the measured length L' is:

$$L' = L \times \cos(\theta)$$

The Cosine Error, introduced by the misalignment, is:

$$\text{Cosine Error} = L - L' = L \times (1 - \cos(\theta)$$

This, at small angles, can be approximated using the Taylor series expansion of cosine:

$$\text{Cosine Error} \approx L \times (\theta 2 / 2)$$

This illustrates how tiny angular misalignments (in the micro- or nanoradian range) can introduce gross errors on the picometer scale. In optical interferometry systems, cosine errors reduce accuracy if the laser beam is not collinear with the direction of travel. Likewise, in coordinate measuring machines (CMMs), minor angular misalignment between the probe and measurement axis causes corresponding projection errors (Podżorny et al., 2013).

5.4 Generalized Straightness Error

Clustering three types of stage motion errors together under a "generalized straightness error" may be convenient for certain applications. Translation of the moving mirrors across the interferometer axis is due to normal straightness errors, misalignment of the direction of motion with the interferometer axis, and stage roll errors. A small angle θ misalignment of the motion axis from the axis of an interferometer will, for example, result in translation of the mirrors perpendicular to the motion direction by Lθ, where L is the displacement. This is equivalent to a straightness error Lθ. Similarly, a roll through the center with amount θR results in the exterior mirrors located at distance r from the center of the plate to move by r κR. The effect is again to translate the mirrors perpendicularly across the interferometer axis, though this is not equivalent to a normal straightness error in the fact that the directions of the translations are reversed for each mirror (Kokuyama et al., 2025).

5.5 Beam Walk

"Beam walk" refers to the movement of the beam over the surfaces of the mirrors. Pitch, yaw, roll, straightness, and alignment with the direction of motion establish this function. In fact, beam walk due to pitch and yaw errors is rendered irrelevant as soon as pitch and yaw are reduced to a suitable level so that no Abbe errors occur; under these circumstances, it is based solely on the overall straightness. Beam walk is the minimal angular or transverse motion of a laser or optical beam on passing through or reflecting off of optical elements such as mirrors, lenses, or beam splitters. At the picometer level, even tiny mechanical variations of optical elements thermal expansion, vibration, or structure drift—will create measurable variations in beam location or direction. Beam walk is a critical problem in high-precision applications like interferometry, atomic force microscopy, or laser alignment systems since beam walk causes measurement and alignment errors. At the picometer level, beam walk is typically a result of subatomic-scale optical mount or substrate deformations or instabilities. A picometer-scale mirror tilt can cause a lateral beam displacement several orders of magnitude larger over long propagation distances. Therefore, beam walk needs to be recognized and accounted for in applications where beam pointing stability and interference accuracy are required (Lakshminarayanan & Richuso, 2024).

5.6 Length Errors Associated with Beam Walk

Beam walk length error is caused by minute angular or lateral misalignments of an optical path that cause a laser beam to shift in position, leading to measurement error. In interferometric and metrological systems of very high precision, even a minute beam displacement, known as beam walk is a critical source of error. Beam walk is a source of huge length errors, especially in the situation of misaligned reflecting surfaces or detectors. Such inaccuracy is becoming critical on the picometer or nanometer level, wherein small mechanical noises, thermal movements, or errors of alignment could shift the path of the beam, effectively modifying the length one wishes to measure. Beam walk leads to angular sensitivity, which makes the optical system highly prone to misalignment, which translates into apparent path length or phase variation and leads to measurement deformities. This has a very pronounced effect in such instruments as laser interferometers, optical encoders, or coordinate measuring machines. Beam walk may cause Abbe-type errors when the detector is moved from the measurement axis, introducing angular misalignment and lateral displacement to produce an aggravated effect of error (Leach, 2015).

6 Applications of Picotechnology

6.1 Bio Medical Science

Despite all these developments made in therapeutical therapies, few medical ailments mainly cancer related issues are growing even higher and higher, and they require full treatment regimens. For curing those complications, the delivery of anticancer drugs appropriately is the optimal treatment and requires promising drug carrier delivery systems in order to approach the tumor cells. HerePico scale particles have appeared to be perfect replacement of previously used nanomaterials in biomedicine and will lead scientist's eyes in to new directions and allow them cross current hurdles. Pico-materials are highly effective in early biomarker and biomolecules detection due to their excellent capacities (Amreddy et al., 2018). Their tiny diameter (10–12) allowed them to penetrate into cells and cross barriers easily and accurately effective target drug delivery of drugs, Nanotechnology has enormous potential applications toward healing most of the health diseases. Ongoing R and D on Nanotechnology not only developed multifunctional benefits in healing the diseases but it also opened the way for Picotechnology. The Picotechnology's physiological platform has made the medical practitioners a viable platform on which they are busy developing new devices and coming up with enhanced treatment strategies with a good safety profile and efficacy compared to the conventional ones. For instance, one of the successful uses of Picotechnology is the novel bio-barcode assay that is specifically very useful in identifying advanced agents of disease even before they exist in trace amounts in living tissues. This highly advanced process of the mixing of gold—picoparticles with DNA amplifiers and recognition particles—is actually verified through Northwestern University (USA) while performing Prostatectomy of prostate cancer biomarker. Tissue engineering and regenerative medicine is biotechnological practice and multidisciplinary discipline that utilizes biological systems with engineering input for one and sole purpose of developing functional substitutes to treat damaged and diseased tissues by enhancing, restoring, or preserving the potential function of biological tissues. Regenerative medicine and tissue engineering strategy integrates three fundamental elements such as growth factors, scaffolds, and stem cells for the fabrication of tissue engineering scaffolds (Mohammadpour et al., 2015). Nanotechnology is now a hallmark technology by the implementation of nanomaterials possessing the characteristics of fantastic drug delivery systems by their ability to cross over to reach various nanostructures inside bones, interacting with the target disease sites, and controlled release of the drug because it provides fine structure scaffolds that are highly beneficial in tissue regeneration, ultimately modifying the tissue engineering interventions. With respect to Picotechnology, the value of the technology can be realized from the fact that it is a deserving candidate to look out for in managing the cellular Micro Tubules (MTs). Fundamentally, MTs are the advanced cellular structures made of tubulins with average size dimension of 25 nm diameter. These filaments are dynamically unstable and are associated with conjugation of end-binding proteins (or plus end-tracking proteins) that possess the ability to surf over dynamic ends of MTs. It should be mentioned that tips of fluorescent proteins shine with green color, and fluorescence is very brightest at microtubules and its intensity keeps decreasing in the direction of minus end of such microtubules. It is approximated that MTs and end-binding proteins can be triggered by an external stimulus source which will in turn increase the cellular movement by simply utilizing the flexible picotechnological science (Mostafavi et al., 2019).

6.2 Environmental Science

The expanding trends of Picotechnology are also establishing themselves in a number of fields other than medicine. In the field of environmental science, Picotechnology holds transformative potential for monitoring, protecting, and restoring monitoring systems with unmatched specificity and sensitivity.

6.2.1 Ultra-Sensitive Environmental Monitoring

One of the major uses of Picotechnology in environmental science is real-time monitoring of the environment. Traditional sensors are usually unable to detect very low levels of toxic substances, particularly when they are spread over large distances or blended with other chemicals. Picotechnology allows for the creation of picosensors, which are instruments that can detect pollutants at the atomic or molecular level. Such sensors can employ quantum dots, single-atom transistors, or molecular recognition units to sense contaminants like heavy metals (mercury, arsenic), volatile organic compounds, and industrial byproducts in water, soil, and air. Such picoscale sensors have a number of benefits: they use less energy, waste very little energy, and perform with unprecedented accuracy. For example, a single molecule of a poisonous compound can cause a measurable signal, so early warning systems in water purification plants or air quality monitors can be implemented. Such accuracy allows preventative environmental interventions, since danger can be recognized and dealt with before it is at dangerous levels 27 (Saleh, 2024).

6.2.2 Water Purification and Treatment

Picotechnology has the potential to greatly enhance water purification techniques by creating pico-engineered filtration membranes and molecular sieves. These filtration devices are engineered with pores of a specific size at the atomic level, allowing them to selectively pass water molecules while rejecting contaminants, salts, or microorganisms. One of the most promising areas relates to the employment of DNA-based nanodevices and molecular machines that behave similarly to biological filters, shifting shape or function upon the detection of special chemical stimuli. These machines can specifically bind to, neutralize, or remove toxic substances from water. Due to their programmatic and adaptive features, they function extremely well in dynamic or complex environments wherein types and concentrations of pollutants can fluctuate (Sanni et al., 2024).

6.2.3 Air Quality Control

In air quality management, Picotechnology provides novel means for the detection and decomposition of harmful airborne substances. For instance, pico-sized catalytic systems may be engineered to interact with specific pollutants, for instance, Nitrogen Oxides (NOx), Sulfur Dioxide (SO_2), or Carbon Monoxide (CO), and reduce them into less toxic substances at the atomic scale. The precise control over surface chemistry and electron interactions offered by Picotechnology increases the efficiency of these catalytic reactions well beyond the capabilities of traditional catalysts. Picosensors integrated into urban infrastructure might continuously measure pollutant concentrations at ultra-high resolution, allowing smart cities to react dynamically to changes in the environment in real time. This strategy facilitates data-driven environmental regulation, optimizing traffic flow, industrial emissions, and public health interventions using live air quality information (Njema & Kibet, 2024).

6.2.4 Soil and Agriculture Application

Agriculture relies to a great extent on fertile soil and pure water, both of which are now under severe threat from chemical fertilizers, pesticides, and industrial waste run-off. Picotechnology would make precision farming possible via picosensors planted in the soil to monitor nutrient content, pH, water content, and contaminants. This will allow need-based and spot-specific application of fertilizers or water, reducing wastage and environmental erosion. Besides, pico-engineered nanocarriers might also be employed for delivering pesticides or nutrients with atomic accuracy, cutting down on environmental footprint and increasing efficiency. The carriers have the ability to deliver their cargo only when they receive specific environmental signals (such as particular plant hormones or microbial metabolism), which increases their specificity and environmental friendliness (Li et al., 2025a, b).

6.2.5 Environmental Mitigation

Picotechnology is also responsible for cleanup of polluted environments. Molecular machines and DNA constructs can be designed to find and bind to particular environmental pollutants, including oil residue, radioactive waste, or persistent organic pollutants. Such devices can then either make the pollutants harmless or allow them to be removed from the environment. In addition, the atomic control provided by Picotechnology can be utilized in order to rebuild compromised ecosystems, e.g., by re-designing damaged soil on the molecular level or by correcting natural microbial balances through selective biochemical intervention (Leiu et al., 2024).

7 Privacy and Ethical Issues

The engineering and use of picomaterials as materials create sub-nanometer sizes raise serious ethics and regulatory questions. Because picomaterials are so very small, their new properties and behaviors are not well understood, which could potentially lead to health and environmental surprises. Picotechnology has the potential to change the field of health measurement through implanting microchips or biosensors within (human) organisms. This could result in the dramatic bettering of human health (and more effective controlling of chronic ailments like diabetes). And yet these new technologies are liable to be enlisted for uses as surveillance and info-gathering instruments. Arguably the most worrying ethical threat on the horizon involves the threat of intrusive monitoring from materials injected with microscopic, if not atomically precise, sensors. These materials could be integrated into everyday objects, clothing, or even implanted in the human body, capable of collecting detailed personal data such as health metrics, location, or behavior, often without an individual's knowledge or consent. This level of data collection poses a serious threat to personal privacy and autonomy. The absence of public awareness regarding the existence and operation of such technologies becomes a concern in terms of informed consent (Gutierrez, 2004). For example, when picomaterials are incorporated into medical implants or diagnostics, patients need to be completely informed regarding the possibilities, risks, and long-term consequences of the materials. From an ethical point of view, researchers and developers have an obligation to provide openness and safeguard personal rights, particularly in vulnerable populations. Another ethical issue pertains to the risk of social inequality, in that access to advanced picotechnological uses might be restricted to wealthier populations, widening gaps in healthcare, security, and education. Also, there exists a misuse of dual use, in which the same Picotechnology created for useful purposes may be repurposed for nefarious applications, including sophisticated weaponry or illicit bio-enhancement. Existing regulatory mechanisms are mostly incapable of addressing the novel challenges of Picotechnology since most standards were intended for larger Nanotechnology. Hence, there is an urgent need for proactive, internationally concerted regulation involving risk assessment, public participation, and formulation of ethical codes of conduct (Mnyusiwalla et al., 2003).

8 Recent Advancements in Picotechnology

In the solution phase and bioactive phases of the body, metal oxide particles may exist at the picomolecular level in nature. The size and functionality of these pico-carrier devices depend on the manufacturing process and physiological conditions of the body. Their prospects, viability, and success rate have not yet been determined. The main differences between nanodevices and pico-devices are Intramolecular and intermolecular thermodynamics in the medium which is based on molecular conformational interaction with molecular assembly in the body's cytoarchitecture. Without interfering with host signal transduction and immune defense, pico-carrier devices can be considered to be an opportunity for detecting the diseases. The design of particle-based pico-drugs, pico-chemicals, and picotargeting molecules could take advantage from the ultrafine size of recently developed picotechnology products, which may also be simpler to operate. The reason for this is related to the type of interactions between the atoms that constitute the gold, interactions that are averaged out into nonexistence in the bulk material, nano gold does not behave like bulk gold (Masciangioli & Zhang, 2003).

To some extent, Nanotechnology is currently proving to be a useful weapon in the fight against climate change. It is crucial in the fight against air pollution. Various pathogenic bacteria at extremely low concentrations, pollutants, and humidity percentages can all be detected with nanosensors. The development of picoscale sensors will be more effective and advanced going forward in order to monitor concentrations, even at the ultra-small level. The research communities will be able to create pico-structured membranes that are more effective than Nanotechnology at separating greenhouse gases because of the currently used nano-structured membranes' strong ability to separate carbon dioxide and methane from exhaust. By controlling molecules and structures that are typically out of control, Picotechnology will make the seemingly impossible feasible. Even with relatively basic appliances, the operational control and functionalisation of equipment at this level will eventually completely change how people interact with one another. The future will see greater recognition for picochips and other microdevices, particularly in environmental and biomedical applications. The most intriguing potential use of Picotechnology could be the ability to repair bodily tissue at the cellular and subcellular levels. By creating picolevel materials, the role of Nanotechnology in increasing fuel consumption efficiency through the use of nanocatalysts will also be strengthened. According to future.fandom.com, effective Picotechnology research and application will enable minds to enter and be sufficient to give life to femto technology (Mostafavi et al., 2019).

The expanding trends of Picotechnology are also establishing themselves in a number of other disciplines outside of medicine. Research is being done to investigate and suitably apply the natural biological interventions of Picotechnology, including quantum mechanics, self-assembly, and self-organization, in the development of novel computing systems. According to research in this area, all plants absorb energy from sunlight during photosynthesis, which is then instantly transmitted to the plants' response houses via quantum mechanical processes with an efficiency of about 100%. This mechanism may eventually be used as a benchmark as a green energy idea to accomplish the goal of creating picosystems for the cost-effective generation and storage of clean solar energy. Because of their increased reactivity and large surface area, picomaterials are useful for creating high-quality catalysts, such as a catalytic converter in cars that dramatically reduces the toxicity of engine fumes. Several types of batteries, fuel cells, and catalysts made using pico-engineering concepts are made with improved reactivity at the picolevel, providing safer, cleaner, and more effective ways to produce and store energy (Baier et al., 1984).

9 Conclusion

Picotechnology, which operates in the picometer range (10^{-12} meters), has evolved as a revolutionary achievement in the field of material science with an unparallelled ability to precisely control matter at atomic and subatomic levels. Where Nanotechnology led to the emergence of novel devices and materials through operation in the nanoscale, Picotechnology makes the next major leap enabling atomic-level position control, electron interaction control, and control of molecular function at an exceptionally high level of accuracy. This ability allows for the creation and development of materials and systems with superior properties more mechanical strength, better electrical or thermal conductivity, greater chemical selectivity, and unparalleled functionality. The application of Picotechnology to material science is unveiling revolutionary applications. Single-atom transistors, for example, illustrate the potential for miniaturization in electronics, opening the door to more powerful, energy-efficient computing. Quantum dots and pic-scale logic gates promise quantum computation and high-resolution sensors, but molecular machines and DNA-based nanodevices with picometric precision are opening the way for advances in drug targeting, biosensing, and even environmental restoration. These fields highlight Picotechnology's potential for transforming the limits and performance of materials in scientific and industrial areas. While offering such promise, Picotechnology is still an evolving discipline, however, with its challenges to be overcome. The material manipulation and fabrication at such a massive scale demand advanced tools such as high-resolution scanning probe microscopes and high-end computational models. Furthermore, ethical concerns of environmental effects, safety, and social implications need to be thoroughly examined as the technology develops. In the future, the integration of Picotechnology with artificial intelligence, quantum mechanics, and biotechnology is likely to expedite its implementation across industries. As atomic-scale design techniques get more advanced and scalable, Picotechnology has the potential to propel the next generation of material innovation—resulting in smart, adaptive, and highly functionalized materials customized for applications in medicine, electronics, clean energy, and more. Altogether, Picotechnology represents the leading edge of material science innovation, holding intriguing potential for precision-designed materials to transform technology and human progress.

References

Altaqui, A., Zhang, X., Wang, M., Diorio, N., Shaar, S., Muthinti, R., Lu, L., Escuti, M., & Oh, C. (2025). Critical dimension metrology and fitting models for polarization volume hologram grating based on liquid crystal polymer. In *Optical architectures for displays and sensing in augmented, virtual, and Mixed Reality (AR, VR, MR) VI* (Vol. 13414, pp. 424–438). SPIE.

Amreddy, N., Babu, A., Muralidharan, R., Panneerselvam, J., Srivastava, A., Ahmed, R., Mehta, M., Munshi, A., & Ramesh, R. (2018). Recent advances in nanoparticle-based cancer drug and gene delivery. *Advanced Cancer Research, 137*, 115–170.

Ariga, K. (2023). Molecular machines and microrobots: Nanoarchitectonics developments and on-water performances. *Micromachines, 14*(1), 25.

Baier, R. E., Meyer, A. E., Natiella, J. R., Natiella, R. R., & Carter, J. M. (1984). Surface properties determine bioadhesive outcomes: Methods and results. *Journal of Biomedical Materials Research, 18*(4), 337–355.

Bolonkin, A. (2009). Converting of matter to nuclear energy by ABGenerator. *American Journal of Engineering and Applied Sciences, 2*, 683–693.

Chen, Z., Grace, I. M., Woltering, S. L., Chen, L., Gee, A., Baugh, J., Briggs, G. A., Bogani, L., Mol, J. A., Lambert, C. J., & Anderson, H. L. (2024). Quantum interference enhances the performance of single-molecule transistors. *Nature Nanotechnology, 19*(7), 986–992.

Durand, M., Wang, Y., & Lawall, J. (2012). Accurate Gouy phase measurement in an astigmatic optical cavity. *Applied Physics B: Lasers and Optics, 108*(4), 749–753.

Ghanbarlou, S., Kahforoushan, D., Abdollahi, H., Zarrintaj, P., Alomar, A., Villanueva, C., & Davachi, S. M. (2025). Advances in quantum dot-based fluorescence sensors for environmental and biomedical detection. *Talanta, 17*, 128176.

Gutierrez, E. (2004). *Privacy implications of nanotechnology*. Electronic privacy information Center. Retrieved March 2, 2021, from http://www.epic.org/privacy/nano/

Kansara, K., Laha, A., Kumar, A., & Bhatia, D. (2025). Programmable DNA-based nanodevices for next-generation clinical and healthcare applications. *Advanced Biology, 9*, 2400593.

Kokuyama, W., Shimoda, T., & Nozato, H. (2025). Primary microvibration calibration of accelerometer with picometer displacement. *Metrologia, 62*(1), 015007.

Kumar, D. (2018). Carbon nanomaterials based nanocomposite as emerg-Ing field for pollution control. In *Nanotechnology environment science* (pp. 125–152). Wiley.

Lakshminarayanan, A., & Richuso, J. (2024). The next generation of DLP Pico technology for augmented reality (AR) displays. In *Optical architectures for displays and sensing in augmented, virtual, and mixed reality (AR, VR, MR)* (Vol. 12913, pp. 141–149). SPIE.

Leach, R. (2015). *Abbe Error/Offset*. Springer. https://doi.org/10.1007/978-3-642-35950-7_16793-1

Leiu, Y. X., Lim, K. M., Chiah, Z. J., Mah, E. S., & Ong, W. J. (2024). Plastic-to-treasure: Innovative advances in photo/electro-catalytic upcycling technologies for commodity chemicals and fuels. *EcoEnergy, 3*, 217.

Li, C., Ma, L., Xue, Z., Li, X., Zhu, S., & Wang, T. (2025a). Pushing the Frontiers: Artificial intelligence (AI)-guided programmable concepts in binary self-assembly of colloidal nanoparticles. *Advanced Science*, 2501000.

Li, P., Zhang, T., Song, Z., Chen, D., Yang, Y., He, J., Lei, X., Tian, H., & Zhang, K. (2025b). Biomass-based mineralized chitin microsheets for high-efficiency radiative cooling. *International Journal of Biological Macromolecules, 1*(308), 142360.

Masciangioli, T., & Zhang, W. (2003). Environmental technologies at the nanoscale. *Environmental Science & Technology, 37*, 102–108.

Mnyusiwalla, A., Daar, A. S., & Singer, P. A. (2003). Mind the gap: Science and ethics in nanotechnology. *Nanotechnology, 14*, R9–R13.

Mohammadpour, M., Jabbarvand, M., Hashemi, H., & Delrish, E. (2015). Prophylactic effect of topical silica nanoparticles as a novel antineovascularization agent for inhibiting corneal neovascularization following chemical burn. *Advanced Biomedical Research, 4*, 124.

Mostafavi, E., Soltantabar, P., & Webster, T. J. (2019). Nanotechnology and Pico technology: A new arena for translational medicine. In *Biomaterials translational medicine* (pp. 191–212). Academic Press.

Nitkin, P. I., Gorshkov, B. G., Nitkin, E. P., & Ksnevich, T. I. (2005). *Sensors and Actuators B: Chemical, 111–112*(11), 500–504.

Njema, G. G., & Kibet, J. K. (2024). A review of novel materials for nano-photocatalytic and optoelectronic applications: Recent perspectives, water splitting and environmental remediation. *Progress in Engineering Science, 1*, 100018.

Podżorny, T., Budzyń, G., & Rzepka, J. (2013). Linearization methods of laser interferometers for pico/nano positioning stages. *Optik, 124*(23), 6345–6348.

Prathyusha, K., Reddy, J. G. M., Venkataswamy, M., & Ramesh, A. (2018). Pico technology: Instruments used and applications in pharmaceutical field. *Research Journal of Pharmaceutical Dosage Forms Technology, 10*, 34.

Saleh, T. A. (2024). Trends in nanomaterial types, synthesis methods, properties and uses: Toxicity, environmental concerns and economic viability. *Nano-Structures & Nano-objects, 37*, 101109.

Sankar, T. V., Patel, D., Samuel, R. S., & Amaranath, G. (2018). Reducing the impacts of greenhouse gases. In K. Sridharan (Ed.), *Emerging trends of nanotechnology in environment and sustainability*. Springer Briefs in Environmental Science.

Sanni, S. E., Oni, B. A., Okoro, E. E., & Pandya, S. (2024). Recent advances in the use of biogenic nanomaterials and photocatalysts for wastewater treatment: Challenges and future prospects. *Frontiers in Nanotechnology, 6*, 1469309.

Sharma, R., Sharma, A., & Chen, C. J. (14 November 2011). "Emerging trends of nanotechnology towards Pico technology: Energy and biomolecules". *Nature Precedings*.

Singh, A., & Prasad, S. M. (2017). Nanotechnology and its role in agro-ecosys-tem: A strategic perspective. *International journal of Environmental Science and Technology, 14*(10), 2277–2300.

Tang, S., & Cao, Z. (2012). Adsorption and dissociation of ammonia on graphene oxides: A first-principles study. *The Journal of Physical Chemistry, 116*, 8778–8791.

Wang, C., Cao, J., Guo, J., Guan, C., Wang, K., Zhang, Q., & Lu, P. (2025). On chip photothermoelectric logic gates with sub-picosecond photothermal response. *Advanced Optical Materials, 13*, 2403528.

Xu, J., Tu, H., Ao, Z., Chen, Y., Danehy, R., & Guo, F. (2019). Acoustic disruption of tumor endothelium and on-demand drug delivery for cancer chemotherapy. *Nano, 15*, 154–1701. https://doi.org/10.1088/1361-6528/aafe4e

Xu, C., Song, J., Liu, X., Lee, C. H., Marinescu, I. D., Hui, J., & Guo, L. (2025). Precision and ultra-precision machining with elastic polishing tools: A review. *Surface Science and Technology, 3*(1), 9.

The Journey Ahead

Future Trends and Challenges in Nanotechnology and Picotechnology

B. Lokeshwari, Shrikaant Kulkarni, U. Subhalakshmi, P. Saranraj, and N. Usha

1 Introduction

Nanotechnology and Picotechnology are the frontiers of scientific progress, enabling control over matter at the atomic and subatomic scales. Nanotechnology in the nanometer scale (1–100 nm) has already revolutionized medicine, electronics, energy, and materials science with the innovations like targeted delivery of medicine, efficient solar cells, and quantum computing devices (Bayda et al., 2020). Picotechnology is the subatomic manipulation of electrons and protons that holds transformative potential for artificial atoms, proton computing, and revolutionary energy solutions (Chen, 2020). These technologies have the significant challenges like fabrication limitation, quantum instability, environmental and health risk, ethical considerations, and concerns over the inequality of economic distribution (Babatunde et al., 2020). The rapid development of Nanotechnology and Picotechnology has the technological boundaries for enabling precise matter control at the atomic and subatomic levels (Stremfelj & Smole, 2021).

While nanotechnology (1–100 nm) has been used in the transforming industries through innovations in healthcare, energy, and computing (Rambaran & Schirhagl, 2022), Picotechnology based on the dimensions to enable transformative in quantum control, materials design, and energy capture (Xie et al., 2004a, b). These developments are accompanied by major challenges, such as fabrication complexity, safety issues, ethical issues, and the regulatory systems (Bleeker et al., 2013).

B. Lokeshwari · U. Subhalakshmi · P. Saranraj (✉) · N. Usha
PG and Research Department of Microbiology, Sacred Heart College (Autonomous), Tirupattur, Tamil Nadu, India

S. Kulkarni
Faculty of Business, Victorian Institute of Technology, Melbourne, VIC, Australia

S. Kulkarni (ed.), *A Journey from Nanotechnology to Picotechnology*, Lecture Notes in Nanoscale Science and Technology 47,
https://doi.org/10.1007/978-3-032-07228-3_15

The scale of dimensions between 1 and 100 nm and offers tremendous potential in industries ranging from medicine and electronics to agriculture and green environmentalism (Martins & Kaczerewska, 2021). As Nanotechnology evolves, it has led to more sophisticated process that deals with matter at the picometer scale (10^{-12} meters), to the subatomic particle and electrons (Xie et al., 2008). Nanotechnology has revolutionized conventional paradigms in medicine, materials science, energy, and computing (Jaramillo-Cabanzo et al., 2021). Nanoparticles enable drug targeting and nanomaterials are being used as the nanoelectronics (Karimi et al., 2012). Technologies like carbon nanotubes, quantum dots, nanoshells, and nanorobots to transform the commercial applications (Eswaraiah et al., 2016).

Nanotechnology has physical, environmental, and ethical benefits as their higher resolution, accuracy, or interaction with quantum dots (Wu et al., 2021). Nanotechnology is manipulating matter at the scale of single atoms, their electrons, and subatomic structures (Obermair et al., 2010). Picotechnology offers ultimate control over material properties, quantum computing elements, and even direct manipulation of DNA or atomic nuclei (Xie et al., 2010). The transition from nano- to Picotechnology has a wide range of issues (Xie et al., 2004a, b). Picometer-precise instruments must be developed that can manipulate subatomic interactions without destabilizing matter (Xie et al., 2008). In addition to technology and ethics, nanotechnology and Picotechnology increasingly involve other disruption areas like the artificial intelligence (AI), biotechnology, robotics, and quantum physics (Chen, 2020). Such interdisciplinarity results in new interdisciplinary inventions, e.g., AI-enabled nanorobots for therapy targeting, nanoscale biosensors for the monitoring of real-time diseases, and atomically precise materials for quantum computing (Wang et al., 2020). These potential benefits are an urgent necessity for the practice of responsible research and innovation (Martins & Kaczerewska, 2021). Environmental and regulatory challenges arise from the use of nanomaterials and the emerging development of picomaterials (Bleeker et al., 2013). Engineered nanosystems can accumulate in the environment, potentially disrupting ecosystems and they must be biocompatible to ensure safety for living organisms, and their disposal requires careful handling to prevent environmental pollution (Babatunde et al., 2020). The issues of intellectual property rights, access, military uses, and technology monopolies need to be resolved under international cooperation and government (Stremfelj & Smole, 2021). The Nanotechnology approaches increased complexity and Picotechnology toward the feasibility of these technologies that are capable of overcoming the development and application of nano- and pico-technologies and ensure their transition into ethical and social issues (Rambaran & Schirhagl, 2022).

2 Future Trends in Nanotechnology

Nanotechnology holds the promise of transforming many disciplines over the decades through advances in materials science, medicine, energy, and computing (Bayda et al., 2020). In medicine, nanomedicine will have levels of unprecedented precision in diagnosis and treatment, such as targeted delivery systems for drugs, nanorobots for invasive surgery, and real-time biosensors for detecting disease at its earliest stages (Li et al., 2021). The creation of smart nanomaterials, such as self-healing polymers and adaptive coatings, will revolutionize industries through increased durability and functionality (Bensalah et al., 2021). For electronics, nanotechnology will expand the limits of miniaturization, with sub-5-nm transistors, quantum dots, and flexible nanoelectronics allowing faster, more efficient, and wearable devices (Huo et al., 2022). Applications in energy will have major advancements, such as high-efficiency perovskite solar cells, nanostructured batteries with ultra-fast charging capability, and nanocatalysts for hydrogen production (Jaramillo-Cabanzo et al., 2021). Environmental nanotechnology has been prominently used in the remediation of pollution, with water purification using nanofilters and real-time air quality monitoring through nanosensors (Cao et al., 2021). Furthermore, the AI and machine learning with nanotechnology with the material discovery and optimization for their ethical and safety issues of the nanoparticle toxicity and regulatory considerations (Rambaran & Schirhagl, 2022). Nanotechnology has the bridge between basic research and practical applications in the small materials that give rise to effects across industries worldwide (Stremfelj & Smole, 2021). Nanotechnology is poised to revolutionize multiple fields, as illustrated in Fig. 1, which highlights key future trends such as advanced nanomaterials, nanomedicine, and nanoelectronics. Emerging areas like 2D materials and metamaterials further underscore the transformative potential of this discipline.

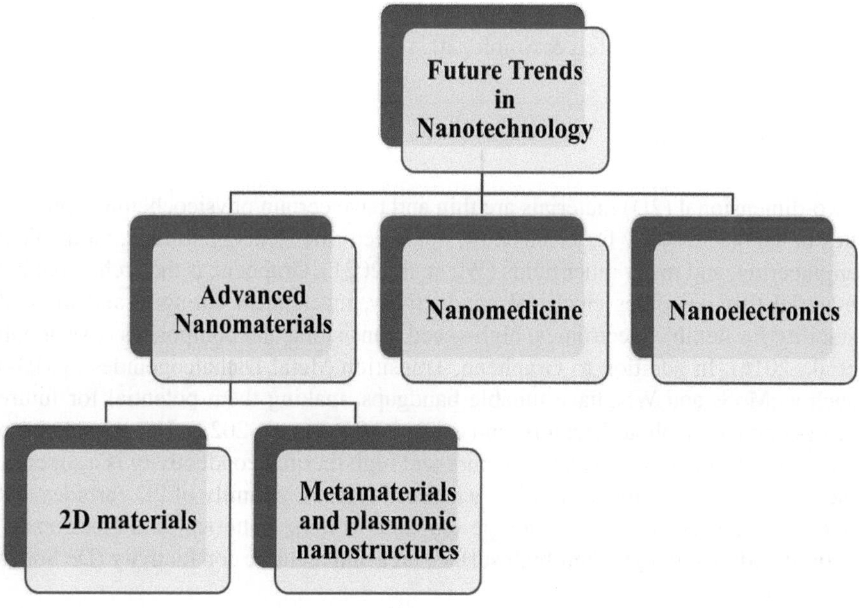

Fig. 1 Future trends in nanotechnology

2.1 Advanced Nanomaterials

Nanotechnology has seen revolutionary advancements with sophisticated nanomaterials that possess several properties as a result of being on the nanoscale in the designed structures (Chen et al., 2019). These materials such as quantum dots, carbon nanotubes, graphene, Metal-Organic Frameworks (MOFs), and metamaterials are transforming industries by their mechanical, electrical, optical, and chemical properties (Wu et al., 2021). In electronics, two-dimensional (2D) materials such as graphene and Transition Metal Dichalcogenides (TMDs) are giving rise to ultra-thin, flexible, high-performance devices, while quantum dots are revolutionizing display technologies for better color purity and efficiency (Eswaraiah et al., 2016). In energy storage and conversion, nanostructured electrodes and solid-state electrolytes are increasing battery capacity and safety, and perovskite nanocrystals in higher standards for the efficiency of solar cells (Jaramillo-Cabanzo et al., 2021).

Biomedical uses sophisticated nanomaterials for targeted drug delivery, imaging, and regenerative medicine, with gold nanoparticles and liposomes used for precise cancer treatment and diagnostic systems (Li et al., 2021). Environmental cleanup uses photocatalytic nanomaterials such as titanium dioxide and porous adsorbents such as MOFs to efficiently break down pollutants and capture greenhouse gases (Cao et al., 2021). The self-healing nanocomposites and lightweight materials with high strength are revolutionizing the aerospace, automotive, and construction industries (Bensalah et al., 2021). Machine learning and computational modeling are used for the identification of nanomaterials in healthcare, energy, and sustainability (Rambaran & Schirhagl, 2022). As Nanotechnology progresses, Nanomaterials will have specific characteristics in future technologies, influencing innovation in the industrial sectors (Stremfelj & Smole, 2021).

2.1.1 2D Materials

Two-dimensional (2D) materials are thin and have certain physicochemical properties in nanotechnology for electronics, optoelectronics, energy storage, biomedical engineering, and many other fields (Wu et al., 2021). Graphene is the archetypal 2D material that possesses electrical conductivity, mechanical strength, and thermal stability for flexible electronics, high-speed transistors, and composites (Eswaraiah et al., 2016). In addition to Graphene, Transition Metal Dichalcogenides (TMDs) such as MoS_2 and WS_2 have tunable bandgaps, making them potential for future semi-conductors, photodetectors, and catalysts (Huo et al., 2022). Hexagonal boron nitride (h-BN) with its insulating nature and high thermal conductivity is a superior dielectric layer for nanoelectronics, whereas MXenes, a family of 2D carbides and nitrides, are used for energy storage (supercapacitors, batteries) and electromagnetic shielding owing to their high surface area and metallic conductivity (De Souza et al., 2019).

The optical characteristics of 2D materials include robust excitonic effects and valley polarization for ultra-thin photonic devices, quantum light emitters, and optoelectronics (Liu et al., 2023). 2D materials are used in biomedicine for biosensing, drug delivery, and antibacterial coatings for their extensive surface area and biocompatibility (Li et al., 2021). The evolution of van der Waals heterostructures, in which distinct 2D materials are aligned with atomic-scale precision, has the modification of electronic and optical properties, allowing for novel quantum phenomena and devices (Chen et al., 2019). Environmental and safety issues of 2D materials include potential toxicity to organisms, requiring careful assessment and management to ensure their sustainable and safe use (Babatunde et al., 2020). The combination of 2D materials with machine learning and artificial intelligence is advancing material discovery and optimization, whereas their combination with other nanomaterials has multifunctional systems for next-generation technologies (Rambaran & Schirhagl, 2022). 2D materials have the potential of nanotechnology, fueling innovation in both fundamental science and industrial applications (Stremfelj & Smole, 2021).

2.1.2 Metamaterials and Plasmonic Nanostructures

Metamaterials and plasmonic nanostructures have light–matter control in photonics, sensing, energy harvesting, and biomedical applications (Yu & Capasso, 2014). Metamaterials, subwavelength-structured engineered composites, possess unusual electromagnetic properties including negative refraction, perfect absorption, and invisibility cloaking by well-designed unit cells that control electric and magnetic responses at microwave, terahertz, and optical frequencies (Luo et al., 2015). Artificial materials have transformed optical devices, such as superlenses that break the diffraction limit, ultra-thin metasurface optics for imaging systems, and electromagnetic cloaks for stealth technology (Boyarsky et al., 2017). Plasmonic nanostructures are the vibration of free electrons (surface plasmons) in noble metals such as silver and gold, focus light to nanoscale volumes much below the diffraction limit, and thus support augmented light–matter interactions for the applications of Surface-Enhanced Raman Spectroscopy (SERS), biosensing on the nanoscale (Khan et al., 2015).

Plasmonic integration with metamaterials resulted in hybrids such as hyperbolic metamaterials for ultra-high states of photonic density for the enhancement of spontaneous emission as well as plasmonic nanocavities for enabling high coupling between quantum emitters and light for processing in the quantum domain (Ginzburg et al., 2017). In energy sectors, solar cells are enhanced with plasmonic nanoparticles through the trapping of light and generation of electrons and metamaterial-based absorbers enable the conversion in solar thermics (Wu & Wang, 2017). Biomedical advancements include hyperlenses for super-resolution imaging based on metamaterials, photothermal therapy by plasmonic nanoparticles, and single-molecule sensitivity label-free biosensors (Khan et al., 2015). Challenges persist in mitigating the loss in plasmonic systems, large-scale fabrication of metamaterials, and active tunability using stimuli-responsive materials such as phase-change materials and 2D materials (Lee et al., 2020). The emergence of metamaterials and plasmonics with machine learning is speeding up the inverse design, allowing nanostructures for their application (Zahra et al., 2021). They also have the potential to revolutionize telecommunications, quantum computing, targeted medicine, and sustainable energy applications, making metamaterials and plasmonic nanostructures for future technologies (Yu & Capasso, 2014).

2.2 *Nanomedicine*

Nanomedicine is an emerging application of nanotechnology for healthcare by implementing nanoscale materials and devices to prevent disease and diagnose the disease at the molecular level (Bayda et al., 2020). Nanomedicine has enabled precise manipulation with biological systems for targeted delivery of drugs, imaging, and regenerative therapy (Li et al., 2021). Nanocarriers such as liposomes, dendrimers, and polymeric nanoparticles may carry therapeutics into target cells directly, minimizing side effects and enhancing efficacy (Logeshwari et al., 2013). The treatment of cancer has been revolutionized using nanomedicine, wherein drugs such as Doxil (liposomal doxorubicin) increase targeting tumors and reduce cardiotoxicity (Li et al., 2021). Moreover, gold nanoparticles and quantum dots are revolutionizing diagnostic imaging by allowing for real-time, high-resolution imaging of molecular processes and tissues (Khan et al., 2015). Nanomedicine plays an important role in the diagnostics and therapy that are combined for personalized medicine (Bayda et al., 2020). Nanomedicine has challenges in biocompatibility, chronic toxicity, and regulatory issues in clinical validation (Bleeker et al., 2013). Recent progress in nanomaterials, advanced fabrication methods, and the development of next-generation treatment technologies (Rambaran & Schirhagl, 2022). Nanomedicine has the potential to medical challenges, ranging from combating infections that are multidrug-resistant to regenerating damaged tissues in precision medicine (Li et al., 2021). The use of machine learning and artificial intelligence enhances nanomedicine's capacity to develop nanotherapeutics as well as predictive diagnostics (Chen, 2020).

2.3 *Nanoelectronics*

Nanoelectronics deals with the design and use of electronic devices and components at the nanoscale (usually 1 to 100 nm), transforming conventional electronics by improving performance, efficiency, and miniaturization (Xie et al., 2004a, b). Utilizing quantum mechanical effects and new nanomaterials, nanoelectronics makes it possible to produce faster, smaller, and more efficient devices (Xie et al.,

2008). Some of the most significant developments in this area involve the application of carbon nanotubes (CNTs), graphene, and quantum dots, which possess outstanding electrical, thermal, and mechanical characteristics, making them suitable for high-speed transistors, flexible electronics, and ultra-high-density memory storage (Eswaraiah et al., 2016). Graphene-based transistors have shown the highest electron mobility, with the potential to substitute silicon in the integrated circuits, while carbon nanotube interconnects for the better conductivity and lower power consumption in nanochips (Huo et al., 2022). The low-power, high-speed data storage and processing, such as in magnetic random-access memory (MRAM) and spin-based transistors (Xie et al., 2010).

Nanoelectronics is also utilized by the Internet of Things (IoT) and wearable electronics, with nanosensors and flexible nanodevices for real-time monitoring of health and sensing the environment (Li et al., 2021). 2D materials (transition metal dichalcogenides) and quantum computing in nanoelectronics from super-efficient processors to biocompatible implants (Chen et al., 2019). The convergence of artificial intelligence and machine learning enhances the design and optimization of nanoelectronic systems, facilitating predictive modeling and defect detection (Rambaran & Schirhagl, 2022). The growing demand for faster, smaller, and eco-friendly electronics by applications in computing, telecommunications, healthcare, and energy (Stremfelj & Smole, 2021). Nanoelectronics intersects with fields like nanophotonics and nanomechanics to advance optoelectronics, NEMS, and ultra-sensitive sensor technologies (Xie et al., 2008).

3 Future Trends in Picotechnology

Picotechnology holds great potential in computers, materials science, medicine, and energy and for atomic and subatomic level control and engineering (Chen, 2020). In contrast, Nanotechnology deals with structures between 1 and 100 nm (Bayda et al., 2020). Picoscale engineering may be able to free quantum processors with the computation power, and decoherence they also facilitate the stabilizing of qubits by adjusting atomic spin for revolutionizing Picotechnology in Quantum computing (Xie et al., 2010). Picoscale material may further be able to make ultra-strong, self-repairing metamaterials whose atomic configurations could be programmed to revolutionizing aerospace, construction, and applications by ultralightweight (Obermair et al., 2010). Picotechnology could enable the atomic-scale devices with the ability to repair DNA damage at the single-nucleotide level, genetic diseases, and reverse cellular aging, thus effectively extending the boundaries of human lifespan in medicine (Xie et al., 2008). Picoscale sensors would also facilitate monitoring chemical reactions in real time, brain function, and even interactions with dark matter (Wang, 2019). The atomic-level weaponization, quantum effects, and environmental factors from picoscale manufacturing, ethical and safety considerations also arise (Babatunde et al., 2020). Picotechnology with artificial intelligence could give rise to self-evolving materials, atomic assembly machines, and picoscale neural

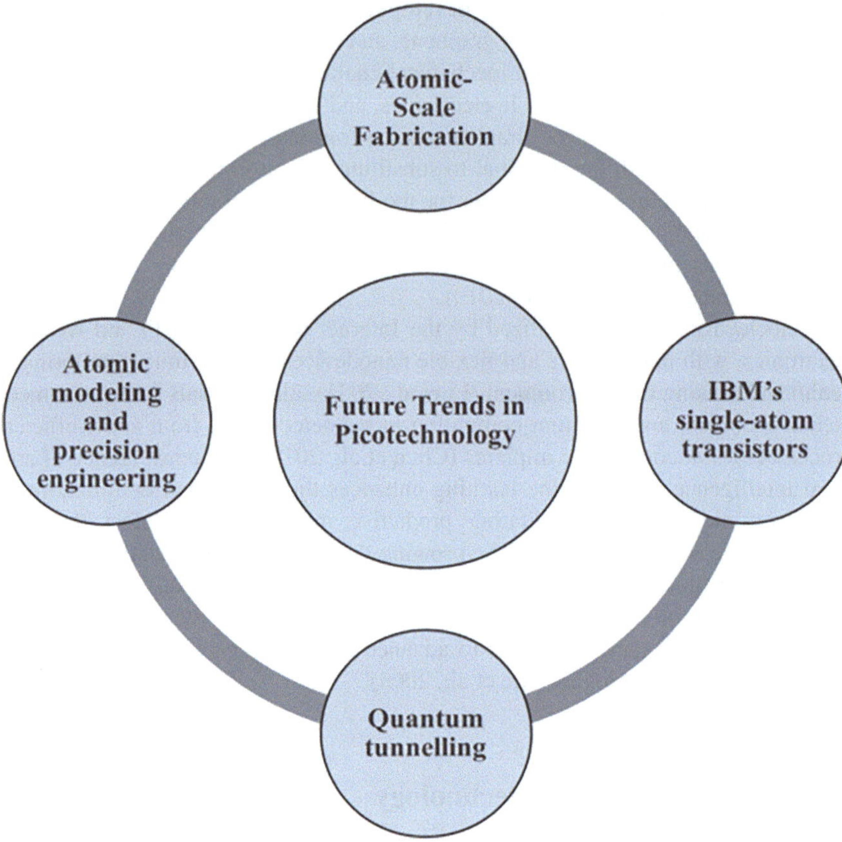

Fig. 2 Future trends in picotechnology

networks (Chen, 2020). Picotechnology has the possibility to the digital revolution at the atomic level (Xie et al., 2004a, b). Picotechnology is advancing rapidly, with breakthroughs like IBM's single-atom transistors and applications of quantum tunneling pushing the boundaries of atomic-scale engineering. As depicted in Fig. 2, these innovations highlight the transformative potential of Picotechnology in enabling unprecedented precision in material design and device fabrication.

3.1 Atomic-Scale Fabrication

Nanotechnology, at the picoscale (10^sup > −12 m), is a frontier of precision engineering that provides unprecedented control over material properties, quantum states, and molecular interactions (Xie et al., 2008). It entails the application of Scanning Probe Microscopy (SPM)-based atom manipulation, which integrates

Scanning Tunneling Microscopy (STM) and Atomic Force Microscopy (AFM) with Laser Technology to single atoms or electrons at surfaces (Xie et al., 2004a, b). This allows for the creation of Artificial Quantum Structures, Ultra-High Density Memory Arrays, and Molecular-Scale Circuits (Xie et al., 2010). STM has established the Picoscale Engineering, with possible uses being Quantum Dot Arrays with Zero Decoherence or Single-Atom Transistors that can operate at terahertz speeds (Obermair et al., 2010).

Picoscale production in electronics may result in silicon processors, fusion catalysts, solid-state batteries with best ion routes, synthetic organelles that boost human metabolism, nanorobots that are able to cure cellular damage at the DNA level, and atomic precision (Xie et al., 2008). Quantum and algorithm optimization under cryogenic fabrication conditions enable automated, atom-by-atom assembly for faster manufacturing (Chen, 2020). Picotechnology would revolutionize human potential to program matter, customize Atomic Medicine, and apply Quantum Artificial Intelligence in Picoscale Networks (Xie et al., 2010). Industrial scale involves parallel atomic assembly, large-scale and power-efficient fabrication methods (Xie et al., 2008).

3.2 IBM'S Single-Atom Transistors

IBM's work on single-atom transistors impacts a shift in Electronics, Quantum Computing, and Material Science (Xie et al., 2010). The single-atom transistors achieve the ultimate limit of Moore's Law by leveraging the Quantum properties of individual atoms to create devices with their efficiency, speed, and density (Obermair et al., 2010). IBM's experiments have evolved advanced methods for isolating and manipulating single-atom switches, particularly with transition metals such as phosphorous-embedded silicon or rare-earth atoms like holmium (Xie et al., 2008). The Quantum Tunneling and Spin-Resolved Electron Transport, Terahertz Frequencies, and Energy Consumption to the desired range are a million times more than Traditional Silicon Transistors (Xie et al., 2004a, b). IBM's transistors consisting of a single atom have immense implications in Quantum Computing that might serve as qubits that are resistant to error because of Nanoscale Confinement (Xie et al., 2010). Fabrication, interconnect complexity, and scalability present major challenges in advanced technology development (Chen, 2020). IBM's research intersects with algorithmic innovation because machine learning models are being utilized to the fore expectation of Optimal Atomic Topologies and Quantum Dots (Xie et al., 2008). Picotechnology using Artificial Intelligence could offer self-tuning systems where transistors reconfigure themselves dynamically to meet computational needs (Xie et al., 2010). IBM's single-atom transistors have the potential to revolutionize industries like medicine, energy, and space exploration (Obermair et al., 2010).

3.3 *Quantum Tunneling*

Quantum Tunneling is a Quantum Mechanical Phenomenon where particles tunnel through Picotechnology energy barriers at 10^-12 m (Xie et al., 2004a, b). Quantum Tunneling has transformed the field of electronics, energy harvesting, and medicine by allowing unparalleled control of matter in subatomic and atomics (Xie et al., 2008). Picoscale devices exploit Single-Electron Tunneling across well-designed Quantum Dots or Molecular Junctions with near-zero power dissipation (Xie et al., 2010). IBM's Single-Atom Transistors on Tunneling between phosphorus atoms in silicon lattices allow logic operations at the Physical Miniaturization (Obermair et al., 2010). Quantum Tunneling also enables Picoscale Biosensors of sensitivity, detecting single biomolecules or weak gravitational waves (Xie et al., 2008).

In energy, Tunneling-enhanced thermionics could have waste heat with their efficiency, and Nuclear tunneling reactions could enable low-energy fusion processes or isotopic separation methods (Chen, 2020). Some challenges are decoherence of quantum and thermal dephase coherent tunneling, where cryogenic or topological protection is needed (Xie et al., 2010). It also encapsulates Tunneling-based quantum computers and Bio-integrated pico-devices for in-time cellular monitoring (Xie et al., 2008). With the development of Picotechnology, Quantum Tunneling may bring various parameters such as Atomic-scale replicators and Zero-power quantum networks revolutionizing our command over the world of sub-atoms (Xie et al., 2004a, b).

3.4 *Atomic Modeling and Precision Engineering*

Picotechnology is Atomic Modeling and Precision Engineering (Xie et al., 2010). It involves the control and management of atoms and subatomic particles, which demands sophisticated computational models, ultra-high-resolution imaging methods, and quantum-level fabrication tools (Xie et al., 2008). Atomic modeling in Picotechnology employs Quantum Mechanical Simulations, Density Functional Theory (DFT), and Molecular Dynamics to predict and analyze matter at the Picoscale (Chen, 2020). Precision engineering entails creating Picoscale manipulators such as Scanning Tunneling Microscopes (STMs) and Atomic Force Microscopes (AFMs) with Quantum Coherence (Xie et al., 2004a, b). Artificial intelligence and machine learning optimize processes for quantum computing, photonics, and energy storage applications (Xie et al., 2010).

Picotechnology has some challenges like quantum decoherence, thermal fluctuations, and van der Waals forces that can perturb atomic stability (Obermair et al., 2010). Prospects involve Picoscale transistors for ultrahigh-efficiency electronics that can reorganize their atomic structure at will, transforming medicine and energy (Xie et al., 2008). Safety and ethics are important issues owing to the possible unintended consequences on the environment and stability at the Picoscale (Babatunde et al., 2020). The evolving technology in Picotechnology may make human ingenuity, atomic-scale 3D printing possible, room temperature superconductors, and the manufacture of man-made exotic factors (Chen, 2020). Picotechnology may result in Picoscale with their interfaces or quantum networks with atomic-level security (Xie et al., 2010).

4 Challenges and Limitations

Picotechnology is operating at the trillionth-of-a-meter scale that has numerous scientific and engineering challenges (Xie et al., 2008). Quantum decoherence, thermal fluctuations, electromagnetic interference, vibrations, and quantum effects pose significant challenges (Xie et al., 2010). These effects can cause misalignment of structures under cryogenic or ultra-high-vacuum conditions (Obermair et al., 2010). The lack of imaging and manipulation with the necessary resolution of the material defects and impurities can alter the desired properties of ultra-pure materials and defect-tolerant designs (Xie et al., 2004a, b). Energy requirements for Picoscale manipulation are also applied to position atoms with such accuracy usually demanding highly extreme conditions (Chen, 2020). Picoscale systems demand large-scale production of Picoscale devices with atomic precision due to slow assembly rates and lack of self-assembly techniques (Xie et al., 2008). Ethics and safety concerns, biosafety, and regulatory concerns remain in the traditional sources of energy also pose challenges (Babatunde et al., 2020). The Integration of Picotechnology with the nanoscale and microscale devices is the transition between Picoscale building blocks and macroscopic devices by several technologies (Xie et al., 2010). Table 1 tabulates Technological and Ethical Challenges in Nanotechnology and Picotechnology.

5 Future Perspectives

The future of Picotechnology and Nanotechnology holds significant potential in science, medicine, energy, and computing (Chen, 2020). Nanotechnology in precision medicine, with drug delivery systems at the nanoscale, potentially has cancer and neurodegenerative diseases (Li et al., 2021). Self-assembling nanomaterials will transform manufacturing to produce ultra-strong, lightweight materials for aerospace and construction, while nanosensors integrated into IoT devices provide real-time environmental monitoring and smart cities (Rambaran & Schirhagl, 2022). Energy storage innovations, such as solid-state nanobatteries and perovskite solar cells, offer ultra-efficient, scalable solutions to global energy needs (Jaramillo-Cabanzo et al., 2021).

Picotechnology at the sub-nanometer level with quantum computing has error-free qubits and room temperature superconductors (Xie et al., 2010). Atomic-scale fabrication could make programmable matter and pico-robots for reversing cellular aging (Xie et al., 2008). Photonics and optoelectronics would be efficient with

Table 1 Technological and ethical challenges in nanotechnology and picotechnology

S.No	Features	Nanotechnology challenges	Picotechnology challenges	References
1.	Limitations	• Precision in nanoscale manufacturing remains difficult • Stability and durability of nanomaterials under varying conditions	• Manipulating matter at the Pico scale (10^{-12} m) is beyond current instrumentation • Theoretical barriers in controlling atomic/subatomic interactions	Chen (2020)
2.	Fabrication & Scalability	• High costs of nanofabrication techniques (e.g., lithography) • Mass production challenges for uniform nanostructures	• No practical tools exist for Pico scale engineering • Quantum effects dominate, making classical engineering approaches ineffective	Xie et al. (2010)
3.	Safety & Ethics	• Unknown long-term health effects of nanoparticles • Environmental risks from nano-waste accumulation	• Potential hazards of manipulating subatomic particles • Ethical concerns over matter reconstruction at fundamental levels	Chen (2020)
4.	Energy requirements	• High energy input for nanomaterial synthesis • Energy inefficiency in self-assembly processes	• Extreme energy demands for Pico scale manipulation • Thermodynamic instability at the Pico scale	Bayda et al. (2020)
5.	Theoretical barriers	• Limitations in quantum tunneling effects for nanoelectronics • Difficulty in predicting nanomaterial behavior	• Lack of theoretical data • Requires breakthroughs in quantum physics and string theory	Yu and Capasso (2014)
6.	Regulatory & Standardization Issues	• Lack of global regulations for nanoparticle usage • Standardization challenges in nanomedicine and manufacturing	• No regulatory framework exists for Pico tech • Potential military/weaponization risks	Babatunde et al. (2020)
7.	Economic feasibility	• High R & D costs limit commercialization • Market adoption barriers in industries like healthcare and energy	• Prohibitively expensive theoretical research • No return on investment in the future	Obermair et al. (2010)

picoscale-engineered metamaterials (Yu & Capasso, 2014). Challenges remain, such as quantum decoherence, thermal noise, and existing imaging technologies (Xie et al., 2010). Additionally, energy demands for Picoscale engineering are expensive, and scalable production is a long-term aspiration (Chen, 2020). The intersection of Nanotechnology and Picotechnology with artificial intelligence will accelerate progress, but it also poses ethical concerns like nanoweaponization, ecological effects, and invasions (Babatunde et al., 2020). Nanotechnology and Picotechnology have the potential for atomically precise manufacturing, ultra-efficient energy systems for climate change, and medical advancements (Xie et al., 2008).

6 Conclusion

Nanotechnology and Picotechnology are two scientific technologies that have unparalleled control over matter at atomic and subatomic levels. Nanotechnology working in the range of 1–10 nm, has revolutionized sectors such as medicine, electronics, energy, and materials science, making possible targeted drug delivery systems, extremely efficient solar cells, and quantum computing hardware. In electronics, nanoelectronics has driven transistor miniaturization, flexible and wearable device creation, and the application of quantum dots in high-performance computing. Picotechnology can also revolutionize this sector further by enabling single-atom transistors, highly efficient quantum processors, and atomic-scale memory storage. Nanostructured batteries, perovskite solar cells, and nanocatalysts for the production of hydrogen are some examples of applications in energy. Environmental nanotechnology is employed to clean pollution, treat water, and track air quality. However, both of these technologies suffer from issues like fabrication precision, material stability, scalability, quantum decoherence, thermal fluctuations, and the lack of useful tools for subatomic manipulation. All these issues are to be solved through interdisciplinary collaboration involving nanotechnology advancements in artificial intelligence, quantum physics, and biotechnology. The convergence of Nanotechnology and Picotechnology with AI and machine learning could have innovation with self-evolving materials, predictive diagnostics, and atomic-scale manufacturing.

References

Babatunde, D. E., Denwigwe, I. H., Babatunde, O. M., Gbadamosi, S. L., Babalola, I. P., & Agboola, O. (2020). Environmental and societal impact of nanotechnology. *IEEE Access, 8*, 4640–4667.

Bayda, S., Adeel, M., Tuccinardi, T., Cordani, M., & Rizzolio, F. (2020). The history of nanoscience and nanotechnology: From chemical-physical applications to nanomedicine. *Molecules, 25*, 112–128.

Bensalah, F., Pézard, J., Haddour, N., Erouel, M., Buret, F., & Khirouni, K. (2021). Carbon Nanofiber/PDMS composite used as corrosion-resistant coating for copper anodes in microbial fuel cells. *Nanomaterials, 11*, 314–324.

Bleeker, E. A., de Jong, W. H., Geertsma, R. E., Groenewold, M., Heugens, E. H., & Koers-Jacquemijns, M. (2013). Considerations on the EU definition of a nanomaterial: Science to support policy making. *Regulatory Toxicology and Pharmacology, 65*, 119–125.

Boyarsky, M., Sleasman, T., Pulido-Mancera, L., Fromenteze, T., Pedross-Engel, A., Watts, C., Imani, M., Reynolds, M., & Smith, D. (2017). Synthetic aperture radar with dynamic metasurface antennas: A conceptual development. *Journal of the Optical Society of America A: Optics and Image Science, and Vision, 34*, 22–36.

Cao, J. J., Huang, Y., & Zhang, Q. (2021). Ambient air purification by nanotechnologies: From theory to application. *Catalysts, 11*, 23–29.

Chen, B. (2020). Nanoscale transistors for biochemical applications. *Nature Nanotechnology, 15*, 831–837.

Chen, H., Zhang, X. L., Zhang, Y. Y., Wang, D., Bao, D. L., Que, Y., Xiao, W., Du, S., Ouyang, M., & Pantelides, S. T. (2019). Atomically precise, custom-design origami graphene nanostructures. *The Sciences, 365*, 1036–1040.

De Souza, F. A. L., Sivaraman, G., Hertkorn, J., Amorim, R. G., Fyta, M., & Scopel, W. L. (2019). Hybrid 2D nanodevices (graphene/h-BN): Selecting NOx gas through the device interface. *Journal of Materials Chemistry A, 7*, 8905–8911.

Eswaraiah, V., Zeng, Q., Long, Y., & Liu, Z. (2016). Black phosphorus nanosheets: Synthesis, characterization and applications. *Small, 12*, 3480–3502.

Ginzburg, P., Roth, D. J., Nasir, M. E., Segovia, P., Krasavin, A. V., Levitt, J., Hirvonen, L. M., Wells, B., Suhling, K., & Richards, D. (2017). Spontaneous emission in non-local materials. *Light: Science & Applications, 6*, 16–23.

Huo, Z., Wei, Y., Wang, Y., Wang, Z. L., & Sun, Q. (2022). Integrated self-powered sensors based on 2D material devices. *Advanced Functional Materials, 32*, 22–26.

Jaramillo-Cabanzo, D. F., Ajayi, B. P., Meduri, P., & Sunkara, M. K. (2021). One-dimensional nanomaterials in lithium-ion batteries. *Journal of Physics D: Applied Physics, 54*, 256–268.

Karimi, A., Fakhroueian, Z., Bahramian, A., Khiabani, N. P., Darabad, J. B., Aniz, R., & Arya, S. (2012). Wettability alteration in carbonates using zirconium oxide nanofluids: EOR implications. *Energy and Fuels, 26*, 1028–1036.

Khan, A. D., Amin, M., Ali, A., Khan, S. D., & Khan, R. (2015). Multiple higher-order Fano resonances in plasmonic hollow cylindrical nanodimer. *Applied Physics A: Materials Science & Processing, 120*, 641–649.

Lee, C.-W., Choi, H. J., & Jeong, H. (2020). Tunable metasurfaces for visible and SWIR applications. *Nano Convergence, 7*, 3–10.

Li, Q., Wu, J. T., Liu, Y., Qi, X. M., Jin, H. G., Yang, C., Liu, J., Li, G. L., & He, Q. G. (2021). Recent advances in black phosphorus-based electrochemical sensors: A review. *Analytica Chimica Acta, 1170*, 34–39.

Liu, G., Tang, Y., Soomro, A. M., Shen, P., Lu, S., Cai, Y., Wang, H., Yang, Q., Chen, H., Shi, Y., Lin, C., Xu, F., Xu, F., Wu, Z., Chen, X., Cai, D., & Kang, J. (2023). Vertically aligned ZnO nanoarray directly orientated on cu paper by h-BN monolayer for flexible and transparent piezoelectric nanogenerator. *Nano Energy, 109*, 19–24.

Logeshwari, P., Silambarasan, P., & Abraham, J. (2013). Ecofriendly synthesis of silver nanoparticles from commercially available plant powder and their anti-bacterial properties. *Scientia Iranica F, 20*, 1049–1054.

Luo, X., Pu, M., Ma, X., & Li, X. (2015). Taming the electromagnetic boundaries *via* Metasurfaces: From theory and fabrication to functional devices. *International Journal of Antennas and Propagation, 201*, 24–35.

Martins, R., & Kaczerewska, O. B. (2021). Green nanotechnology: The latest innovations, knowledge gaps, and future perspectives. *Applied Sciences, 11*, 4513–4526.

Obermair, C., Xie, F. Q., & Schimmel, T. (2010). The single-atom transistor: Perspectives for quantum electronics on the atomic-scale. *Europhysics News, 41*, 25–28.
Rambaran, T., & Schirhagl, R. (2022). Nanotechnology from lab to industry—A look at current trends. *Nanoscale Advances, 4*, 3664–3675.
Stremfelj, J., & Smole, F. (2021). Nanotechnology and nanoscience—from past breakthroughs to future prospects. *Informacije Midem-Journal of Microelectronics Electronic Components and Materials, 51*, 25–48.
Wang, N. (2019). Picosecond chemical reactions monitored by nanoscale sensors. *Science Advances, 5*, 45–52.
Wang, X., Wang, H., Jian, J., Rutherford, B. X., Gao, X., Xu, X., Zhang, X., & Wang, H. (2020). Metal-free oxide-nitride Heterostructure as a tunable hyperbolic metamaterial platform. *Nano Letters, 20*, 6614–6622.
Wu, J., & Wang, Y. (2017). Optical absorption and thermal effects of Plasmonic nanostructures. In *Nanoplasmonics: Fundamentals and applications* (pp. 157–171). Intech Open.
Wu, Y., Yuan, W., Xu, M., Bai, S., Chen, Y., Tang, Z., Wang, C., Yang, Y., Zhang, X., Yuan, Y., Chen, M., Zhang, X., Liu, B., & Jiang, L. (2021). Two-dimensional black phosphorus: Properties, fabrication and application for flexible supercapacitors. *Chemical Engineering Journal, 412*, 125–134.
Xie, F. Q., Nittler, L., Obermair, C., & Schimmel, T. (2004a). Gate-controlled atomic quantum switch. *Physical Review Letters, 93*, 128–132.
Xie, F. Q., Obermair, C., & Schimmel, T. (2004b). Switching an electrical current with atoms: The reproducible operation of a multi-atom relay. *Solid State Communications, 132*, 437–442.
Xie, F. Q., Maul, R., Augenstein, A., Obermair, C., & Starikov, E. B. (2008). Independently switchable atomic quantum transistors by reversible contact reconstruction. *Nano Letters, 8*, 4493–4497.
Xie, F., Maul, R., Obermair, C., Wenzel, W., Schön, G., & Schimmel, T. (2010). Multilevel atomic-scale transistors based on metallic quantum point contacts. *Advanced Materials, 22*, 2033–2036.
Yu, N., & Capasso, F. (2014). Flat optics with designer Metasurfaces. *Nature Materials, 13*, 139–150.
Zahra, S., Ma, L., Wang, W., Li, J., Chen, D., Liu, Y., Zhou, Y., Li, N., Huang, Y., & Wen, G. (2021). Electromagnetic Metasurfaces and reconfigurable Metasurfaces: A review. *Frontiers of Physics, 8*, 15.

Index

A
Advanced nanomaterials, 183, 359–362
Antimicrobial, 35, 117, 120, 124, 128
Applications, 3, 28, 45, 93, 105, 134, 165, 231, 255, 276, 301, 314, 335, 358
Applications of nanotechnology, 8, 14–17, 19, 35–36, 39, 106, 127, 135–140, 153, 182–185
Atomic force microscopy (AFM), 7, 8, 28, 29, 37, 81, 94, 97, 98, 102, 103, 106, 169–173, 243, 244, 259, 260, 300, 315, 317–320, 323, 336, 340, 345, 365
Atomic-scale manipulation, 235

B
Bioimaging, 5, 313–331
Biological science, 118, 299–309
Biomarkers, 107, 113, 116, 122, 125, 128, 195, 222, 306, 314, 322–327, 329–331, 339, 346, 347

C
Carbon capture, 40, 73, 151–154, 262–263, 270
Catalysis, 5, 6, 10, 29, 46, 71, 100, 106, 139–140, 148, 262, 339
Characterization, 3, 7, 12, 15, 28, 78–80, 93–103, 171, 200, 243–245, 259–261, 278, 317
Characterization techniques, 12–14, 19, 79–80, 93, 94, 102, 270, 317
Chemical composition analysis, 57, 94
Computing system, 121, 176, 188, 279–281, 284, 351
Cryo-electron microscopy (Cryo-EM), 38, 315, 318–321, 324, 331

D
Diagnostics, 14–15, 18, 82, 107, 108, 112, 113, 116, 117, 121, 122, 124, 125, 128, 129, 171, 191, 192, 195, 197, 203, 212, 213, 216, 223–225, 242, 246, 248, 287, 300, 307, 313–317, 320, 323–326, 328–331, 336, 340–342, 350, 360, 362, 369
DNA, 47, 61, 62, 64, 71, 72, 76, 98, 102, 113, 126, 184, 191–201, 203–210, 212, 216, 218–221, 223–225, 306, 315, 324, 339, 342, 347, 349, 358, 363, 365
Drug delivery, 5, 6, 9, 10, 14–15, 18, 28, 29, 40, 41, 64, 71, 72, 78, 80–82, 108–125, 128, 129, 167, 171, 197, 223, 232, 233, 235–238, 246, 248, 249, 300–302, 305, 306, 308, 315, 326, 336, 337, 340, 342, 346, 347, 360, 361, 367, 369

E
Energy efficiency (EE), 123, 134, 144, 154, 176, 177, 185, 187, 188, 266, 271, 280, 291

S. Kulkarni (ed.), *A Journey from Nanotechnology to Picotechnology*, Lecture Notes in Nanoscale Science and Technology 47,
https://doi.org/10.1007/978-3-032-07228-3

Energy storage, 5, 28, 73, 80–81, 99, 103, 134–140, 144–148, 154, 160, 161, 167–169, 173, 232, 233, 235–237, 239, 245, 246, 248, 257–258, 264–271, 291, 360, 367
Environmental remediation, 12, 16, 19, 28, 80, 148, 262
Ethical consideration, 18, 41, 126–127, 157, 256, 328, 331, 357

F
Fabrication, 11, 15, 30, 32, 39, 45–82, 94, 140, 166, 231–250, 256, 260, 327, 336, 344, 347, 352, 357, 362, 364, 367–369
Fabrication techniques, 30–33, 45–82, 158, 168–169, 174, 175, 231–250, 258–261, 269
Fabrication tools, 366

H
High-performance materials, 232
High-resolution imaging, 7, 8, 38–39, 94, 107, 247, 249, 315, 318–320, 362
Hydrogen fuel cells, 136, 144, 239

I
Internet of medical things (IoMT), 285–288

M
Material science, 7, 10, 14, 36, 38, 106, 233, 235, 238, 314, 316, 335–352, 365
Membrane, 32, 33, 63, 71, 81, 109, 111, 139–140, 149, 152, 156, 191–194, 196, 197, 200–203, 212, 238, 259, 262, 270, 286, 308, 319, 324, 348, 350, 351
Molecular manipulation, 39
Molecular sequencing (MS), 191–226
Multi-omics approaches, 326–327

N
Nanocarriers, 72, 109, 115, 128, 248, 349, 362
Nanoelectronics, 5, 15, 41, 47, 50, 107, 237–240, 243, 260, 281, 358–360, 362–363, 368, 369
Nanomaterials, 4, 27, 45, 93, 105, 134, 167, 233, 261, 301, 337, 358
Nanomedicine, 15, 18, 45, 107, 108, 115, 121, 171, 192, 223, 236, 358, 359, 362, 368
Nanopores, 191–226, 267
Nano-scale, 66, 78
Nanosensors, 17, 18, 41, 82, 107, 125, 128, 143, 147, 154, 183, 184, 187, 248, 308, 338, 350, 359, 363, 367
Nanotechnology, 3, 27, 45, 93, 105, 134, 165, 233, 255, 275, 299, 313, 335, 357, 358

P
Personalized medicine, 195, 216, 226, 248, 249, 325, 326, 328, 329, 362
Picometer scale, 36–38, 40, 160, 234, 235, 237, 239, 241–245, 249, 300–302, 315–317, 319, 320, 322–325, 330, 336, 339–341, 345, 358
Picometer scale probes, 39
Picotechnology, 28, 160, 231–250, 255–271, 275–293, 299–309, 313–331, 335–352, 357–369
Picotechnology design, 29, 237, 265, 364
Pollution control, 16, 40, 134, 149–150, 270, 302, 337
Precision medicine, 128, 129, 205, 224–226, 249, 315, 326–327, 331, 362, 367

Q
Quantum dot, 5, 6, 8, 15, 19, 29, 37, 39, 55, 100, 107, 113, 117, 135, 168, 231, 237, 238, 241, 244, 246–248, 256, 258–262, 264, 270, 284, 314, 316–318, 320, 321, 336, 340–342, 348, 352, 358–360, 362, 363, 369
Quantum dot technologies, 174, 320, 324, 331
Quantum tunneling, 8, 28, 37–39, 244, 340, 364–366, 368

R
Raman spectroscopy, 94, 100–103, 329, 361
Regenerative medicine, 74, 76, 114, 128, 129, 248, 301, 302, 306, 337, 347, 360
Renewable energy, 16, 41, 73, 134–137, 139–147, 153, 154, 157, 160, 161, 235, 263
RNA, 120, 193–200, 203–209, 212, 219, 221, 223–225

S
Scanning electron microscopy (SEM), 7, 12, 19, 79, 81, 94–95, 102, 103
Scanning tunneling microscopy (STM), 3, 7, 8, 19, 28, 37–39, 106, 165, 169–173, 243, 244, 260, 315–317, 336, 338, 340, 342, 365
Sequencing platforms, 212–215
Single-atom devices, 260
6G ultra-dense cellular network, 290
Solar cells, 5, 15, 30, 51, 53, 73, 76, 80, 134–136, 139–142, 148, 153, 156, 159, 160, 236, 237, 258, 280, 357, 359–361, 367, 369
Surface topography, 8, 94, 97, 102, 170, 260
Sustainable development, 11, 18, 73, 153–157, 161, 257
Sustainable energy, 135, 362
Synthesis, 10–13, 19, 27, 28, 30–33, 47, 51–53, 55–59, 61, 63, 65, 66, 69–73, 75, 78, 94–97, 100, 101, 119, 158, 168, 237–239, 245, 262, 270, 271, 277, 301, 303, 305, 308, 309, 337, 368

T
Theranostics, 128, 325–326, 331
Tissue engineering, 29, 64, 71, 76, 78, 108, 113, 114, 192, 237, 249, 300, 305, 306, 308, 347
Transmission electron microscopy (TEM), 7, 12, 14, 19, 38, 80, 94–97, 102, 103, 243, 259, 260, 317

V
Vaccine delivery, 128
Vapor phase deposition, 51–55

X
X-ray photoelectron spectroscopy (XPS), 94, 98–99, 101, 103

MIX
Papier aus verantwortungsvollen Quellen
Paper from responsible sources
FSC® C105338

If you have any concerns about our products,
you can contact us on
ProductSafety@springernature.com

In case Publisher is established outside the EU,
the EU authorized representative is:
Springer Nature Customer Service Center GmbH
Europaplatz 3, 69115 Heidelberg, Germany

Printed by Libri Plureos GmbH
in Hamburg, Germany